Sustainable Energy Management

Sustainable Energy Management

Mirjana Golušin

Siniša Dodić

Stevan Popov

AMSTERDAM • BOSTON • HEIDELBERG • LONDON
NEW YORK • OXFORD • PARIS • SAN DIEGO
SAN FRANCISCO • SINGAPORE • SYDNEY • TOKYO

Academic Press is an Imprint of Elsevier

Academic Press is an imprint of Elsevier
225 Wyman Street, Waltham, MA 02451, USA
The Boulevard, Langford Lane, Kidlington, Oxford OX5 1GB, UK

Notices
Knowledge and best practice in this field are constantly changing. As new research and experience broaden our understanding, changes in research methods, professional practices, or medical treatment may become necessary.

Practitioners and researchers must always rely on their own experience and knowledge in evaluating and using any information, methods, compounds, or experiments described herein. In using such information or methods they should be mindful of their own safety and the safety of others, including parties for whom they have a professional responsibility.

To the fullest extent of the law, neither the Publisher nor the authors, contributors, or editors, assume any liability for any injury and/or damage to persons or property as a matter of products liability, negligence or otherwise, or from any use or operation of any methods, products, instructions, or ideas contained in the material herein.

Library of Congress Cataloging-in-Publication Data
Golušin, Mirjana.
  Sustainable energy management / Mirjana Golušin, Stevan Popov, Siniša Dodić.
    pages cm
  Includes bibliographical references.
  ISBN 978-0-12-415978-5 (hardback)
  1. Energy development.   2. Renewable energy sources.   3. Sustainable engineering.   I. Popov, Stevan.   II. Dodic, Sinasa.   III. Title.
  TJ163.2.G657   2013
  333.79'4—dc23
                                        2012026621

British Library Cataloguing-in-Publication Data
A catalogue record for this book is available from the British Library.

For information on all Academic Press publications visit our website at http://store.elsevier.com

Printed in the United States of America
13 14 15 16 9 8 7 6 5 4 3 2 1

Contents

Acknowledgments

This book is part of the following five projects: 1) Project No. III 47 009 – *European integrations and social and economic changes in Serbian economy on the way to the EU*, which is supported by the Ministry of Science and Technological Development of Serbia in the period 2011-2014. 2) Project No. 179015 – *Challenges and prospects of structural changes in Serbia: strategic directions for economic development and harmonization with EU requirements*, which is supported by the Ministry of Science and Technological Development of Serbia in the period 2011-2014. 3) Project No. TR-31002 – *Development of bioethanol production from products of sugar beet processing*, which is supported by the Ministry of Science and Technological Development of Serbia. 4) Project No. 114/451-2089/2011 – *Development of bioethanol production on small farms*, which is supported by the Provincial Secretariat of Science and Technological Development of Voivodina. 5) EUREKA project No E15832 – *Development of technology for efficient and economical production of bioethanol fuel at small farms*, which is supported by the Ministry of Science and Technological Development of Serbia.

This book was translated by Mrs. Sanela Šipragic Đokić.

Chapter 1

Introduction

Sustainable development is a comprehensive concept of development that was adopted in order to preserve the planet's resources to the extent that will satisfy the needs of present generations without compromising the ability of the next generation to meet the same needs. The concept is comprehensive, multidimensional, and multidisciplinary, and it can be applied widely.

The study of sustainable development involves considering the whole range of relevant issues, mechanisms, solutions, and critical attitudes. Currently there is no single concept of sustainable development that can be universally applied in all areas of human activity throughout the world, and it is necessary to examine it in some sectors that are of special interest for a certain range of issues.

Since the creation of contemporary mankind, especially with the progressive development of human activities, there has been a need for the provision and consumption of certain types and amounts of energy. Every activity is related to the pending charges of energy that are created, transferred, transformed and consumed.

Energy production and consumption is a special problem in the modern world and it needs to be understood on several grounds. First of all, with a proportional increase in population there is an increasing need for the provision of sufficient energy. Given the nature of population growth and inability to stop this growing trend, the problem of providing energy for all human needs becomes global and constant [63].

Energy needs are very different in different parts of the world. The highest energy consumption, since the beginning of the Industrial Revolution in the late eighteenth century, has been found in the most developed countries [65]. Consumption of large amounts of energy has brought these countries a significant economic advantage and created the need to face a constant requirement of providing sufficient energy for future economic and social development.

Energy consumption in the world can be defined and measured on several grounds, but the most suitable is the energy consumption in the total amount of time, in some regions, as well as energy consumption from particular sources.

The total energy consumption in the world undergoes complex monitoring and control, which has multiple objectives. First of all, monitoring of energy-related indicators in the world is necessary because it provides insight into the

Sustainable Energy Management. http://dx.doi.org/10.1016/B978-0-12-415978-5.00001-1

general situation of the energy sector, helps identify the differences among particular regions, as well as understand the trends and predict future tendencies. Tracking energy-related indicators requires extremely complex measurements, so that accurate data can be obtained only after several years. The framework overview of energy consumption in the world until 2005 is shown in Figure 1.1.

Energy consumption has been measured in an organized and precise way since 1965. Power consumption in the world is characterized by three periods. In the base year 1965, energy consumption in the world was less than 5 Terawats. Out of the 5 Terawats, the largest portion was provided by exploitation of traditional sources (oil and coal), almost in equal ratios. Half that amount of energy was obtained from gas, and the least amount was obtained by exploitation of water energy. Even then, 1% of energy was obtained from nuclear power plants [64].

In the second period, which started in 1970, the energy consumption of petroleum was doubled and reached its maximum in 1980. Consumption of coal was also growing, but it was much slower and more balanced. Energy consumption increased steadily from gas, which can be partly explained by population growth and technology. The huge increase in consumption of petroleum can be understood only because of the increase in traffic. In this period (from 1980 to 1990), the consumption of water resources remained at the same levels, and the consumption of energy from nuclear sources grew slowly but not significantly.

The third period began in 1990 and is characterized by the beginning of a stable and balanced growth in energy consumption from traditional sources. Consumption of oil and gas continued to grow almost evenly, while the consumption of coal was stabilized and even slightly reduced by the end of the period. Production of energy from water and nuclear sources reached its peak in 1990 and there were no significant changes until 2000.

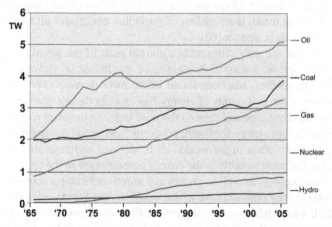

FIGURE 1.1   Energy consumption in the world until the year 2005

In the beginning of the new millennium, by a shift in the third distinct period, energy consumption from oil and gas continued to grow at the same pace, while the consumption of energy obtained from coal, after a decade of stabilization, began to grow again. At the end of the third period, in 2005, for which the precise data is available, the total world energy consumption was less than 15 TW, three times more than what was recorded 40 years earlier [75].

The reasons for increased consumption are numerous and complex and include the increase in the number of people in the world, improvement of the needs of industry and transport, inefficient consumption, and poor energy efficiency indicators. Whatever the reasons, this large consumption of energy, apart from the exhaustion of the world energy resources, has led to significant environmental consequences. All of which raises the need for a strategic redirecting in the field of energy management at all levels, with the knowledge that energy consumption is uneven in some parts of the world, as shown in Figure 1.2.

Along with the monitoring of the total consumption of energy in the world, the monitoring of energy consumption in some regions of the world is connected with a number of problems, so final estimates cannot be considered absolutely accurate, although they provide valuable information about the approximate values of energy consumption.

The highest energy consumption can be seen in the United States, Canada, Norway, and Saudi Arabia, which can be largely explained by the high level of technological development and living standards, as well as the intensity of traffic. Somewhat weaker consumption has been recorded in Russia, Scandinavia, and Australia. Compared to the first group of countries, with the highest energy consumption (indicated above), the European countries, Japan, South Africa, and Argentina, spend half that energy, followed by the Eastern European countries.

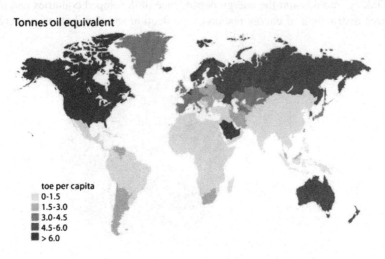

Tonnes oil equivalent

toe per capita
0-1.5
1.5-3.0
3.0-4.5
4.5-6.0
> 6.0

FIGURE 1.2   Regional distribution of energy consumption in the world

Even lower power consumption has been recorded in Brazil, and the lowest in Africa and the Far East.

Energy consumption in some regions depends on many factors, but it is evident that it is related to the question of whether the region also produces the energy. Countries with the largest energy reserves are also the biggest consumers of energy, which can not necessarily be considered justified. The European countries are mainly well-developed countries, but they do not use too much energy, which is mainly due to the great efforts for rational management of energy [7].

Consumption of different types of energy also has its particularities, which are shown in Figure 1.3 [77].

From the beginning of the industrial revolution (technical and technological improvements and expansion of transport) energy from oil is mostly used, and one-third of the world's energy needs is provided in that way. Apart from this, the trend of oil consumption is obvious. Over the years, oil consumption has continued to grow and that trend is expected to continue. Somewhat less energy is obtained by exploitation of coal (25%), but that trend has decreased, which can be explained by specific technological processes and the use of automobiles that cannot use coal as fuel. About one-fifth of world energy demand is obtained by the exploitation of gas [12].

The remaining quantities of energy are provided in the same proportion of energy from the biomass (11%), nuclear (6.4%) energy, and water courses (21%). Exploitation of nuclear energy is limited to countries where such production is allowed. The least quantities of energy are obtained from alternative sources (2.2%). The average general values have been presented, but it should be emphasized that there are significant differences in certain countries and regions [6].

Taking into account the energy dependence of developed countries and the uneven distribution of energy resources, production and distribution of energy

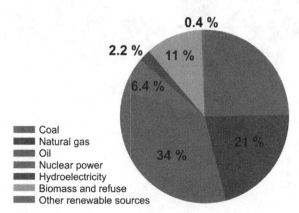

FIGURE 1.3   **Energy consumption in the world by the type of energy**

has become a problem of particular international political and economic importance. Energy in all its forms has become a subject of international trade, numerous disputes, negotiations, and military conflicts. Since the demand for energy
on the planet is constantly growing, it will occupy a more important place in
global economic and political changes in the future.

With the development of awareness and responsibility of individuals and
businesses and countries with the objective to respond to the growing number
of environmental problems, the concept of sustainable development has been
defined to seek options for successful solutions to the environmental problems,
including energy exploitation and consumption, which are connected with substantial negative environmental impact. The problem of providing sufficient
energy is one of the main challenges of sustainable development and clearly
breaks the trend of uncontrolled energy consumption and implicitly imposes the
need for changes in this field.

Sustainable development involves, among other things, the gradual implementation of specific measures to drastically change the current approach to
energy production and consumption, because it implies developing new technologies, utilization of new energy sources, development and implementation
of comprehensive measures of saving, and the development and implementation of numerous legal regulations, all with the aim of raising levels of energy
efficiency, i.e., to stop the trend of uncontrolled consumption of energy that
inevitably leads to the rapid and total depletion of existing energy resources
[78]. These measures are binding for all countries that accept the concept of
sustainable development and are a part of many international agreements and
protocols that govern this area.

The energy problem, with the need for economic development on the one
hand and the need to keep energy resources and reduce pollution on the other,
requires definition of a special mechanism to manage this area, which is essential to the process of planning at all levels, in companies, across regions and
countries, and in the international community as a whole [79]. Sustainable
energy management, with all the elements of modern management science,
along with the integration of requirements that are set by implementing the
concept of sustainable development, is an effective mechanism that allows long-
term planning of sustainable development. Sustainable energy management is
designed so that its steps are clearly defined and measurable and its goals are
specified and measurable. Only this model of management, which is based on
the impact of the factors of sustainability, with clearly defined goals, provides
quality monitoring of the progress of the whole process of sustainable development. However, in sustainable energy management the impact that energy
exploitation and utilization of energy resources may have on other indicators of
development can be monitored, measured, and kept under control.

Sustainable development consists of four subsystems that are interconnected
although different in nature and intensity. The economic, environmental, social,
and institutional subsystems are a kind of sustainable development unit, whose

harmonious development represents the best possible option for long-term, stable development and survival of mankind. Energy management is a key problem of sustainable development because energy consumption reduces the values of the environmental subsystem but has positive effects on the values of the economic subsystem.

Sustainable energy management enables the monitoring of economic and ecological parameters of development because it includes the exact methods for assessing the effectiveness of implementation, and the control is based on a method for determining the degree of sustainable development through measuring its indicators. Defined in this way, the sustainable energy management model can provide real insight into the level of achieved sustainable development as a whole, taking into account that determining the balance between the economic and ecological development is particularly important [38].

Sustainable energy management is designed to provide a management model that can be accepted not only by individual companies but by countries or regions that create a geographical, natural, and economic unit. The sustainable energy management model is based solely on determining the most appropriate form of management of the system of future energy production and consumption, indicating the most important objectives in the field of sustainable energy development, pointing out how these goals can be achieved, and controlling the complex process by applying the methodology of assessment of the achieved degree of sustainable development. Only in this way is it possible to plan, organize, monitor, and control the complex and complicated process of energy management in the modern world.

Chapter 2

Energy and Sustainable Development

Sustainable development as a theory, concept, and idea is a comprehensive framework for the development of humankind in the future, i.e., the attempt to plan future development based on past experience and predicted future needs. The concept of sustainable development implies the need for reviewing and understanding its complex and multidisciplinary, multidimensional, and heterogeneous structure, making it one of the most complex concepts of development since the creation of humankind.

Any human activity can be observed from different points of view and evaluated on several grounds. Sustainable development, however, being the starting assumption and the basic criterion for the acceptance of the idea, requires that the observed activity, event, or asset at any stage of its implementation, development, and consumption not have any detrimental effect on the environment. Broadly, sustainable development is the need to encourage and allow the performance of only those activities that do not have a detrimental impact on the quality of life of future generations.

Taking into account the different types of human activities, diversity of resources, and ways of their exploitation and consumption, it is necessary to study certain aspects of sustainable development (economic, environmental and social) as a separate phenomena, but also consider its relationship with all other parameters of sustainability.

2.1 DEFINITION OF SUSTAINABLE DEVELOPMENT

Sustainable development is a recently accepted concept of development that has emerged as a result of the fact that the development of civilization has exhausted natural resources to the extent the earth has become unsustainable, thus challenging the prospects of development and survival of future generations. The awareness of the need for preservation and restoration of natural resources has always existed, as indicated by written documents and oral traditions. In the early periods of human development, people were much more aware that they depended on the earth and they treated it with respect. With the increase in population, geographical discoveries, the exploitation of

Sustainable Energy Management. http://dx.doi.org/10.1016/B978-0-12-415978-5.00002-3

colonies, and the Industrial Revolution, people suddenly stopped being interested in the preservation of the human connection with the earth, and its resources have been exhausted relentlessly in the last two centuries. Only after World War II, especially after the Chernobyl nuclear accident (Ukraine, 1986), was public awareness raised again about existing and future environmental problems [11].

The first public reaction prompted authorities to start to legally regulate pollution, and in the 1980s environmental issues were raised to a higher level, and the emphasis shifted from sanctioning polluters to prevention of the problem.

Laws and technology have prescribed and found a variety of solutions that prevent environmental incidents or decrease their likelihood. The scope of interest has also extended to the field of ecology, and the need for conservation, rational exploitation of natural resources, and replacement of non-renewable resources have become imperatives of environmental protection.

Sporadic solutions and intentions of certain countries to make progress in the field of ecology soon encountered the barriers that exist in the form of borders, social organizations, and conflicting economic interests, but global environmental problems have been increasing and the international community is aware that it has to cooperate in this field. The need to define a new concept of growth and development of the planet is the imperative that would determine further development of humankind.

The concept of sustainable development was first officially used as a possible model of development in 1987 at the 42nd United Nations General Assembly in the report of the Commission for Environment and Development called "Our Common Future." The Commission was established in the 38th session of the General Assembly in 1983, and the report was widely known as "The Report of the Brundtland Commission," named after Gro Harlem Brundtland, the Norwegian Prime Minister who chaired the historic session. The most commonly cited definition of sustainable development is the definition given in the Brundtland report: "Sustainable Development should meet the needs of present generations without compromising the ability of future generations to meet their needs." If development is defined as an increase in welfare, then sustainable development can be understood as preservation of welfare over time, the Brundtland Commission report Our Common Future (Oxford: Oxford University Press, 1987).

The Brundtland Commission promoted the politically correct idea of sustainable development, and its report is similar to the Club of Rome report of 1972, which was published under the title "The Limits to Growth." In contrast to Brundtland's report, "The Limits to Growth" was sharply criticized, especially by economists, and did not have a significant influence on international and national environmental policy. This report was seen as highlighting the importance of redistribution, while Brundtland's report was interpreted as a desirable continuation of economic growth. Brundtland's report became notable due to four important observations, as follows [28]:

1. First, it proposed the concept of sustainable development, defining it as meeting current needs without compromising the ability of future generations to meet their needs as well.
2. Second, it emphasized that international cooperation is essential but very difficult to achieve. Brundtland's report points out the problem succinctly: "The earth is one, but the world is not." The policies of different countries are often driven by local and regional interests that do not comply with global environmental expectations.
3. Third, it recommended adoption of the United Nations program on sustainable development and organization of the International Conference on Environment and Development.
4. Fourth, it suggested the strengthening of national environmental agencies, institutions, and organizations.

Thus the concept of sustainable development was widely adopted quickly. During the United Nations Conference on Environment and Development, popularly called the Earth Summit, held in June 1992 in Rio de Janeiro, Agenda 21 was defined and adopted as a comprehensive plan of action for achieving sustainable development in the twenty-first century. In 2003, the World Summit on Sustainable Development was organized (Rio 10+), and the importance of sustainable development was reaffirmed by individual countries and globally. It was also acknowledged that the implementation of the idea of sustainable development had not advanced significantly and therefore it was necessary to take concrete actions and measures to fight poverty, change unsustainable models of production and consumption, and protect rational provision of natural resources.

One of the first definitions of sustainable development was given by R. Repetto, who said that the idea of sustainability is basically grounded in the belief that decisions made today should not jeopardize the prospects for the preservation or improvement of living standards in the future [76]. This definition implies that economic systems should be managed by using dividends on available resources, where the resource base should be kept and improved, so that next generations can live as well as the previous ones. This view has many similarities with the ideal concept of income considered by J. R. Hicks more than half a century ago. According to Hicks, the calculation of income makes it possible to determine the maximum amount that can be spent in the current period, while at the same time not diminishing the prospects for consumption in the future [55].

More recently, R. Solow defined the concept of sustainability – concept of elasticity. According to Solow, sustainability does not imply an obligation to spend natural resources according to sustainable needs. It is necessary to substitute all non-renewable natural resources with renewable alternatives. Also, the pattern of production consumption should be re-considered and re-defined in order to spend optimal quantity of resources and to produce less possible waste.

The principle of sustainability highlights the existence of freedom to use resources that future generations will be deprived of as long as their living conditions remain the same as the conditions provided for present generations. In other words, inability of future generations to use resources would mean a departure from the criteria of sustainability, if that worsens the living conditions of future generations as compared to the previous ones. Although sustainable development may be understood and interpreted differently, the problem occurs when it is necessary to formulate a model of sustainable economy.

Ten years after the Brundtland's report, determination of sustainable development became globally recognized and the subject of discussion on the highest level started together with its global challenge (Conference in Rio de Janeiro in 1992). Therefore, sustainable development started to be connected with the need for certain changes at all levels (local, regional and global), plus intensification of international cooperation, strongest commitment and awareness of responsibility for wellbeing of future generations.

Although this concept was accepted in 1987, there is no proposal for its operational implementation.

Only in recent debates on sustainable development has the idea of a long-term preservation of nature been discussed. In the past, nature was seen as something needed to meet current and future needs, and the idea of sustainability was not environmental, but primarily economic.

As a whole, the concept of sustainable development emphasizes the international aspect of economic activities and their impact on the environment and future resources. Therefore, the opportunity (direct and environmental cost) of providing a certain future are in the center of the concept. Sustainable development is connected with the concept of optimal growth. It aims to establish a balance between economic growth and environmental degradation. The costs of environmental pollution and utilization of resources have an important role in making current economic decisions, since they represent the cost of the damage inflicted on the environment and natural resources that would be paid by future generations.

The concept of sustainable development includes, apart from economic issues, social aspects such as poverty, social disorders, and other problems of social and political stability. It also includes degradation, environmental pollution, and depletion of natural resources. Therefore, sustainable economic development could mean an increase in real GDP (Gross Domestic Product), basic indicator of economic development per capita, which is not the result of environmental degradation, exhaustion of resources, and social disorders. In this interpretation, the environment and nature are designated as a certain restriction on economic development. The concept of sustainable development seeks to prevent negative economic growth that may increase future costs.

The normative content of economically sustainable development is defined in the requirements to preserve the natural conditions and environment for future generations. The implicit requirements of the concept of sustainability are the following [31]:

- Sustainable intergenerational and economic wealth of people;
- Ensuring the survival of humankind as long as possible;
- Striving for flexibility in production and economic systems and/or stability of their characteristics (their ability to recover when exposed to shocks);
- Maintenance of biodiversity;
- Ensuring sustainability of the community; and
- Stabilization of the biosphere.

The complexity of environmental issues, which are primarily a consequence of the rapid development of western industrialized countries, has influenced the acceptance of the notion that traditional economies cannot resolve these questions adequately. If the volume of economic output increases (being the result of the increased consumption of energy and raw materials, as well as the growing number of people), this would necessarily lead to greater use of natural resources and environmental services.

2.2 SUSTAINABLE DEVELOPMENT PRINCIPLES

While some authors believe that the concept of sustainable development is a new theory or doctrine of development, most believe this is a new concept whose wider implementation would change patterns of production, consumption, and everyday life, because it is partly based on the new generation of normative ethical theories. This concept is a landmark ("asymptotical ideal"), and the paradigm of sustainable development is never directly applicable. In order to be realized, the sustainable development concept requires undertaking of new activities as follows:

- New legal and institutional arrangements;
- Adequate financial and economic arrangements;
- New technology and technical solutions;
- Promotion and education;
- New methods of public communication and interaction, with an emphasis on openness, participation, and transparency; and
- New coalitions for sustainable development.

However, these social, economic, political, ecological, spatial, and intergenerational principles and criteria are still general and therefore are not directly applicable. They require operationalization, which has been undertaken in most countries facing this problem. The specific strategy or policy of sustainable development combines a number of general and specific principles and criteria. Depending on the number, content, and specific problem they relate to, the following general principles and criteria of sustainability are usually partly operationalized:

- The principle of strict conservation and the principle of precaution. This principle gives priority to environmental protection. This is especially applicable in situations where scientific and other objective views do not provide sufficient knowledge on potential consequences of the anticipated economic activities, so the rigorous conservation of natural resources should be exercised even at the cost of limiting economic growth. However, applications of these principles should not prevent the use of innovative procedures that can be easily predicted to have negative impacts on the environment.
- In the analysis of environmental impact and costs and benefits, the central place should be given to the principle of alternative solutions, it being one of general principles. When more intensive exploitation of natural resources is inevitable, it should: 1) be minimized, 2) use non-renewable resources as little as possible and renewable resources as much as possible, and 3) be applied with respect to high and rigorous ecological and spatial standards. Intensive economic growth that would be accomplished at the cost of significant exploitation and endangering of natural resources, especially the non-renewable ones, is not acceptable [30].
- The principle of rational use of raw materials and other resources and inputs.
- The principle of risk reduction is applied when there is an estimate and economic activity can be understood as an objective danger and risk to the environment.
- Applying the principle of durability (of products and services) allows the production of quality output, with reasonable costs, that are socially and market acceptable.
- The principle of "polluter pays" and the principle of "user pays" are applied in all cases where economic activities threaten the ecological capacity and end thresholds.
- Depending on the type of activities and types of areas, the following principles are applied individually or in combination: 1) planning principles; 2) market principles; and 3) principles of regulatory control (bans, permissions, etc.).

Application of the principle of re-usability/using and recycling should lead to expanding the use of resources that can be reused or that are biologically degradable. Decisions on projects, programs, and plans for economic development and protection should be made based on the analysis of their impact on the environment (*EIA – Environmental Impact Assessment, SIA – Strategic Impact Assessment, SEA – Strategic Environmental Assessment*) [54]. In addition to the analysis of direct costs and benefits, the main impact of indirect costs and benefits should be evaluated. Market prices as parameters in evaluating the impact should be used if the assumption on the smooth operation of ("perfect") market mechanism is justifiable. In cases where this assumption cannot be justified, the parameters from the class of "shadow price" should be used, etc.

Economic decisions should be based as far as possible on sound systematic insights into the ecological-spatial capacity and extreme thresholds, as a general reference and basis for evaluation.

The principles of public openness, democracy, and participation should be the basis for the preparation, adoption, and implementation of decisions on sustainable economic development and environmental protection, with appropriate IT support.

Practice has shown that general principles and categories of sustainability are not directly usable in preparation, adoption, and implementation of business decisions and all decisions related to sustainability. These principles, therefore, need to be operationalized and generalized so that they express the criteria and the contents of a specific market and geographic space. The disputes over the various meanings of sustainability originate here.

While there is a kind of consensus about general principles and criteria in terms of selected interpretations of sustainability, different interpretations can be found in cases when concrete goals have to be determined, as well as the means for their implementation. In such cases, certain aspects of sustainability are emphasized and some other aspects are ignored. In such division, the aspects of interest play the most important role. On the other hand, this is a consequence of the fact that sustainability is a multilayered phenomenon that has many meanings that can be mutually incompatible. In practice, there are very few situations where it is possible to implement all layers of sustainability, which indicates the necessity to make a "trade-off." Given that sustainability has several meanings and contexts, for community understanding the following defines sustainable:

- Something that provides continuous economic growth and development;
- Something that is socially acceptable;
- Something that harmonizes the ratio and the measure of economic and social development according to ecological capacities; and
- Something that is politically acceptable (can be the subject of agreement in official institutions).

Among the various meanings and contexts of sustainability, economic, social, and political sustainability emphasize the importance of so-called intragenerational equality and fairness, while intergenerational equality and fairness are of greater importance in the ecological understanding of sustainability.

2.3 ENERGY SUSTAINABILITY AS CRITERIA FOR DEVELOPMENT

A particular problem today and a challenge in the field of its development is how to define the terms sustainability and acceptability in terms of sustainable development. Sustainability can be understood as a term, a concept, a philosophy, an attitude, a request, or a need. The current concept of sustainability has emerged as a result of decades of defining activities and events that are acceptable or unacceptable in terms of opportunities to meet the needs of present and future generations. A particular problem is the need to define the concept of sustainability when it comes to energy development and security in this area [10].

In the early 1970s the first significant energy problems struck the world's social scene, and the first ecological requirements were generated. Later, ecological regulations and recommendations were standardized and continue to be developed today. With the development of ecology as a multidisciplinary science and an increase in people's awareness of the need to preserve and improve the quality of the environment, the term "sustainable" has begun to be widely used.

At first, ecology, and thus sustainability, was seen narrowly as a set of rules that exist just to prevent environmental pollution. In the 1970s the first regulations in the field of ecology appeared, but all remained limited to the regulation of physical, chemical, mechanical, and microbiological impacts of some agents on the immediate environment.

During this time, with the rise of awareness regarding energy problems and the increasing number of companies looking to gain a profit at the environment's expense, sustainability took on a new dimension. It was at this time that government agencies started making reports about the deterioration of the environment and the depletion of non-renewable energy resources. The public became familiar with the problems in the environment and demands emerged for conflict resolution. It was then that ecology was given a new dimension, and many intensive studies about the degree of exploitation of renewable and non-renewable natural resources began.

This awareness resulted in the adoption of the concept of sustainable development at the highest level of decision making as the only prerequisite for long-term quality survival. Because of this, sustainability has been given a new dimension [45].

In the early 1980s, it was suggested that a sustainable energy process is one based on the accepted concept of sustainable global development and the development of such business and any other strategy that adopts the criteria of the concept. During this early period, the impact of political structure and citizens was intensified, and relevant laws, regulations, recommendations, proposals, and numerous other initiatives were adopted. The period of the early 1980s can be seen as a turning point in understanding and respecting the criteria of sustainability at all levels. Developed countries were leading in this movement. Previously adopted regulations on the prevention of pollution of nature were tightened and new ones were created. Due to the development of other sciences, such as chemistry, biology, and medicine, new findings were made on the harmful effect of certain substances in nature, and the list of hazardous and potentially hazardous materials and activities was expanded. Sustainable became only that which was not on the list of activities that threaten the environment. Later, the rejection of certain energy processes started taking place, which later led to the closing of whole power plants that were not environmentally suitable.

Sustainability of energy at that time moved out of the domain of simple monitoring and sanctioning the work of power plants. New strategies were adopted to preserve the quality of environment. Long-term sustainability became an important factor in decision-making processes when planning new power plants or infrastructure used for the transfer of energy. Ecological movements were strengthened along with political parties that observe ecology as a priority. In such an environment, energy sustainability was given new importance, as the living environment began to be seen as unique (global) and people began to accept themselves as inseparable parts of nature.

Human activities that did not adversely affect nature and man himself became sustainable, which led to the expansion of savings of all types of materials, the study of alternative forms of energy, and the problem of nuclear power plants was raised. Companies that were not ecologically suitable came under attack, paid huge fines, and were boycotted or closed forever.

Nothing can be observed separately from the living environment as a whole. Science is increasingly trying to study and explain the complexity of interactive processes in nature, and man understands that any activity that he undertakes will, after a certain period of time, affect his health and quality of life in general.

It is quite clear that the energy sustainability of a particular process or activity can no longer be regarded only as a simple set of numbers and standards. Most people cannot (and do not have to) explain the harmful impact of a process but can clearly see the consequences. People want not only to avoid such impacts but to prevent them. Awareness about limited energy sources and need for their replacement for the first time becomes part of strategies for development of companies, development of the region, and the global community as a whole.

In the 1990s energy sustainability ceased to be viewed only as a precondition for the preservation of the environment. Instead, modern man accepted the fact that economic suitability should be observed in the broadest sense possible. Sustainability ceased to be seen as a qualitative characteristic of a process, but became a parameter of evaluation in all areas of life and work. Environmental requirements in the broadest sense became something that must be respected in order to survive and make progress [81].

Also in the 1990s sustainability began to be widely applied in the description and evaluation of not only economic activities but also in the field of politics, culture, education, sports, and entertainment. The environmental impact of any human activity was examined, and at the end of the last millennium developed countries raised awareness of the need for ecological evaluation and exclusion of anything that can negatively affect humankind. The principle of responsibility towards the environment, towards us, and others became an integral part of cultural values, especially in the most developed countries.

The new millennium brought about new challenges and problems, but the desire of the international community to make ecological improvements whenever possible is clear. Sustainability, and therefore the sustainability of energy, has become a philosophy of life, and each individual is educated to support the requirements to live a healthy and quality life, and to be aware that generations to come have the right to have such a life as well. Sustainability is not easy to achieve – it has to be a part of laws and a generally accepted human value. Energy sustainability is no longer desired, it is a must. Of course, most developed countries have already achieved this level of sustainability, but the situation is much different in underdeveloped countries [80].

2.4 DIMENSIONS OF ENERGY SUSTAINABILITY

In order to more easily accept energy sustainability as a business philosophy, it is first necessary to determine what is desirable in terms of environmental protection and what should be strived for so as to direct the activities in that way. Defining the concept and the essence of energy sustainability is a particular challenge because the list of ecologically suitable or unsuitable activities or phenomena can never be final.

Energy sustainability, especially in developed countries, has been broadly analyzed, so that it can freely be said that energy sustainability can be sought in any phenomenon and activity that takes place in the sphere of production, transmission, and consumption of energy. Energy sustainability, contrary to many qualitative measures, has a kind of temporal, spatial, political economic, and civilization dimension. It is hard to say that the dimensions of energy sustainability are determined once and for all and not subject to changes. Time will inevitably bring many new requirements, but it is still necessary to accurately determine the specific features of these dimensions.

The *time dimension* of energy sustainability implies that the time factor is very important in ecology and includes several methods of observation of the phenomena in the field of energy, which can be described as sustainable or unsustainable. First, an observation is made, as well as the study and proper assessment of events from the distant and recent past, in order to determine their impacts on the current situation and learn useful lessons.

Monitoring and controlling all current phenomena relies directly on the past and most directly on the future. Naturally, energy sustainability means conducting appropriate activities now and understanding how they will affect the state of energy in the future because it includes what is most important in sustainable development – the timely prevention of all ecologically unacceptable events and activities. Future may be limited only partially and for practical reasons, because the whole concept of sustainable development is based in the long term on the principle of intergenerational justice.

The *spatial dimension* is based on the generally accepted fact that energy sustainability can and must be observed in any space. Each individual can and should find sustainable and environmental-friendly behaviors and actions in their immediate vicinity and according to that initiate their wishes and requirements. Ecological eligibility is observed, found, and presented at the level of settlements, areas, countries, regions, and so on. Of special importance is the understanding that energy sustainability should be viewed primarily from the global point of view.

The *political dimension* has become particularly important in the past two decades. Numerous changes in the political arena and misunderstandings and conflicts are associated with the growing insecurity in the field of energy security of countries and regions. All developed countries in the world estimate sustainability of energy, both in their own country and the region. Because of that, intensification of political interference can be expected in determining energy sustainability and all it implies in the near or distant future. Respecting the rules of sustainable development has become one of the prerequisites of integration in world trends.

The *economic dimension* of energy sustainability is of particular importance, given that successful economic development and stability are of immediate relevance for each individual and are closely linked with the capabilities to meet the needs for energy. Today, businesses face great challenges as well as opportunities to respond to consumer demands that send clear messages that they do not want to support any unsustainable energy or environmentally harmful technologies and products. In most developed countries there is no special requirement for educating citizens in that direction, given that the area is, according to public pressure, fully embedded in the institutions of the system.

The laws are strict, efficient, and apply equally to all. Business processes are reviewed in developed countries even before they have started to determine their ecological suitability. Substantial financial resources are allocated for that purpose. Companies that plan to operate in the future have to accept such rules, not only because of legal regulations, but even more because of the demands of consumers.

The *civilization dimension* of sustainability is a kind of summary of all the above and includes much more, so it is particularly difficult to define and almost impossible to restrict. Thus, energy sustainability must be considered together with the sustainable development of all other activities. Modern man has a right to live in a proper living environment with water, air, soil, flora and fauna. However, most importantly, the human environment includes other people and their relationships. Physical and chemical parameters are relatively easy to determine, and it can easily be identified what is ecologically suitable and what is not and then make decisions as to what is ecologically suitable.

Inseparability of human nature and the inevitable development of human relations are the two main factors that determine sustainability in an entirely new way, which means that the actions of each person must be evaluated in terms of ecological suitability. Man is in constant interaction with nature, with other people, and with himself. Through thousands of years of history of humankind, experiences have been accumulated along with achievements that have become an integral part of every human being.

Human problems, misunderstandings, wars, and intolerance have strongly influenced the development of the consciousness of modern man. A civilized man should not base his existence on the disappearance and destruction of his environment to meet his own, often short and limited objectives. It is unrealistic to expect that man would give up a certain quality of life for ecology, but with a little more effort and cooperation, energy development in harmony with nature can be achieved, which will have only long-term positive effects on the mental and physical health of each individual.

2.5 BASIC CONCEPT OF ENERGY SUSTAINABILITY

The international community as a whole is a special challenge. Differences in the degree of economic development, political system, culture, religion, and customs among different countries and regions have led to different ideas about energy sustainability. Therefore, in order to study these issues it is necessary to briefly examine the historical development in certain parts of the world, given that the events in the past directly influence the current level of development of environmental awareness, and therefore the understanding of energy sustainability.

Today, the most developed countries have the most elaborate system regarding managing the quality of the environment, as a direct response to strong public pressure. Thanks to a high degree of economic development, advanced education systems, and the achieved level of responsibility and political maturity of society, consumers openly express their demands, seek answers, and often opt for environmentally friendly technologies, products, or services. Legislation in the field of ecology is developed and functional, so it can be said that the systems in developed countries almost entirely support the concept of sustainable development. The European Union countries, Scandinavian countries, United States, Canada, and Japan are leading in terms of pointing out the importance of ecological suitability in all areas of life and work. Energy sustainability in these countries includes economical use of energy resources, energy-efficient operation, production of energy-efficient products, and use of renewable energy sources.

On the other hand, there are a number of countries at the medium level of economic development, concentrated in southern and eastern Europe, that lag behind in terms of attention to ecology in general. The history of the development of these countries has not been favorable for the development of a healthy economic system, so ecology has been pushed into the background. Thanks to broad education, citizens are familiar with environmental problems, they support the concept of sustainable development, and from time to time highlight their demands, but generally poor conditions in the environment do not create a favorable ground for advancement in the field of ecology in underdeveloped countries. A society that is burdened with many problems is not able to fully address the problem of ecology and sustainable energy development, but companies need to be aware that their future survival, and above all possibilities of exports, depends on business operations that are in line with global environmental standards.

The largest group of countries consists of non-developed and underdeveloped countries at a very low level of economic and social development. They are the victims of historical events and not strong enough to successfully handle their own existential problems, let alone pass appropriate laws and educate their citizens in the direction of ecological and energy sustainability. It is a paradox that these countries were examples of untouched nature and are rich in flora and fauna, as well as energy and mineral resources, where man lived on, in our terms, a low level of development. Today, these are the countries whose resources are maximally exploited with little or no environmental protection. The industrial plants of the so-called dirty technology and huge amounts of waste are moved to third-world countries. Inhumane working conditions, exploitation of child labor, poor health care, and the unavailability of broader education are just some of the indicators of the quality of life.

Sustainable energy development is the need to harmonize the energy development of humankind in accordance with the possibilities – natural energy resources. Regardless of the comprehensiveness of the problems and diverse situations in the world today and that are expected in the future, several basic concepts of sustainable development have been defined, as follows:

- The concept of non-exhaustible energy resources includes conservation of both natural and gained total natural energy resources;
- The concept of non-exhaustible natural energy resources results from the previous concept, and implies that natural energy resources are held constant, considered as a whole and per capita, and should be especially considered because of constant population growth;
- The concept of elasticity, which explains the reduction of specific natural energy wealth, is necessary and inevitable, and occurs as a logical consequence of population growth and growth of their demands.

It should be emphasized that these concepts were generated over time after the adoption of Agenda 21 at the UN conference in Rio de Janeiro in 1992. But certain assumptions about the basic concept of sustainable development existed even before, starting when the energy crisis of the 1970s initiated the change in the approach and the mode of exploitation of nature by man.

In the very beginning, there was an assumption, actually a requirement, to fully preserve energy resources as they were found. Consequently, this meant the cessation of further exploitation of energy resources and their complete conservation. In certain countries, and in some cases, the idea of total cessation of exploitation of energy resources was implemented, but it was sporadic and negligible. Given the rapid population growth and growing energy needs, the concept of non-exhausting and acquired natural wealth was abandoned.

Taking into account the shortcomings of the concept of non-exhausting and acquired natural resources, the concept of non-exhausting wealth was adopted and took into consideration the number of inhabitants on the planet. In fact, this concept implicitly implied that each inhabitant of the planet should be given a certain amount of energy wealth. This concept tried to eliminate the problem of the place of birth of the inhabitant. Actually, this concept implies that all natural resources belong to all, regardless of their place of birth or residence. Regardless of the civilization progress, the concept was defined in such a way that could not be sustained in the long term, primarily because of the uneven distribution of energy resources on the planet and the lack of intention to introduce any changes in the field of energy production and consumption. Energy resources have already become an important element of world trade and the way of acquiring economic and social power.

Currently the most acceptable concept is the one of elasticity, which takes into account the basic shortcomings of the two previously developed and abandoned concepts of sustainable energy development. The concept takes into consideration the global condition on the planet and above all the increase in the number of inhabitants and the need for a certain quality of life that the inhabitants are not willing to abandon. But there is a need to revise the current power management mode, both globally and locally. Changes in the traditional way of thinking, doing business, and dealing with energy at all levels are needed.

In short, the concept of sustainable energy development, i.e., its basic principle from which all other principles are built, is based on the principle of intergenerational justice, which includes meeting the energy needs of present generations without compromising the ability to meet the power needs of future generations. In other words, the current generation has an obligation to leave to the next one the energy situation it had itself.

The basic concept of sustainable energy development can be developed based on different criteria, but it can be generally summarized as follows:

- The principle of conservation of the existing non-renewable resources, where, depending on the type, quantity, and quality of available resources as well as the needs and opportunities in specific areas, the most acceptable way of conservation of existing resources is considered, or permanent reduction of the intensity of their exploitation;
- The principle of exploitation of renewable energy resources, which to some extent replaces the use of non-renewable resources and allows the use of energy in areas that are insufficiently rich in traditional energy resources, or in hardly accessible areas;
- The principle of energy efficiency, which implies an efficient and economical utilization of energy in all phases of its existence, from the energy that is accumulated in the resources through efficient production, distribution, and consumption, to promotion and support, to production of goods that use less energy than the same or similar goods;
- The principle of intergenerational justice, which incorporates the rule of energy management in all plans of energy development that would enable future generations to meet their own needs for energy;

- The principle of harmonization of economic development and energy consumption, which determines the development discussed above, is an especially sensitive principle whose implementation is linked with many problems, because it suggests the need to change traditional ways of thinking. In fact, economic growth was seen as the only measure of the progress of each country, and only indicators of economic growth were significant and available in determining the situation in each country and its position in the international community. The need for sustainable energy development imposes the acceptance of the new principle, by which it is necessary to cease economic growth if it leads to excessive depletion of energy and other resources;
- The principle of paying damages caused by excessive and inappropriate use of energy resources, which creates the possibility that all countries in the world and all business entities understand the importance of adequate energy management, and will take on obligations resulting from that and responsibilities if they do not fulfill these obligations;
- The principle of measurability is derived from the need for sustainable strategic management, which is possible only if during the process of planning, implementation, and evaluation we deal exclusively with goals and data that are strictly and exactly measurable. Arbitrary and universal definitions, plans, and objectives do not provide a true picture and ensure proper application of sustainable energy management tools that are available; and
- The principle of promotion and education, which imposes the need for continuous promotion of sustainable development of energy. Only with thorough education that begins at an early age can conditions be created for the long-term responsible treatment of energy and creating a future that is not limited.

These basic principles of energy sustainability highlight the basic framework necessary for planning and implementation of sustainable energy management and present a framework within which energy management can be implemented. Only through the respect of the basic principles can energy stability and long-term energy sustainability be provided.

2.6 BASIC PROBLEMS OF FUTURE ENERGY DEVELOPMENT

The future of the planet carries a number of assumptions about future environmental problems which the current generation faces and which future generations will face, too. Therefore, the main challenges to the concept of sustainable energy development are defined globally, as follows [66]:

- Demographic growth is one of the factors against which it is not possible to act globally with any measure. There are examples of countries that seek to encourage or restrict population growth by application of certain measures, but these are examples that cannot have a global impact. From the very beginning, the world's population has been steadily and progressively increasing. It would not take long to double the number of inhabitants on the planet. At the beginning of the 21st century, it took almost a thousand years to double the population, while current estimates indicate that the number of inhabitants on the planet will double in 50 years. Realistic estimates suggest that by 2025 about 8.4 billion people will be living on the planet. The significant part of the current population suffers from hunger, thirst, poverty, and unemployment. Demographic problems are characterized by the following:

- Global overpopulation, because obviously more people live on the planet than is optimal, and prospects in terms of that are not encouraging because the number of people in the future will increase;
- Uneven population density, as a characteristic of the modern world, which creates many problems. Globally observed, the population on the planet is very unevenly distributed as a result of development of transport and migration of population in search of better living conditions, as well as the consequences of uncontrolled high birth rates. There are areas in which extremely large numbers of people live in difficult economic and health conditions (certain regions of Africa, Latin America, and Southeast Asia). In these regions the majority of the population is subject to minimum living conditions. On the other hand, there are developed countries where more people live above normal living conditions. Life in such large agglomerates (cities, metropolises) offers its residents much better living conditions, but it creates a number of ecological, sociological, psychological, and other problems whose consequences are felt by all residents, as well as services that are obliged to resolve such problems. Despite the fact that the planet is overpopulated globally, there are entire regions inhabited by fewer people than the resources of that region can withstand (parts of Canada, Australia, and Russia). In each country uneven population density is a problem that creates a series of environmental problems that will continue into the future;

Uneven accessibility to vital resources is associated with uneven distribution of resources on the planet, as well as with population density. In some regions people have at their disposal an abundance of arable land, water, and forests, while other regions are ruled by catastrophic drought and famine due to lack of water and food (arable land). This problem is very difficult to solve, for the simple reason that resources cannot be relocated and made available to people in need;

Existential problems of world population are another characteristic of modern humankind and affect most of the world's population. It is believed that only one sixth of the total number of people on the planet has satisfactory living conditions (dwelling place, sufficient food and access to water, employment, education, and medical treatment). Poverty, unemployment, and hunger are problems of sustainable development that pose great challenges for the world and the ways to solve these problems are uncertain;

The fact that most conventional energy resources are not renewable emphasizes the global problem. Conventional energy resources (fossil fuels, coal) are extremely limited and their use is intensive. Almost all industry in today's world is based on the use of energy that comes from these resources, and they will inevitably cease to be exploited one day. Most experts agree that reserves will suffice until 2020, but after 2050 huge problems may appear in this field. Only rational energy management and prevention of any huge spending can make the situation easier to handle;

Renewability of only small energy resources as an alternative to the traditional sources implies that there are resources (wood, energy, and water – if water is used rationally and there is no major drought) that can be renewed. This type of energy supply is limited and meets only minor needs of the population. In order to preserve these resources responsible actions of authorities and society as a whole is required in terms of their conservation and restoration;

Limited use of alternative energy resources as a way to meet specific needs. Wind energy, solar, biomass energy, and the like are used in a limited way because these ways of obtaining energy are still rather expensive. It is realistic to expect increased efforts in the future to use alternative energy resources more and develop solutions that are appropriate for more users;

Uneven distribution of natural energy resources is the problem civilization has always been faced with, and the first civilizations were created in areas generally rich in resources. With the increased needs for energy that emerged in the period of rapid industrialization, most developed countries became forced to purchase large amounts of energy in the market, i.e., from imports. Identification of the position of certain energy supplies leads to gradual and then ever increasing polarization among the countries that have large amounts of energy resources and the countries dependent on them. With the emergence and intensive use of alternative energy sources, this problem can be alleviated to some extent, but it certainly cannot be completely removed. Uneven distribution of energy resources is one of the important factors of general instability, and the inability to solve this problem imposes the need to develop more advanced forms of global cooperation;

Concentration of large energy consumers, as another aspect of global energy imbalance, directs the flow of energy and the problems associated with it. Namely, as in most human activities, in the area of economic growth, which is closely related to the use of energy, there is a relatively sharp division between countries that consume the biggest quantities of the world energy supplies and the countries whose consumption is average or even symbolic. Strong economic development in certain parts of the world (Europe, United States, Canada, Japan, China, and Russia) created the need for directing the largest amount of energy in the direction of these countries and regions;

Energy as a subject of world trade, a phenomenon caused by the energy dependence, or the inability to achieve economic growth without energy exploitation in various forms. Energy resources that are concentrated in certain parts of the world became an extremely complex and important subject of world trade during the last decades of the twentieth century. Some countries are very dependent on the import of energy, while for some countries energy is the main export product. This arrangement opens up opportunities for a number of discrepancies, inconsistencies and conditioning.

The need for uniform legislation and harmonious development is of particular importance for sustainable energy development since the energy problem has long crossed local borders and become a problem of general, global interest. In some countries of the world there are numerous legal regulations and measures that promote sustainable development in general. In a great part of the world rules have been applied that encourage sustainable energy development, but in this respect there is no general consensus of all countries, rather there are great differences between individual regions, as well as a wide gap between rich and poor countries.

The problems stated above are the most obvious challenges of survival in the future and are therefore subjects of great interest. Given that the concept of sustainable energy development implies the involvement of all organizations and individuals, it is essential that everyone, depending on the activity, redirects their activities so as to contribute to reducing the intensity of these problems.

2.7 LEGISLATION

Global sustainable development strategy is based on the application of a certain number of documents, both binding and non-binding. Following the selection of sustainable growth as the only development path (UN Conference in Rio de Janeiro, 1992), a number of regulations have been passed in this area, particularly in the most developed countries in the world, which are also the biggest polluters.

Energy management related resolutions are of a special global interest since energy production and consumption in the world brings about numerous problems, conflicts, and economic crises. Exaggerated use of energy substances leads to energy resources exhaustion and pollution of the atmosphere, which has been the main reason for adopting the Kyoto Convention, Kyoto Protocol, and a number of directives mandatory for the European Union member states [46].

The Kyoto Protocol, together with the UN Framework Convention on Climate Change, is an international agreement adopted in December 1997 in Kyoto, Japan. The main difference between the protocol and the convention is that the convention urges industrially developed countries to reduce emission of greenhouse gases, while the protocol makes it mandatory. The European Union is highly dependent on imported energy and, at the same time, a huge emitter of harmful gases, and has taken the following measures aimed at finding resolution to the issue:

- Introduction of energy saving and rational use by application of energy efficiency measures, achieving a total savings of the final energy up by 20-40% through transport, distribution, and consumption;
- Energy saving and rational use measures by application of energy efficiency measures in building, community energy, industrial energy, transport, and other places;
- Replacing fossil fuels with renewable energy sources, achieving lower levels of import dependence, decreased emissions, and environmental protection, including new vacancies.

The Kyoto Protocol came into force on February 16, 2005. By the end of 2008, 183 states ratified the protocol. Its main goal is reduction of global anthropogenic emission of gases with the greenhouse effect by at least 5% in comparison with the reference year of 1990, in the course of the first mandatory period between 2008 and 2012. Annex A of the Protocol requires reduction of six gases with the greenhouse effect (GHG): carbon dioxide (CO_2), methane (CH4), nitrous oxide (N2O), and industrial gases, such as hydrofluorocarbon (HFC), Perfluoroisobutene (PFC), and the sulfur hexafluoride (SF6) group.

Annex B contains a list of state signatories and quantified responsibilities pertaining to decreased emission of GHG in percentages in comparison with the reference year. In other words, quantified responsibility of reduced emission of GHG has been envisioned for 38 industrially developed countries, including 11 central and east European countries with transitional economies. In addition to this, the protocol allows states to decide by themselves which of the listed six gases will be included in their national strategy for reduction of gases emission.

It is important to note that the industrialized states with quantified responsibilities listed in Annex B of the Kyoto Protocol are actually developed countries. State signatories that are not in the Annex list do not have the responsibility of a quantified decrease of GHG emission within the Kyoto Protocol. However, they have to fulfill general liabilities envisioned by the convention and the protocol. Other significant directives pertaining to the energy policy of the European Union are as follows.

Directive 2001/77/EC

Directive on promotion of electricity from renewable energy sources in international energy market

Promotion of renewable energy sources (RES) implies implementation of a number of activities aiming at long-term reinforcing and maintaining of a certain level of energy production from renewable sources, which is carried out by passing adequate development strategies in every country, identifying national goals, and creating a support system and origin guarantees – green certificate, short and simple administrative procedures, defining conditions and tariffs to join the network, etc.

Directive 2009/28/EC

Directive on promotion of the use of biofuels and other renewable fuels for transport

This directive refers to increased use of biofuels in the market by 10% of the total fuel quantity for transport by the end of 2010.

Directive 2001/80/EC

Directive on emission limits in the air from huge combustion systems

This directive refers to huge fueling systems of thermal power higher than 50 MW, for which it stipulates strict emission limits in the air for new and existing power plants that are to be achieved by application of the best available technologies. Its full implementation is supposed to be completed by December 2017 due to the significant investment required from the energy sector in order to maintain the standards stipulated by this directive.

Directive 1999/32/EC

Directive on reduction of sulfur content in liquid fuels

Implementation of this directive is envisioned to be completed by December 31, 2011 by the Annex to the Energy Community Treaty and pertains to the reduction of emissions and setting limits to sulfur dioxide. In the period of 2000-2010, the total emission of sulfur dioxide in the European Union member states went down by 60%, thanks to application of directives on huge systems [32].

Directive 96/61/EC

Integrated Pollution Prevention and Control (IPPC Directive)

In essence, the IPPC directive is about achievement of integrated prevention and supervision of pollution caused by different activities (comprising the whole energy sector). The IPPC directive stipulates measures for the prevention or reduction of emission in water, air, and land, including measures pertaining to waste, aimed at reaching a high level of environmental protection.

This Directive is highly important for the energy sector since thermal energy plants with heating input above 50 MW and oil, mineral oil, and gas refineries need to obtain an integrated permit for the future work of such systems. The year 2015 is the deadline for obtaining the integrated permit issued by authorities of existing plants and activities. The permit stipulates conditions guaranteeing a high level of environmental protection as a whole.

2.8 CASE STUDY – EUROPEAN UNION

Energy policy monitoring and control in European Union

Abstract

The main objective of this study is determination of energy-economy degree, as part of energy policy monitoring, using two methods — *Energy Intensity*, a well-known indicator, and *Index of Energy Intensity Costs*, suggested by the authors. The results show significant differences in the final level of energy-economy degree determined by using energy consumption expressed in both physical and monetary values. Monitoring of energy-economy degree in the EU27 region, by using two chosen indicators, shows opposite final results. Measuring of Energy Intensity, the energy-economy degree in the EU27 region showed positive trends. Measuring of *Index of Energy Costs Intensity*, the energy-economy degree in the EU27 region, showed negative trends. Finally, this study determines strengths, weaknesses, and the differences between these indices. Both indicators of energy-economy degree are useful tools, but the authors suggest their use for different purposes. *Energy Intensity* is more suitable for strategic planning purposes, while *Energy Costs Intensity* is an indicator more reliable for determination of economic-related monitoring.

Keywords: monitoring; energy intensity; energy costs intensity

1 INTRODUCTION

Monitoring of the energy-economy balance became one of the crucial problems in the area of sustainable development monitoring and measurement. GDP, as a traditional economic indicator of development, requires changes and adjustments in compliance with the European Union (EU).

Energy consumption is one of the most important factors that determines and limits possibilities for economic development. Even in the case of slow economic development caused by a global crisis, energy consumption continues to grow. Limited energy resources, energy disruptions, changes in energy markets, problems with supply, and all other problems related to energy security make EU27 a region of high energy dependence. The EU can provide only about 50% of needed energy from its own resources. RES - renewable energy sources related investments are also under consideration.

The EU Sustainable Development Strategy (EU SDS) as well as other related documents, underlines sustainable energy-economic growth as one of the strategic pillars of the EU's future. Strategic plans for EU development determine the main objectives and activities needed in order to achieve indicated goals. Strategy implementation can be adequate only if a precise system of control, as a separate strategic field, is used. Monitoring of sustainable EU development is regulated by numerous documents that cover different fields of development.

The EU Energy Roadmap 2050 defines, as one of its priorities, implementation of all activities that lead to energy security in the EU region. The proposal provides solutions for monitoring and measurements of progress in this area, and it is in compliance with the basic EU Strategy on Sustainable and Inclusive Growth and the Kyoto Protocol.

The EU Strategy 2020 contains three basic management stages: planning, realization, and control. The control stage monitors and measures the level of achieved results. Monitoring is performed by using indicators (Eurostat – The European Statistical System):

- Greenhouse emissions, base year 1990;
- Share of renewables in gross final energy consumption; and
- Energy intensity of the economy (proxy indicator for energy savings, under development).

Efficient monitoring of EU sustainable development demands determination and use of efficient indicators. The characteristics of efficient indicators of sustainability are:

- Relevance in relation to aim: they show important characteristics of a monitored subsystem;
- Comprehensibility: they are comprehensible to the public, not only to the experts of the monitored areas;
- Reliability: the pieces of information incorporated in the indicator are accurate; and
- Availability of the data: the data are adjusted to the national statistical system for processing the data and information.

The area of energy intensity is still under development. The traditional index of *Energy Intensity* shows the quantity of energy required to perform a particular activity (service), expressed as energy per unit of output, in this case, per unit of GDP. This indicator has several very important characteristics, which make it suitable, reliable, and comparable. First of all, energy intensity is expressed in physical values and shows total energy consumption. On the basis of this, each country can calculate its own needs and, according to desirable economic development and residential trends, can plan future supply and storage. This main characteristic emphasizes the fact that this indicator should be used mainly for decision making and strategic planning purposes. In this group of activities, this indicator shows all characteristics of efficient indicators indicated above.

The negative characteristic of *Energy Intensity*, as an indicator, is that it is less suitable for economic-related analysis. This indicator does not cover costs derived from carbon emissions, as external costs, which is a direct consequence of energy exploitation. Besides this, *Energy Intensity* does not value the importance of energy prices. The price of energy can play a very important role in energy policy and development as a whole. In a region such as the EU, where each country determines energy prices, the influence of energy price can be significant.

On the other hand, the *Energy Intensity* indicator is less suitable for economic purposes. *Energy Intensity* is an indicator expressed in physical values, which does not include main economic instruments, but is primarily energy prices. Energy consumption, expressed in monetary values, is more suitable for economic purposes and for EU monitoring linked to the economy. The third group of indicators, defined by Eurostat, emphasizes the need to measure the energy-economy degree in an economic way – in monetary values. From this point of view, *Energy Intensity* does not satisfy requirements determined as essential for efficient indicators.

The *Index of Energy Intensity Costs*, suggested by the authors as an energy-related indicator of economic development, is expressed in monetary values. Both scientific community and decision makers agree that the GDP has to be modified. It shows the amount of energy-related costs spent for creation of the GDP. This index emphasizes the fact that the GDP must be decreased by the value of these costs, because they are already involved in the GDP through energy incorporated in goods and services. Total energy-related costs include costs of energy consumption and costs of carbon emissions. Ninety-six percent of carbon emission is caused by energy exploitation, so its costs should be included.

As indicator developed for determination of indicator called "Energy intensity of the economy" (Eurostat), the Index of Energy Intensity Costs shows characteristic described for efficient indicators. It is "relevance in relation to aim: they show important characteristics of a monitored subsystem," because its main objective is determination of energy costs of economic development, expressed in monetary values, which is a basic value in an economy. Indicators in this area should determine and monitor the level of energy-economy degree according to different scenarios [36].

2 METHODOLOGY OF THE RESEARCH

The main objective of the research:

- Comparative analysis of the energy-economy degree determined by using two different indicators of energy intensity.

The secondary objectives of the research:

- Determination of energy intensity in EU27 region;
- Determination of energy costs intensity in EU27 region;
- Determination of energy-economy degree in EU27 region using *Energy Intensity* indicator; and
- Determination of energy-economy degree in EU27 region using *Energy Costs Intensity* indicator.

Research sample:
Research covers data for EU27 region

Time of research:

Research uses data related to the period between 2000 (first year) and 2010 (last year)

Input parameters:

• Research is defined as a procedure of energy-economy degree determination by using two indicators. After calculation of indicated degree, both obtained results will be compared in order to define similarities and differences. In order to provide comparability, the GDP has been used as unit of measurement.

2.1 Determination of Energy Intensity

Energy Intensity, as a standard indicator for determination of the energy-economy degree, is determined using existing data.

Total Energy Consumption – data available on European Environment Agency, Report, 2000 – 2011.

GDP – data available on Eurostat.

Final *Energy Intensity* is determined as:

$$EI = \frac{EC}{GDP}$$

where:

EI – Energy Intensity

EC – Energy Consumption

GDP – Gross Domestic Product

2.2 Determination of Energy Costs Intensity

Energy Costs Intensity includes all costs related to energy exploitation, including both direct (costs of consumed energy) and indirect (external, carbon emission) costs. It is calculated as:

$$ECI = \frac{GPD}{GDP - (ECC + CEC)}$$

where:

ECI – Energy Costs Intensity

ECC – Energy Consumption Costs

CEC – Carbon Emission Costs

GDP – Gross Domestic Product

The *Index of Energy Costs Intensity* considers the fact that energy-related costs are costs that have a negative influence on national wealth. The GDP, as a measure of economic development, must be reconsidered and minimized for the value of energy-related costs that were involved in its creation.

Carbon emissions, caused mainly by energy exploitation, should be considered carefully, as well as other emissions and damages.

Total Energy Consumption is expressed in millions of tons of oil equivalent. This value is converted into KWh units and multiplied by energy prices related to separate countries in the indicated year (European Environment Agency, Reports, 2000-2011).

Carbon Emission Costs will be determined by multiplying the volume by 30 EUR per ton of carbon [2].

3 RESULTS OF THE RESEARCH

3.1 Input data

The EU27 region is characterized by a significant level of economic development, measured by all traditional economic indicators, including the GDP. The level of the GDP in the indicated region is presented in Table 2.1.

The EU27 region is a region with a significant level of GDP in total, but differences between countries are substantial. In the first year of research, the highest levels of GDP were recorded in Germany, France, Italy, and the UK - the indicated four countries together cover about half of the total EU GDP value. The same situation is clear throughout the ten-year-long period.

Some countries recorded the significant changes in GDP. These countries are eastern countries that are new members of the EU region – Estonia, Latvia, and Lithuania. Significant economic development was recorded in Slovenia, Poland, and the Czech Republic. This data is useful for further research, because countries with such intensive economic development most likely will record similar higher levels of energy consumption and air pollution. Data related to total energy consumption (in Mtoe and KWh) are shown in Tables 2.2 and 2.3.

In order to provide comparability between indicators, total energy consumption is given in KWh. The results are shown in Table 3.

The period between 2000 and 2010 in the EU27 region is characterized by significant energy consumption. Of course, the four most developed countries consume half of all energy in the region. However, the level of energy consumption varies during the ten-year period. The four largest countries increased their energy consumption, but not by more than 1%. A similar situation can be seen in Bulgaria, The Netherlands, Slovakia, and the Czech Republic, regardless of the fact that these countries, in the same period of time, recorded a higher GDP. A significantly higher level of energy consumption was seen in Greece, Lithuania, and Latvia. In these cases, economic development is in direct connection with energy consumption. The only country where energy consumption in 2010 was lower than in 2000 was Sweden.

In order to obtain monetary values suitable for calculation of the final outcome, all previously determined values have been multiplied by an average energy price in the sample countries. Energy prices in the EU27 region for the considered period are given in Table 2.4.

TABLE 2.1 GDP in EU27 region (2000-2010) [33]

Country	2000	2001	2002	2003	2004	2005	2006	2007	2008	2009	2010
GDP PPP (millions of EUR)											
Austria	153,650	154,450	156,990	158,250	166,270	172,250	178,680	182,570	175,470	178,910	178,910
Belgium	191,360	192,860	195,500	197,030	206,890	212,460	218,670	220,870	214,790	219,470	219,470
Bulgaria	22,130	23,050	24,120	25,450	27,170	28,890	30,770	32,760	34,800	32,880	32,940
Czech Rep	103,670	106,220	108,230	112,130	117,160	124,560	133,040	141,190	144,670	138,670	141,890
Denmark	133,260	134,190	134,820	135,340	138,450	141,830	146,650	148,970	147,300	139,620	142,540
Germany	2,704,050	2,737,580	2,737,580	2,731,630	2,764,620	2,785,440	2,879,270	2,955,800	2,985,000	2,844,120	2,947,410
Estonia	9,400	10,200	11,010	11,840	12,700	13,890	15,360	16,420	15,590	13,420	13,660
Ireland	154,780	163,600	174,310	182,000	190,370	201,820	212,560	224,520	216,560	200,140	198,060
Greece	136,400	132,450	147,160	189,170	217,690	223,010	236,770	272,010	292,360	275,680	258,700
Spain	961,530	996,600	1,023,550	1,055,250	1,089,720	1,129,110	1,174,480	1,216,460	1,226,920	1,181,250	1,179,550
France	1,083,940	1,103,840	1,114,100	1,124,120	1,152,720	1,173,780	1,202,730	123,0220	1,229,220	1,195,670	1,213,360
Italy	1,699,340	1,730,230	1,738,090	1,737,800	1,764,420	1,775,990	1,812,150	1,839,010	1,814,670	1,719,990	1,742,280
Cyprus	14,520	15,100	15,420	15,710	16,370	17,020	17,300	18,510	19,170	18,980	19,180
Latvia	10,820	11,690	12,450	13,350	14,500	16,040	18,000	19,800	18,960	15,560	15,500

(Continued)

TABLE 2.1 Continued

Country	2000	2001	2002	2003	2004	2005	2006	2007	2008	2009	2010
GDP PPP (Millions of EUR)											
Lithuania	17,850	19,050	20,360	22,450	24,100	25,980	28,010	30,770	31,670	27,000	27,360
Luxembourg	31,540	32,340	33,660	34,180	35,690	37,630	39,500	42,120	42,730	41,170	42,620
Hungary	89,360	93,030	97,120	101,300	106,060	110,200	111,030	111,890	112,810	105,260	106,500
Malta	5,640	5,550	5,700	5,680	5,730	5,960	6,180	6,400	6,570	6,430	6,630
Netherlands	597,390	608,890	609,360	611,400	625,080	637,870	659,520	685,380	698,270	670,930	682,790
Poland	261,100	264,240	268,060	278,420	293,300	303,910	322,840	344,740	362,420	368,400	382,460
Portugal	183,470	187,080	188,410	186,650	189,560	191,000	193,740	198,370	198,350	193,410	195,980
Romania	75,010	79,280	83,330	87,660	95,020	98,990	106,810	113,210	123,890	113,360	114,430
Slovenia	29,830	30,680	31,900	32,800	34,210	35,750	37,820	40,390	41,810	38,540	39,000
Slovakia	53,040	54,880	57,400	60,150	63,170	67,380	73,110	80,850	85,840	80,520	80,920
Finland	94,330	96,490	98,240	100,210	104,330	107,370	112,110	118,090	119,180	109,400	112,820
Sweden	324,510	328,610	336,770	344,630	359,230	370,580	386,510	399,320	396,870	375,700	396,500
UK	2,015,620	2,065,230	2,108,540	2,167,750	2,231,730	2,280,220	2,343,790	2,406,720	2,405,150	2,287,900	2,316,520
TOTAL	1,115,7540	1,137,7410	11,532,180	1,172,2350	12,046,260	12,288,930	12,857,400	13,097,360	13,161,040	12,592,380	12,807,980

Source: World Bank World Development Indicators, International Financial Statistics of the IMF

TABLE 2.2 Total energy consumption in EU27 region (2000-2010) – Mtoe

Country	2000	2001	2002	2003	2004	2005	2006	2007	2008	2009	2010
Total energy consumption (million tons of oil equivalent)											
Austria	23.7	25.0	25.3	26.6	26.9	28.3	27.6	27.4	27.6	26.3	26.1
Belgium	37.4	37.9	36.1	38.3	37.7	36.6	36.1	34.6	37.5	34.5	39.4
Bulgaria	8.8	8.8	8.9	9.5	9.4	9.9	10.4	10.1	9.9	8.6	8.9
Czech Rep	24.7	25.2	24.4	25.7	26.2	26.0	26.4	25.8	25.5	24.4	25.6
Denmark	14.7	15.1	14.6	15.1	15.4	15.5	15.7	15.7	15.5	14.8	15.1
Germany	219.1	222.7	219.2	230.8	230.9	229.6	233.3	215.7	224.2	213.3	222.3
Estonia	2.4	2.7	2.6	2.7	2.8	2.9	2.9	3.1	3.1	2.8	2.9
Ireland	10.7	11.1	11.2	11.5	11.9	12.5	13.2	13.2	13.2	11.8	12.2
Greece	18.6	19.2	19.5	20.5	20.3	20.8	21.4	21.9	21.3	20.5	23.8
Spain	79.4	83.4	84.7	90.0	94.3	97.4	96.1	98.8	95.8	89.0	91.0
France	154.5	161.0	157.6	161.2	162.8	162.3	161.5	158.6	160.7	155.5	158.2
Italy	124.7	126.0	125.5	131.0	132.5	134.4	132.3	129.3	128.3	120.9	128.1
Cyprus	1.6	1.7	1.7	1.8	1.8	1.8	1.8	1.9	2.0	1.9	1.9
Latvia	3.2	2.6	3.6	3.8	3.9	4.0	4.2	4.4	4.2	3.9	3.9
Lithuania	3.7	3.9	4.0	4.1	4.3	4.5	4.8	5.0	4.9	4.4	4.8
Luxem-bourg	3.5	3.7	3.7	3.9	4.4	4.5	4.4	4.3	4.4	4.1	4.2
Hungary	16.1	16.9	16.9	17.6	17.5	18.2	17.9	16.9	17.1	16.4	16.9
Malta	0.4	0.4	0.4	0.5	0.5	0.5	0.4	0.4	0.5	0.4	0.4
Nether-lands	50.5	51.3	51.3	52.0	52.8	52.3	50.9	49.8	51.1	50.4	50.1
Poland	55.6	56.0	54.5	56.0	58.1	58.2	60.8	61.7	62.2	60.9	61.0
Portugal	17.7	18.0	18.4	18.4	18.9	19.0	18.7	19.0	18.5	18.2	18.5
Romania	22.5	22.9	22.9	24.0	24.4	24.7	24.9	24.1	24.6	22.1	23.0
Slovenia	4.4	4.6	4.6	4.7	4.8	4.9	4.9	4.9	5.3	4.7	4.9
Slovakia	11.0	11.5	11.6	11.2	11.1	11.5	11.4	11.2	11.5	10.7	11.1
Finland	23.9	24.3	25.3	26.0	26.4	25.5	26.9	26.7	25.9	24.0	25.0
Sweden	34.9	34.3	34.1	34.0	33.9	33.6	33.2	33.3	32.6	31.6	32.5
UK	152.4	153.3	148.8	150.5	152.3	153.3	151.2	148.7	148.2	137.5	149.0
TOTAL	1010.1	1143.5	1131.4	1171.4	1186.2	1192.7	1193.3	1166.5	1175.6	1113.6	1160.8

Source: EEA – European Environment Agency, 2011.

TABLE 2.3 Total energy consumption in EU27 region (2000-2010) – KWh

Country	2000	2001	2002	2003	2004	2005	2006	2007	2008	2009	2010
Total energy consumption (millions of KWh)											
Austria	275,631	290,750	294,239	309,358	312,847	329,129	320,988	318,662	320,988	305,869	303,543
Belgium	434,962	440,777	419,843	445,429	438,451	425,658	419,843	402,398	436,125	401,235	458,222
Bulgaria	102,344	102,344	103,507	110,485	109,322	115,137	120,952	117,463	115,137	100,018	103,507
Czech Rep	287,261	293,076	283,772	298,891	304,706	302,380	307,032	300,054	296,565	283,772	297,728
Denmark	170,961	175,613	169,798	175,613	179,102	180,265	182,591	182,591	180,265	172,124	175,613
Germany	2,548,133	2,590,001	2,549,296	2,684,204	2,685,367	2,670,248	2,713,279	2,508,591	2,607,446	2,480,679	2,585,349
Estonia	27,912	31,401	30,238	31,401	32,564	33,727	33,727	36,053	36,053	32,564	33,727
Ireland	124,441	129,093	130,256	133,745	138,397	145,375	153,516	153,516	153,516	137,234	141,886
Greece	216,318	223,296	226,785	238,415	236,089	241,904	248,882	254,697	247,719	238,415	276,794
Spain	923,422	969,942	985,061	1,046,700	1,096,709	1,132,762	1,117,643	1,149,044	1,114,154	1,035,070	1,058,330
France	1,796,835	1,872,430	1,832,888	1,874,756	1,893,364	1,887,549	1,878,245	1,844,518	1,868,941	1,808,465	1,839,866
Italy	1,450,261	1,465,380	1,459,565	1,523,530	1,540,975	1,563,072	1,538,649	1,503,759	1,492,129	1,406,067	1,489,803
Cyprus	18,608	19,771	19,771	20,934	20,934	20,934	20,934	22,097	2,023,260	22,097	22,097
Latvia	37,216	30,238	41,868	44,194	45,357	46,520	48,846	51,172	48,846	45,357	45,357
Lithuania	43,031	45,357	46,520	47,683	50,009	52,335	55,824	58,150	56,987	51,172	55,824

(Continued)

TABLE 2.3 Continued

Total energy consumption (millions of KWh)

Country	2000	2001	2002	2003	2004	2005	2006	2007	2008	2009	2010
Luxembourg	40,705	31,401	31,401	45,357	51,172	52,335	51,172	50,009	51,172	47,683	48,846
Hungary	187,243	196,547	196,547	204,688	203,525	211,666	208,177	196,547	198,873	190,732	196,547
Malta	4,652	4,652	4,652	5,815	5,815	5,815	4,652	4,652	5,815	4,652	4,652
Netherlands	587,315	596,619	596,619	604,760	614,064	608,249	591,967	579,174	594,293	586,152	582,663
Poland	646,628	651,280	633,835	651,280	675,703	676,866	707,104	717,571	723,386	708,267	709,430
Portugal	205,851	210,503	213,992	213,992	219,807	220,970	217,481	220,970	215,155	211,666	215,155
Romania	261,675	266,327	266,327	279,120	283,772	287,261	289,587	280,283	286,098	257,023	267,490
Slovenia	51,172	53,498	53,498	54,661	55,824	56,987	56,987	56,987	61,639	54,661	56,987
Slovakia	127,930	133,745	134,908	130,256	129,093	133,745	132,582	130,256	133,745	124,441	129,093
Finland	277,957	282,609	294,239	302,380	302,380	296,565	312,847	310,521	301,217	279,120	290,750
Sweden	405,887	398,909	396,583	395,420	394,257	390,768	386,116	387,279	379,138	367,508	377,975
UK	1,772,412	1,782,879	1,730,544	1,750,315	1,771,249	1,782,879	1,758,456	1,729,381	1,723,566	1,599,125	1,732,870
TOTAL	13,026,763	13,288,438	13,146,552	13,623,382	13,790,854	13,871,101	13,878,079	13,566,395	13,648,970	12,951,168	13,500,104

Author's results

TABLE 2.4 Energy prices in EU27 region (2000–2010)

Country	2000	2001	2002	2003	2004	2005	2006	2007	2008	2009	2010
Energy prices (EUR per KWh – excise tax included)											
Austria	0.0949	0.0945	0.0932	0.0926	0.0981	0.0964	0.0894	0.1050	0.1271	0.1380	0.1427
Belgium	0.1171	0.1184	0.1137	0.1120	0.1145	0.1116	0.1123	0.1229	0.1500	0.1431	0.1449
Bulgaria	0.0402	0.0419	0.0443	0.0466	0.0486	0.0537	0.0552	0.0547	0.0593	0.0685	0.0675
Czech Rep	0.0475	0.0538	0.0642	0.0654	0.0660	0.0729	0.0829	0.0898	0.1060	0.1102	0.1108
Denmark	0.0718	0.0781	0.0865	0.0947	0.0915	0.0927	0.0997	0.1170	0.1203	0.1239	0.1168
Germany	0.1191	0.1220	0.1261	0.1267	0.1259	0.1334	0.1374	0.1433	0.1299	0.1401	0.1381
Estonia	0.0401	0.0422	0.0457	0.0550	0.0550	0.0576	0.0620	0.0635	0.0639	0.0712	0.0695
Ireland	0.0795	0.0795	0.0883	0.1006	0.1055	0.1197	0.1285	0.1465	0.1559	0.1789	0.1589
Greece	0.0564	0.0564	0.0580	0.0606	0.0621	0.0637	0.0643	0.0661	0.0957	0.1055	0.0975
Spain	0.0895	0.0859	0.0859	0.0872	0.0885	0.0900	0.0940	0.1004	0.1124	0.1294	0.1417
France	0.0928	0.0914	0.0923	0.0890	0.0905	0.0905	0.0905	0.0921	0.0914	0.0908	0.0940
Italy	0.1500	0.1567	0.1390	0.1449	0.1434	0.1440	0.1548	0.1658	0.1699	0.1711	0.1752
Cyprus	0.0845	0.0990	0.0845	0.0915	0.0928	0.0915	0.1225	0.1177	0.1528	0.1336	0.1597
Latvia	0.0388	0.0401	0.0499	0.0552	0.0487	0.0702	0.0702	0.0583	0.0802	0.0957	0.0954
Lithuania	0.0409	0.0451	0.0498	0.0512	0.0535	0.0609	0.0609	0.0658	0.0729	0.0799	0.0955

(Continued)

TABLE 2.4 Continued

Country	2000	2001	2002	2003	2004	2005	2006	2007	2008	2009	2010
Energy prices (EUR per KWh – excise tax included)											
Luxembourg	0.1056	0.1120	0.1148	0.1191	0.1215	0.1288	0.1390	0.1509	0.1442	0.1619	0.1433
Hungary	0.0622	0.0634	0.0723	0.0733	0.0794	0.0851	0.0896	0.1019	0.1277	0.1227	0.1349
Malta	0.0609	0.0617	0.0631	0.0652	0.0636	0.0727	0.0904	0.0940	0.0945	0.1627	0.1615
Netherlands	0.0938	0.0978	0.0923	0.0970	0.1031	0.1102	0.1207	0.1400	0.1270	0.1440	0.1266
Poland	0.0688	0.0710	0.0818	0.0775	0.0699	0.0823	0.0923	0.0945	0.0965	0.0883	0.1049
Portugal	0.1194	0.1200	0.1223	0.1257	0.1283	0.1313	0.1340	0.1420	0.1074	0.1264	0.1093
Romania	0.0504	0.0552	0.0594	0.0633	0.0648	0.0655	0.0792	0.0855	0.0885	0.0814	0.0856
Slovenia	0.0830	0.0837	0.0858	0.0833	0.0841	0.0861	0.0874	0.0887	0.0911	0.1056	0.1057
Slovakia	0.1002	0.1007	0.1008	0.1012	0.1024	0.1123	0.1216	0.1292	0.1148	0.1294	0.1277
Finland	0.0645	0.0637	0.0697	0.0738	0.0810	0.0792	0.0809	0.0877	0.0915	0.0974	0.0998
Sweden	0.0637	0.0629	0.0701	0.0838	0.0898	0.0846	0.0876	0.1088	0.1085	0.1040	0.1195
UK	0.1056	0.0996	0.1031	0.0959	0.0837	0.0836	0.0971	0.1254	0.1394	0.1399	0.1321

Source: EEA – European Environment Agency, 2011.

In general, prices of energy were higher in 2010 than in 2000. The average price of energy was 0.1056 EUR/KWh in 2000 and 0.1321 EUR/KWh in 2010. The highest prices in the EU27 region are in three island countries – Ireland, Cyprus, Malta, and Italy. Energy prices are the lowest in eastern EU countries – Bulgaria, Greece, Latvia, Lithuania — and Finland (see Table 2.5) [3].

The indicated results lead to the conclusion that monetary values of energy consumption depend on energy prices in each individual country. Countries that recorded higher levels of energy consumption expressed in physical values (Latvia and Lithuania) recorded lower energy consumption when expressed in monetary values. Significant differences have been recorded in the case of Italy, the UK, and France.

The standard methodology for calculation of *Energy Intensity* does not cover pollution caused by energy exploitation as a measure of damage which leads to a reduction of natural wealth. In order to define *Energy Costs Intensity*, it was necessary to determine total carbon emission and related costs. The results are presented in Tables 2.6 and 2.7.

The EU27 region marks significant changes in the level of carbon emissions among the countries and during the ten years studied. In total, emission of carbon declined, and in 2010 the level of carbon emission was lower than in 2000. The highest levels of emission were recorded in 2005 and 2006. Also, the highest emitters are countries with the highest GDP.

There is no regularity between energy consumption and related carbon emissions. In most EU27 countries, the level of carbon emission declined and in 2010 it was lower than in 2000. The higher level of carbon emission was noted in Estonia, Spain, Latvia, Lithuania, Netherlands, and Slovenia. These are countries characterized by substantial economic development in the same period of time. Other countries recorded GDP growth and higher levels of energy consumption, but total carbon emissions declined.

Carbon dioxide is a gas whose emission has a certain value, which is an external cost created by energy consumption. Carbon emission costs were calculated according to the current price determined by the European Trading Scheme – 30 EUR per ton of CO_2. The results are presented in Table 2.7.

As expected, higher levels of carbon emission costs were seen in Germany, Italy, the UK, and France. High levels were also noted in Poland and Spain.

On the basis of the determined *Energy Consumption Costs* and *Carbon Emission Costs*, the total costs related to energy consumption were calculated and are presented in Table 2.8.

The total value of energy-related costs indicates that the costs were highest in Germany, France, Italy, and the UK. High levels were also noted in Poland and the Netherlands. Unlike the carbon emission indicator, which declined slowly in all countries, the indicators of values of all energy-related costs in all countries constantly increased. This is surely a result of the inclusion of external costs into the total costs related to energy consumption.

TABLE 2.5 Total energy consumption in EU27 region (2000-2010) – EUR [57]

Country	2000	2001	2002	2003	2004	2005	2006	2007	2008	2009	2010
Total energy consumption (millions of EUR)											
Austria	26,157	27,476	27,423	28,647	30,690	31,728	28,696	33,460	40,798	42,210	43,316
Belgium	50,934	52,188	47,736	49,888	50,203	47,503	47,148	49,455	65,419	57,417	66,396
Bulgaria	4,114	4,288	4,585	5,149	5,313	6,183	6,677	6,425	6,828	6,851	6,987
Czech Rep	13,645	15,767	18,218	19,547	20,111	22,044	25,453	26,945	31,436	31,272	32,988
Denmark	12,275	13,715	14,688	16,631	16,388	16,711	18,204	21,363	21,686	21,326	20,512
Germany	303,483	315,980	321,466	340,089	338,088	356,211	372,805	359,481	338,707	347,543	357,037
Estonia	1,119	1,325	1,382	1,727	1,791	1,943	2,091	2,289	2,304	2,319	2,344
Ireland	9,893	10,263	11,502	13,455	14,601	17,401	19,727	22,490	23,933	24,551	22,546
Greece	12,200	12,594	13,154	14,448	14,661	15,409	16,003	16,835	23,707	25,153	26,987
Spain	82,646	83,318	84,617	91,272	97,059	101,949	105,058	115,364	125,231	133,938	149,965
France	166,746	171,140	169,176	166,853	171,349	170,823	169,981	169,880	170,821	164,209	172,947
Italy	217,539	229,625	202,880	220,760	220,976	225,082	238,183	249,323	2,535,13	240,578	261,014
Cyprus	1,572	1,957	1,671	1,915	1,943	1,915	2,564	2,601	0	2,952	3,529
Latvia	1,444	1,213	2,089	2,440	2,209	3,266	3,429	2,983	3,917	4,341	4,327
Lithuania	1,760	2,046	2,317	2,441	2,675	3,187	3,400	3,826	4,154	4,089	5,331

(Continued)

TABLE 2.5 Continued

Country	2000	2001	2002	2003	2004	2005	2006	2007	2008	2009	2010
Total energy consumption (millions of EUR)											
Luxembourg	4,298	3,517	3,605	5,402	6,217	6,741	7,113	7,546	7,379	7,720	7,000
Hungary	11,647	12,461	14,210	15,004	16,160	18,013	18,653	20,028	25,396	23,403	26,514
Malta	283	287	294	379	370	423	421	437	550	757	751
Netherlands	55,090	58,349	55,068	58,662	63,310	67,029	71,450	81,084	75,475	84,406	73,765
Poland	44,488	46,241	51,848	50,474	47,232	55,706	65,266	67,810	69,807	62,540	74,419
Portugal	24,579	25,260	26,171	26,899	28,201	29,013	29,142	31,378	23,108	26,755	23,516
Romania	13,188	14,701	15,820	17,668	18,388	18,816	22,935	23,964	25,320	20,922	22,897
Slovenia	4,247	4,478	4,590	4,553	4,695	4,907	4,981	5055	5,615	5,772	6,024
Slovakia	12,819	13,468	13,599	13,182	13,219	15,020	16,122	16,829	15,354	16,103	16,485
Finland	17,928	18,002	20,508	22,316	24,493	23,488	25,309	27,233	27,561	27,186	29,017
Sweden	25,855	25,091	27,800	33,136	35,404	33,059	33,824	42,136	41,136	38,221	45,168
UK	187,167	177,575	178,419	167,855	148,254	149,049	170,746	216,864	240,265	223,718	228,912
TOTAL	1,307,118	1,342,327	1,334,834	1,390,791	1,393,999	1,442,618	1,525,381	1,623,087	1,669,420	1,646,249	1,730,695

Author's results

TABLE 2.6 Total carbon emission in EU27 region (2000-2010) [32]

Country	2000	2001	2002	2003	2004	2005	2006	2007	2008	2009	2010
Total CO_2 emissions (million tons)											
Austria	66.0	71.0	72.0	78.1	77.6	79.0	77.6	74.2	71.0	69.9	67.7
Belgium	123.7	124.1	123.0	127.0	126.6	123.4	119.5	114.5	110.1	108.2	104.8
Bulgaria	51.0	50.5	52.1	49.3	53.3	54.0	55.1	58.9	54.4	52.9	50.5
Czech Rep	127.1	128.7	125.0	126.3	127.0	126.4	128.6	129.9	121.9	117.7	116.9
Denmark	52.9	54.7	54.3	59.4	53.9	50.2	58.1	53.2	50.9	48.3	46.0
Germany	883.7	901.3	886.4	889.0	879.9	851.7	867.0	841.2	840.2	799.2	786.7
Estonia	15.6	15.9	15.4	17.2	17.4	16.8	16.3	19.1	18.8	18.5	18.3
Ireland	44.7	47.2	45.8	45.0	45.9	47.6	47.2	47.5	46.1	45.5	43.6
Greece	103.4	105.6	105.3	109.5	109.7	111.0	109.6	113.6	101.8	98.9	97.8
Spain	307.7	311.4	330.1	334.0	351.4	368.0	358.4	366.4	340.5	331.3	329.3
France	406.4	411.9	401.1	409.6	414.0	416.7	407.0	397.1	380.8	379.7	376.9
Italy	462.7	468.4	470.6	486.0	489.0	490.1	485.8	475.3	450.6	466.6	445.1
Cyprus	7.7	7.6	7.5	7.5	7.7	8.0	8.2	8/3	8.5	8.5	8.6
Latvia	7.1	7.5	7.5	7.6	7.7	7.8	8.3	8.6	8.0	7.7	7.6

(Continued)

TABLE 2.6 Continued

Country	2000	2001	2002	2003	2004	2005	2006	2007	2008	2009	2010
Total CO_2 emissions (million tons)											
Lithuania	12.0	12.8	13.0	13.0	13.6	14.4	14.6	15.9	15.5	15.4	15.1
Luxembourg	8.9	9.2	10.3	10.8	12.2	12.3	12.2	11.8	11.0	10.8	10.5
Hungary	58.5	59.9	58.5	61.6	59.9	61.1	59.8	57.8	55.5	54.9	54.6
Malta	2.3	2.4	2.5	2.6	2.6	2.7	2.6	2.7	2.7	2.6	2.6
Netherlands	169.6	175.2	175.7	179.6	181.0	175.8	172.5	127.7	154.5	163.3	173.8
Poland	320.6	317.2	305.7	317.0	317.3	318.2	329.6	328.3	320.2	318.8	316.1
Portugal	63.8	65.1	69.3	64.6	67.0	69.7	65.2	62.8	60.7	58.9	56.3
Romania	95.3	100.3	106.3	111.4	112.2	105.9	111.1	110.9	99.1	87.2	94.7
Slovenia	15.2	16.1	16.3	16.0	16.4	16.7	16.9	17.0	17.1	17.1	17.2
Slovakia	40.3	41.7	40.0	41.4	41.1	40.7	40.0	38.1	38.0	37.9	37.6
Finland	56.7	62.0	64.4	72.0	68.1	56.3	67.7	66.1	60.3	59.1	56.5
Sweden	53.4	54.1	55.1	55.9	55.3	52.9	52.7	51.6	50.9	49.9	49.1
UK	550.0	561.8	544.8	556.2	556.6	554.0	551.8	543.2	533.3	529.4	522.9
TOTAL	4106.3	4183.6	4158.0	4247.6	4264.4	4231.4	4243.4	4141.7	4022.4	3958.2	3906

Source: EEA – European Environment Agency, 2011.

TABLE 2.7 Total carbon emission costs in EU 27 region (2000-2010)

Country	2000	2001	2002	2003	2004	2005	2006	2007	2008	2009	2010
Total CO_2 emissions (millions of EUR)											
Austria	1,980	2,130	2,160	2,343	2,328	2,370	2,328	2,226	2,130	2,097	2,031
Belgium	3,711	3,723	3,690	3,810	3,798	3,702	3,585	3,435	3,303	3,246	3,144
Bulgaria	1,530	1,515	1,563	1,479	1,599	1,620	1,653	1,767	1,632	1,587	1,515
Czech Rep	3,813	3,861	3,750	3,789	3,810	3,792	3,858	3,897	3,657	3,531	3,507
Denmark	1,587	1,641	1,629	1,782	1,617	1,506	1,743	1,596	1,527	1,449	1,380
Germany	26,511	27,039	26,592	26,670	26,397	25,551	26,010	25,236	25,206	23,276	23,601
Estonia	468	477	462	516	522	504	489	573	564	555	549
Ireland	1,341	1,416	1,374	1,350	1,377	1,428	1,416	1,425	1,383	1,365	1,308
Greece	3,102	3,168	3,159	3,285	3,291	3,330	3,288	3,408	3,054	2,967	2,934
Spain	9,231	9,342	9,909	10,020	10,542	11,040	10,752	10,992	10,215	9,939	9,879
France	12,192	12,357	12,033	12,288	12,420	12,501	12,210	11,913	11,424	11,391	11,307
Italy	13,881	14,052	14,118	14,580	14,670	14,703	14,574	14,259	13,518	13,998	13,353
Cyprus	231	228	225	225	231	240	246	249	255	255	258
Latvia	213	225	225	228	231	234	249	258	240	231	228
Lithuania	360	384	390	390	408	432	438	477	465	462	453

TABLE 2.7 Continued

Total CO$_2$ emissions (millions of EUR)

Country	2000	2001	2002	2003	2004	2005	2006	2007	2008	2009	2010
Luxembourg	267	276	309	324	366	369	366	354	330	324	315
Hungary	1,755	1,797	1,755	1,848	1,797	1,833	1,794	1,734	1,665	1,647	1,638
Malta	69	72	75	78	78	81	78	81	81	78	78
Netherlands	5,088	5,256	5,271	5,388	5,430	5,274	5,175	3,831	4,635	4,899	5,214
Poland	9,618	9,516	9,171	9,510	9,519	9,546	9,888	9,849	9,606	9,564	9,483
Portugal	1,914	1,953	2,079	1,938	2,010	2,091	1,956	1,884	1,821	1,767	1,689
Romania	2,859	3,009	3,189	3,342	3,366	3,177	3,333	3,327	2,973	2,616	2,841
Slovenia	456	483	489	480	492	501	507	510	513	513	516
Slovakia	1,209	1,251	1,200	1,242	1,233	1,221	1,200	1,143	1,140	1,137	1,128
Finland	1,701	1,860	1,932	2,160	2,043	1,689	2,031	1,983	1,809	1,773	1,695
Sweden	1,602	1,623	1,653	1,677	1,659	1,587	1,581	1,548	1,527	1,497	1,473
UK	16,500	16,854	16,344	16,686	16,698	16,620	16,554	16,296	15,999	15,882	15,687
TOTAL	123,189	125,508	124,740	127,428	127,932	126,942	127,302	124,251	120,672	118,746	117,204

Author's results [42]

TABLE 2.8 Total energy-related costs (ERC) in EU27 region (2000-2010) [43]

Country	2000	2001	2002	2003	2004	2005	2006	2007	2008	2009	2010
Total ERC = energy consumption costs + carbon emission costs (millions of EUR)											
Austria	28,137	29,606	29,583	30,990	33,018	34,098	31,024	35,686	42,928	44,307	45,347
Belgium	54,645	55,911	51,426	53,698	54,001	51,205	50,733	52,890	68,722	60,663	69,540
Bulgaria	5,644	5,803	6,148	6,628	6,912	7,803	8,330	8,192	8,460	8,438	8,502
Czech Rep	17,458	19,628	21,968	23,336	23,921	25,836	29,311	30,842	35,093	34,803	36,495
Denmark	13,862	15,356	16,317	18,413	18,005	18,217	19,947	22,959	23,213	22,775	21,892
Germany	329,994	343,019	348,058	366,759	364,485	381,762	398,815	384,717	363,913	370,819	380,638
Estonia	1,587	1,802	1,844	2,243	2,313	2,447	2,580	2,862	2,868	2,874	2,893
Ireland	11,234	11,679	12,876	14,805	15,978	18,829	21,143	23,915	25,316	25,916	23,854
Greece	15,302	15,762	16,313	17,733	17,952	18,739	19,291	20,243	26,761	28,120	29,921
Spain	91,877	92,660	94,526	101,292	107,601	112,989	115,810	126,356	135,446	143,877	159,844
France	178,938	183,497	181,209	179,141	183,769	183,324	182,191	181,793	182,245	175,600	184,254
Italy	231,420	243,677	216,998	235,340	235,646	239,785	252,757	263,582	267,031	254,576	274,367
Cyprus	1,803	2,185	1,896	2,140	2,174	2,155	2,810	2,850	255	3,207	3,787
Latvia	1,657	1,438	2,314	2,668	2,440	3,500	3,678	3,241	4,157	4,572	4,555
Lithuania	2,120	2,430	2,707	2,831	3,083	3,619	3,838	4,303	4,619	4,551	5,784

(Continued)

TABLE 2.8 Continued

Country	2000	2001	2002	2003	2004	2005	2006	2007	2008	2009	2010
Total ERC = energy consumption costs + carbon emission costs (millions of EUR)											
Luxembourg	4,565	3,793	3,914	5,726	6,583	7,110	7,479	7,900	7,709	8,044	7,315
Hungary	13,402	14,258	15,965	16,852	17,957	19,846	20,447	21,762	27,061	25,050	2,8152
Malta	352	359	369	457	448	504	499	518	631	835	829
Netherlands	60,178	63,605	60,339	64,050	68,740	72,303	76,625	84,915	80,110	89,305	78,979
Poland	54,106	55,757	61,019	59,984	56,751	65,252	75,154	77,659	79,413	72,104	83,902
Portugal	26,493	27,213	28,250	28,837	30,211	31,104	31,098	33,262	24,929	28,522	25,205
Romania	16,047	17,710	19,009	21,010	21,754	21,993	26,268	27,291	28,293	23,538	25,738
Slovenia	4,703	4,961	5,079	5,033	5,187	5,408	5,488	5,565	6,128	6,285	6,540
Slovakia	14,028	14,719	14,799	14,424	14,452	16,241	17,322	17,972	16,494	17,240	17,613
Finland	19,629	19,862	22,440	24,476	26,536	25,177	27,340	29,216	29,370	28,959	30,712
Sweden	27,457	26,714	29,453	34,813	37,063	34,646	35,405	43,684	42,663	39,718	46,641
UK	203,667	194,429	194,763	184,541	164,952	165,669	187,300	233,160	256,264	239,600	244,599
TOTAL	1,430,305	1,467,833	1,459,582	1,518,220	1,521,932	1,569,561	1,652,683	1,747,335	1,790,092	1,764,298	1,847,898

Author's results

A doubling of total costs related to energy in comparison to the year 2000 is seen in all countries of the region, except in Slovakia and Slovenia, where the change was very small. The only country where a decrease of energy-related costs was registered is Portugal. The four biggest energy consumers (Germany, Italy, France, and the UK) also showed an increase, between 15% and 25%.

3.2 Energy Intensity in the EU27 region

Energy Intensity was determined on the basis of previously defined fundamental parameters, GDP and *Total Energy Consumption*, expressed in KWh. The values of this index for the EU27 region are shown in the Table 2.9.

Generally, the results show a reduction of energy consumption measured in physical values in the EU27 region in total. Only Belgium and Italy noted small increases in energy consumption. The level of energy consumption increased between 2000 and 2003, and then it started to drop. The highest level of energy per GDP unit was recorded in Bulgaria – 3.14 KWh per GDP unit. In seven EU countries, energy consumption per GDP unit was between 2 and 3. Energy consumption between 1 and 2 was present in 12 EU countries. In the remaining seven countries, energy consumption was less than 1 KWh per GDP unit. In Germany, the UK, and Italy, the largest energy consumers in the EU region, less than 1 KWh per GDP unit was seen. The lowest level of energy intensity was in Malta.

The level of energy consumption, measured by *Energy Intensity*, is presented in Figure 2.1.

According to the data, the average level of energy-economy degree in the EU27 region, measured by *Energy Intensity*, shows positive results.

3.3 Index of Energy Costs Intensity in EU27 region

The *Index of Energy Costs Intensity* was determined on the basis of previously defined data as an indicator of the relationship between the value of consumed energy (including direct and indirect costs) and GDP units. For the determination, monetary values of the consumed energy and emitted carbon were used. The final values of the *Index of Energy Costs Intensity* are shown in Table 2.10.

The *Index of Energy Costs Intensity*, as a parameter expressed in monetary values, shows a different picture in the EU27 region. First of all, the total index in 2010 was higher than in 2000, which is not a case when the level of energy-economy degree was calculated using the traditional *Energy Intensity* indicator. Furthermore, in this case there were no significant differences between individual years.

In relation to the first year (which makes a big difference in relation to the measurement by *Index of Energy Intensity*) there is no decrease in values. Moreover, in most countries, the index is higher in 2010 than in 2000. This fact leads to the conclusion that in all countries costs related to energy are higher

TABLE 2.9 Energy intensity in EU27 region (2000–2010)

Country	2000	2001	2002	2003	2004	2005	2006	2007	2008	2009	2010
Energy intensity = energy consumption/GDP											
Austria	1.79	1.88	1.87	1.95	1.88	1.91	1.80	1.75	1.83	1.71	1.70
Belgium	2.27	2.29	2.15	2.26	2.12	2.00	1.92	1.82	2.03	1.83	2.09
Bulgaria	4.62	4.44	4.29	4.34	4.02	3.99	3.93	3.59	3.31	3.04	3.14
Czech Rep	2.77	2.76	2.62	2.67	2.60	2.43	2.31	2.13	2.05	2.05	2.10
Denmark	1.28	1.31	1.26	1.30	1.29	1.27	1.25	1.23	1.22	1.23	1.23
Germany	0.94	0.95	0.93	0.98	0.97	0.96	0.94	0.85	0.87	0.87	0.88
Estonia	2.97	3.08	2.75	2.65	2.56	2.43	2.20	2.20	2.31	2.43	2.47
Ireland	0.80	0.79	0.75	0.73	0.73	0.72	0.72	0.68	0.71	0.69	0.72
Greece	1.59	1.69	1.54	1.26	1.08	1.08	1.05	0.94	0.85	0.86	1.07
Spain	0.96	0.97	0.96	0.99	1.01	1.00	0.95	0.94	0.91	0.88	0.90
France	1.66	1.70	1.65	1.67	1.64	1.61	1.56	1.50	1.52	1.51	1.52
Italy	0.85	0.85	0.84	0.88	0.87	0.88	0.85	0.82	0.82	0.82	0.86
Cyprus	1.28	1.31	1.28	1.33	1.28	1.23	0.12	1.19	0.00	1.16	1.15
Latvia	3.44	2.59	3.36	3.31	3.13	2.90	2.71	2.58	2.58	2.91	2.93
Lithuania	2.41	2.38	2.28	2.12	2.08	2.01	1.99	1.89	1.80	1.90	2.04

(Continued)

TABLE 2.9 Continued

Country	2000	2001	2002	2003	2004	2005	2006	2007	2008	2009	2010
Energy intensity = energy consumption/GDP											
Luxembourg	1.29	0.97	0.93	1.33	1.43	1.39	1.30	1.19	1.20	1.16	1.15
Hungary	2.10	2.11	2.02	2.02	1.92	1.92	1.87	1.76	1.76	1.81	1.85
Malta	0.82	0.84	0.82	1.02	1.01	0.98	0.75	0.73	0.89	0.72	0.70
Netherlands	0.98	0.98	0.98	0.99	0.98	0.95	0.90	0.85	0.85	0.87	0.85
Poland	2.48	2.46	2.36	2.34	2.30	2.23	2.19	2.08	2.00	1.92	1.85
Portugal	1.12	1.13	1.14	1.15	1.16	1.16	1.12	1.11	1.08	1.09	1.10
Romania	3.49	3.36	3.20	3.18	2.99	2.90	2.71	2.48	2.31	2.27	2.34
Slovenia	1.72	1.74	1.68	1.67	1.63	1.59	1.51	1.41	1.47	1.42	1.46
Slovakia	2.41	2.44	2.35	2.17	2.04	1.98	1.81	1.61	1.56	1.55	1.60
Finland	2.95	2.93	3.00	3.02	2.90	2.76	2.79	2.63	2.53	2.55	2.58
Sweden	1.25	1.21	1.18	1.15	1.10	1.05	1.00	0.97	0.96	0.98	0.95
UK	0.88	0.86	0.82	0.81	0.79	0.78	0.75	0.72	0.72	0.70	0.75
TOTAL	1.17	1.17	1.14	1.16	1.14	1.13	1.08	1.04	1.04	1.03	1.05

Author's results

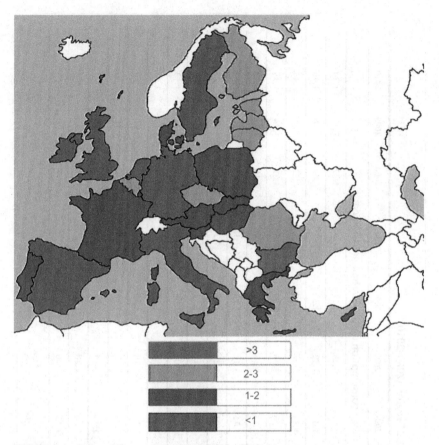

	>3
	2-3
	1-2
	<1

FIGURE 2.1   Index of Energy Intensity in EU27 – 2010

every year. The main reason for this is changes in energy prices. According to this, the energy-economy degree in the EU27 region becomes more worrisome.

Most countries in the EU27 region have an *Index of Energy Costs* intensity smaller than 1. They include the largest consumers — Germany, France, and the UK. Seven countries recorded an index between 1.21 and 1.30. The highest indices were noted in Austria, Bulgaria, Belgium, the Czech Republic, and Finland, with the highest in Latvia.

The final values of the *Index of Energy Costs Intensity* for 2010 are presented at Figure 2.2.

The energy-economy degree, measured by the *Index of Energy Costs Intensity*, shows the opposite results. All the values recorded in 2010 were higher than in 2000, which leads to the conclusion that each year the EU27 region invests more money resources in energy in order to provide certain economic development.

TABLE 2.10 Index of energy costs intensity in EU27 region (2000-2010)

Country	2000	2001	2002	2003	2004	2005	2006	2007	2008	2009	2010
Index of energy costs intensity = GDP/GDP – (energy consumption costs + carbon emission costs)											
Austria	1.22	1.24	1.23	1.24	1.25	1.25	1.21	1.24	1.32	1.33	1.34
Belgium	1.40	1.41	1.36	1.37	1.35	1.32	1.30	1.31	1.47	1.38	1.46
Bulgaria	1.34	1.34	1.34	1.35	1.34	1.37	1.37	1.33	1.32	1.35	1.35
Czech Rep	1.20	1.23	1.25	1.26	1.26	1.26	1.28	1.28	1.32	1.34	1.35
Denmark	1.12	1.13	1.14	1.16	1.15	1.15	1.16	1.18	1.19	1.19	1.18
Germany	1.14	1.14	1.15	1.16	1.15	1.16	1.16	1.15	1.14	1.15	1.15
Estonia	1.20	1.21	1.20	1.23	1.22	1.21	1.20	1.21	1.23	1.27	1.27
Ireland	1.08	1.08	1.08	1.09	1.09	1.10	1.11	1.12	1.13	1.15	1.14
Greece	1.13	1.14	1.12	1.10	1.09	1.09	1.09	1.08	1.10	1.11	1.13
Spain	1.11	1.10	1.10	1.11	1.11	1.11	1.11	1.12	1.12	1.14	1.16
France	1.20	1.20	1.19	1.19	1.19	1.19	1.18	1.17	1.17	1.17	1.18
Italy	1.16	1.16	1.14	1.16	1.15	1.16	1.16	1.17	1.17	1.17	1.19
Cyprus	1.14	1.17	1.14	1.25	1.20	1.14	1.02	1.18	1.01	1.20	1.25
Latvia	1.18	1.14	1.23	1.25	1.20	1.28	1.26	1.20	1.28	1.42	1.42
Lithuania	1.13	1.15	1.15	1.14	1.15	1.16	1.16	1.16	1.17	1.20	1.27

TABLE 2.10 Continued

Country	2000	2001	2002	2003	2004	2005	2006	2007	2008	2009	2010
Index of energy costs intensity = GDP/GDP – (energy consumption costs + carbon emission costs)											
Luxembourg	1.17	1.13	1.13	1.20	1.23	1.23	1.23	1.23	1.22	1.24	1.21
Hungary	1.18	1.18	1.20	1.20	1.20	1.22	1.23	1.24	1.32	1.31	1.36
Malta	1.07	1.07	1.07	1.09	1.08	1.09	1.09	1.09	1.11	1.15	1.14
Netherlands	1.11	1.12	1.11	1.12	1.12	1.13	1.13	1.14	1.13	1.15	1.13
Poland	1.26	1.27	1.29	1.27	1.24	1.27	1.30	1.29	1.28	1.24	1.28
Portugal	1.17	1.17	1.18	1.18	1.19	1.19	1.19	1.20	1.14	1.17	1.15
Romania	1.27	1.29	1.30	1.32	1.30	1.29	1.33	1.32	1.30	1.26	1.29
Slovenia	1.19	1.19	1.19	1.18	1.18	1.18	1.17	1.16	1.17	1.19	1.20
Slovakia	1.36	1.37	1.35	1.32	1.30	1.32	1.31	1.29	1.24	1.27	1.28
Finland	1.26	1.26	1.30	1.32	1.34	1.31	1.32	1.33	1.33	1.36	1.37
Sweden	1.09	1.09	1.10	1.11	1.12	1.10	1.10	1.12	1.12	1.12	1.13
UK	1.11	1.10	1.10	1.09	1.08	1.08	1.09	1.11	1.12	1.12	1.12
TOTAL	1.15	1.15	1.14	1.15	1.14	1.15	1.15	1.15	1.16	1.16	1.17

Author's results

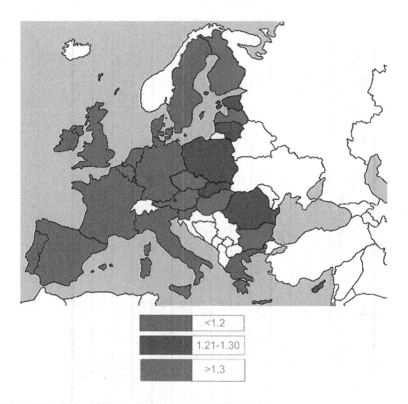

FIGURE 2.2   Index of Energy Costs Intensity in EU27 – 2010

4 DISCUSSION

Analysis of energy-related cost indicators in the EU27 region shows certain specific characteristics and inconsistencies, discussed as follows.

GDP Changes

Primarily, the region has constant and gradual GDP growth, with a peak in 2008, after which the GDP slightly falls. However, this fall was mainly less than 5%. The EU region continues with gradual economic growth, in compliance with the EU STRATEGY 2020, which stipulates growth according to sustainable principles. For that purpose, the countries of the EU27 region (as well as the countries' candidates to be full members of the EU) must direct their activities toward development based on less natural resources consumption. Energy scarcity in the EU and high energy import dependency imposes the need for special attention on energy consumption. The monitoring process through certain indicators is a tool for successful monitoring and supply of data valuable in further development process. Further, economic growth is assumed in the EU region, therefore transitions to less energy-demanding activities, higher use of RES, and a higher degree of energy efficiency are the key development directions.

Total Energy Consumption

During the considered period, total energy consumption increased constantly and was mainly parallel with the increase in the GDP rate growth. However, the increase in total energy consumption was not proportional to the increase in the GDP rate growth. The only countries with a decrease in energy consumption were Sweden and the UK. In most other countries, the GDP growth was not related to the same degree of energy consumption. All this indicates that GDP growth is possible with a reduction in energy consumption as well. The EU27 region had the highest energy consumption in 2008, after which consumption slowly decreased.

CO$_2$ Emission

The most favorable situation was in terms of CO$_2$ emission, which is monitored as an external cost. The EU27 region released the most CO$_2$ into the atmosphere in 2006. This is two years prior to the year when the highest GDP was registered. This leads to the conclusion that CO$_2$ emission is not exclusively related to the degree of economic activities, and apart from that it shows that the EU region undertakes substantial measures for low-carbon technologies implementation. The final result shows that the CO$_2$ emission in 2010 was approximately 5% less than in 2000, and the GDP in the same ten-year period increased by 15%.

Energy Prices

The EU27 region is a unique territory, with free flow of goods, capital, and people. At the same time each country member determines energy prices for its territory. Energy prices gradually increase every year. However, differences among countries are even more important. Energy prices varied substantially throughout the EU27 region, from 0.0675 EUR/KWh in Bulgaria to 0.1752 EUR/KWh in Italy (data for 2010), which is almost double. The lowest energy prices were in the east, countries that most recently became EU members. There was also a substantial difference in energy prices among the most developed countries. In this group of countries, by far the lowest was in France – 0.0940 EUR/KWh — while in Germany and the UK prices were a little over 0.13 EUR/KWh.

Energy Intensity Changes

The level of energy-economy degree, measured by *Energy Intensity*, showed more positive results for each country and for the EU27 region as a whole. The Total Energy Intensity for the region was smaller in 2010 than in 2000, which implies a positive trend. In general, the EU27 regions spent less energy (approximately 10%) to create a GDP unit in 2010 than in 2000. The worst situation was in Bulgaria, which needs three times more energy to create a GDP unit than an average EU country. A similar disconcerting situation was in Belgium, the Czech Republic, Estonia, Latvia, Lithuania, and Romania, where *Energy Intensity* was between 2 and 3. In relation to the year 2000, in 2010 all countries showed a decrease in Energy Intensity, a positive parameter.

Energy Costs Intensity Changes

The energy-economy degree, measured by the *Index of Energy Costs Intensity*, shows different results. In general, there were no significant changes between parameters for 2000 and 2010. Regardless of the energy savings and smaller carbon emissions, energy-related costs were still high and caused a high price of economic development.

Comparison of Final Results Obtained Using Two Methods

Monitoring of the energy-economy degree in the EU27 region by using two chosen indicators shows opposite final results.

Measuring with *Energy Intensity*, the energy-economy degree in the EU27 region showed positive trends.

Measuring with the *Index of Energy Costs Intensity*, the energy-economy degree in the EU27 region showed negative trends.

5 CONLUSIONS

The EU STRATEGY 2020 poses unique aims for all country members; however, their capacities, development levels, and ways of strategy realization are different. Therefore, a unique monitoring system is an absolute priority. Data acquired through this study were acquired for each individual country and should serve as an efficiency measure of implementation of the EU sustainable development strategy, which is a fundamental indicator to further direct implementation in the right direction.

Measurement of the energy-economy degree in the EU27 region in this research showed different final results, depending on the method used. In *Energy Intensity* measurements, the final results pointed to positive trends. The quantity of energy needed for certain economic development decreased each year and carbon emissions were also lower.

On the other hand, measurement of the energy-economy degree using the *Index of Energy Costs Intensity* showed a different picture. With each passing year, costs of energy that were included in economic development became higher. Energy prices, as a very important instrument of economy, cannot be neglected. This index considers it to be an important parameter. Energy-related costs (costs of energy and carbon costs) minimize the GDP by a certain value. Energy consumption and carbon emissions have economic values, which have a negative impact on economic development. Countries that recorded lower levels managed to achieve certain economic development with a lower level of energy-related costs.

Both indices, *Energy Intensity* and *Index of Energy Costs Intensity*, should be monitored in the future. The authors suggest changes in the purpose of their implementation. *Energy Intensity*, as an index that includes physical values, should be used for strategic planning in energy supply, energy transport, and storage areas. On the other hand, the *Index of Energy Costs Intensity* is more suitable for economic planning and monitoring of energy-economy degree. Future improvements of the indices require, above all, the development of methodologies for calculation and monetization of all external costs related to energy exploitation.

Implementation of the *Index of Energy Consumption* Costs can provide several important outcomes, linked to energy resources exploitation, among which the most important are determining the environmental costs related to energy exploitation and precise separation (decoupling) of the economic and environmental costs. It can also be used as an indicator of sustainable development, because it includes energy, environmental, and economic components. The suggested index can be used as a method for comparison between countries and between different periods of time in order to monitor levels of sustainable energy development and for consideration of energy scarcity costs, as a future parameter. The suggested index can be applicable throughout the European Union and beyond.

Chapter 3

Energy Management – Planning

3.1 TRADITIONAL CONCEPT OF ENERGY MANAGEMENT

Energy management is in most cases planned in theory and implemented in practice in a traditional way by implementing the already established principles of management that can be applied to almost any other human activity. Classical management science has developed valuable and effective mechanisms for management in all sectors of human labor, in which, depending on the needs and specific nature of work, certain adjustments can be made but are pre-determined and largely known in advance. Thus developed management, based on the learning of James A. F. Stoner, Ichak Adzes, Peter F. Dracker, and other founders of modern management science, has greatly improved business in the second half of the twentieth century and increased efficiency and effectiveness of human labor in all areas where it has been applied.

Traditional management has been adapted to modern business operations under the conditions of globalization and controlled flows of people, goods, and capital, and it greatly enables the persistence of certain system and socio-economic relations in the world. With changes in social consciousness and the first criticisms of the existing relations in the global economic and political scene, there emerges the need for certain changes in many spheres of interest for modern mankind. This imposes the need to make changes and adjustments in the field of management that would empower this science to respond to challenges of doing business in the future.

Energy management is a particularly demanding challenge in all areas, especially in terms of the need to manage energy in a manner required by the modern world, which is not a simple application of the process of management in the field of energy. With the increasing intensity of energy use in the world, there has been a need to develop adequate energy management at all levels, where it is necessary to develop mechanisms that will react and adapt the whole process of management, depending on the character and intensity of changes taking place in the environment.

The process of energy management can be seen in an extremely simplified way as a process that takes place from the moment of extraction of energy raw material to its final consumption and disposal of the waste energy. This process

Sustainable Energy Management. http://dx.doi.org/10.1016/B978-0-12-415978-5.00003-5

takes place in an extremely complex environment, is subject to the impact of numerous factors, and represents certain challenges in countries and regions where very stable conditions prevail and parameter changes are reduced to minimum. A particular problem is power management in the areas where frequent changes occur regarding type and intensity. Moreover, particularly challenging is the attempt to develop a unique model for energy management globally.

With the introduction of the principle of sustainability, as the only principle of the development of mankind in the future, the issues of energy management are given a new dimension, which greatly differs from traditional management and imposes the need for the adoption of new solutions. In the most widely accepted context, energy management involves the use of a sustainable model of strategic management in the field of monitoring of energy flows in terms of production, transport, transformation, and consumption.

Given the complexity of energy management, it is necessary to use the available approaches and modify them in accordance with the requirements of sustainability. Creating an adequate model of sustainable energy management is only possible with the integration of sustainable development requirements, particularities of the energy sector, and the science of strategic management. By integrating these criteria a new interdisciplinary approach to energy management has been created that has its own characteristics and is subject to changes depending on many factors. Therefore, it is primarily necessary to give a brief overview of the development of the science of strategy and strategic management in general.

A single definition that would include all the features and aspects of the strategy would be long and complicated, so each author who contributes to development of modern management emphasizes one or more aspects.

Perhaps the earliest definition of the strategy originates from the ancient Greek writer Xenophon: "The strategy is the knowledge of the business which you propose to be achieved" [27]. Xenophon clearly indicates the direct responsibility of those who manage (managers) and emphasizes the link between leadership and strategy formulation.

The important contributors to management science, Igor Ansoff and Anthony McDonnell point out that strategic management is a systematic approach to managing change [5]. By this definition the authors make a distinction between strategy and policy, where the policy is a universal decision that is always made in the same way if the same circumstances occur, while the strategy applies similar principles, allowing for different decisions to be made depending on the concrete circumstances.

On the other hand, Ichak Adizes points out the following: "Instead of the story of a manager who plans, organizes and so on, we should speak of the management team that performs these functions. The roles of producers, managers, entrepreneurs and integrators must be performed by the complementary team, because no single person can be responsible for all of them [1]."

Kenechi Ohmae's definition of strategy explicitly emphasizes the importance of meeting the expectations of customers' needs as the driving force of the strategy and points out the competitive aspect of strategy. Ohmae defines strategy as follows: "The way in which corporation seeks to differentiate itself in a positive way compared to its competitors, using its relevant forces to better meet customers' needs [70]."

In the definition by Kenneth Andrews strategy is seen as the relationship between the goal and possibilities of the realization of that goal, making the strategy sensitive to values and culture as much as to chances and opportunities arising in the environment. Andrews gives a broad definition of strategy: "Modeling of the main aims, purposes and achievements and the important policies or plans to achieve these goals, which are stated in such a way that they define in which business the company is or it will be, as well as the current type of company or what the company should be [4]."

Authors Samuel Certo and Paul Peter point out that strategic management is a continuous, iterative process targeted at the maintenance of organization completely appropriate to respond to its environment [9].

There are several traditional scientific approaches to strategic management, and all are in use in all areas of human activity around the world. Authors Certo and Peter define the process of strategic management as a continuous, iterative process that involves several steps:

1. Analysis of the environment
2. Directing the organization (mission and objectives)
3. Formulation of strategy
4. Implementation of strategy
5. Strategic control

The process of strategic management in this case is defined as the process that is going on in an infinite number of repetitions, as shown in Figure 3.1.

Thus the defined process of strategic management has been widely accepted in practice because of its simplicity. Complexity of a particular process does not necessarily lead to its improvement, but such a simplified approach to strategic management is not quite acceptable for energy management, which must respect the requirements of sustainability. The main objection to the presented model is primarily the narrow scope of defining the strategy, i.e., the ways to achieve

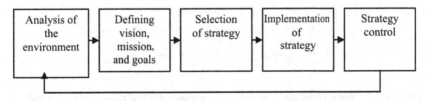

FIGURE 3.1   The process of strategic management according to Certo and Peter

the desired goals. In addition, the process of control does not take into consideration legal limits but is only based on examination of the achieved objectives, which are often not properly set, measurable, or given in an exact way.

Authors Gerry Johanson and Kevan Scholes developed a similar and yet redefined model, because the process of strategic management, which could serve as the starting point for developing the model of sustainable management of energy, is observed in three steps. According to these authors, the process of strategic management consists of the following:[59]

1. Strategic analysis, which consists of analysis of the environment, analysis of the expectation of goals and strengths, and analysis of the resources.
2. Strategic choice, which consists of generating options, estimation of options, and selection of strategies.
3. Strategic implementation, which consists of planning resources, organizational structure, and people and systems.

Contrary to the previously shown and proposed basic model, the process of strategic management of energy is based on the relationship between the three basic steps, which can be summarized in Figure 3.2.

The model presented above suggests to some extent the application of new solutions that can be regarded as acceptable for the sustainable management of energy, primarily because of the emphasis on the connection of the three basic elements, especially implying a particular framework for strategic choice. Namely, the process of energy management is very complex and it is therefore necessary that in the planning stage there is the possibility of defining and redefining the choice of strategy to be used for the realization of the set goals. The basic elements of this model presented here have their own characteristics and subsystems.

Strategic analysis includes processes of analysis and prediction of environment, analysis of the resource capabilities, and expectations and the strength of the main social groups. These analyses should result in understanding of the strategic position of the region, opportunities and threats, and definition of the main goals of the process of energy management.

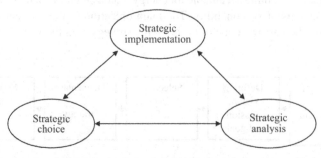

FIGURE 3.2   Model of the elements of strategic management according to Johanson and Scholes

Strategic choice also consists of three elements. It includes identification, description, and evaluation of strategic options and choice of strategy that could be implemented in the process of energy management. According to the theory and science of management, the determination for further actions should be defined at this point, i.e., the selection of several options: the stabilization of energy production and consumption, expansion, reduction or pause, and a more or less radical strategic shift in which the process of energy management has to undergo significant changes. Respecting the very concept of sustainable management of energy, the strategic approach to management can be considered more acceptable than the one previously discussed (according to Certo and Peter) because it opens the possibility for more strategic choices and allows the implementation of the strategic shift as an option that takes into account the need for significant changes in dealing with energy resources and energy consumption [29].

Strategic implementation refers to issues that determine the fulfillment of requirements for the implementation of selected strategic options. For this reason, it includes planning input, defining the appropriate organizational structure and selection of management style, and corporate culture and recruitment.

Authors Higgins and Vincze have a similar approach, and in the process of strategic management they start from formulation of the vision, mission, and developmental goals and preparation for implementation [56].

1. Formulation of vision, mission, and goals
2. Setting strategic goals
3. Development of strategy
4. Implementation of the selected strategy
5. Assessment and control of strategy

The presented model of strategic management requires greater efforts in the field of planning, which has been realized in the first two strategic steps, when the vision and mission have been defined. Particular attention has been paid to setting strategic goals. Although one step has been divided into two steps, the model has to be defined and analyzed as a separate model of management, as shown in Figure 3.3.

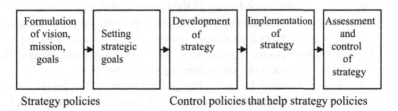

FIGURE 3.3   Model of the process of strategic management according to Higgins and Vincze

The model of strategic management shown in Figure 3.3 is partly redefined the first strategic model, shown in Figure 3.2, where the main motive was the attempt to surpass possible shortcomings of the previous models. Special emphasis in this case is given to the process of defining the strategic goals that determine all further actions. Therefore, the wrong definition of the strategic goals in the early stage of planning can have far-reaching good and bad consequences in terms of implementation and success of the entire strategy. This is also the greatest weakness of the proposed model, since the process of sustainable energy management is a process that has consequences of a proper or improper planning and definition of the objectives will be felt after many decades, which questions the purpose of the control.

Taking into account all the strengths and weaknesses of the described strategic models, we should present the model that can be regarded as the most appropriate one for the initial creation of the model of sustainable development of energy, which consists of eight strategic steps:

1. *Defining vision and mission of the company,* defining the main attitudes that direct the attitudes of the company towards the future (defining vision), and determining the role of the company in a given system of doing business (defining mission).

2. *Analyzing and forecasting the direct and indirect environment* should provide a) diagnosis of the current strategic position of the company and b) assess opportunities and threats that will be created with the projected development of the environmental factors.

3. *Analyzing and forecasting internal possibilities* of the company and the conditions for their provision.

4. *Defining objectives* through quantitative and qualitative determination of the type of business performance the company strives for, and in relation to that it measures its effectiveness and efficiency in the implementation of its mission and strategy of the company.

5. *Identifying the critical factors of business success* and determination of activities that a company must undertake to operate in the selected field of activities. Then concrete critical factors are searched for so as to gain competitive and comparative advantages.

6. *Identifying and evaluating strategic options* (alternatives) and choice of appropriate strategies, i.e., searching for potential ways and instruments for the selection of goals and selection of the business strategy.

7. *Implementation,* i.e., creation of the appropriate plan of business culture and management style, the plan to identify resources, the selection of the organizational structure, systems of motivation, and estimation of the value of the work so as to implement the determined business strategy.

8. *Monitoring and auditing,* i.e., measurement and evaluation of the plan (control) to know whether the plans have been achieved in accordance with the possibilities of companies (audit).

The proposed method of management consists of several more steps compared with the previous ones, which to some extent complicates its implementation, but on the other hand, provides an opportunity for detailed planning and monitoring, which is imperative in the process of sustainable energy management. This process is influenced by a number of factors, lasts long, and has no precisely defined end. Additionally, its implementation can significantly influence economic, ecological, and any other development in general. The process of strategic management could in this case be as presented in Figure 3.4.

The proposed initial model has its advantages and disadvantages. The main disadvantage of the proposed model is its complexity, which can lead to problems in the planning of resources, people, and time [56]. The main advantage of the initial model is the need to more carefully plan all the steps of management, because the implementation of the model is in this case the last, eighth step. All other stages involve preparation with particular emphasis on the analysis of the internal and external environment. For the purposes of energy management

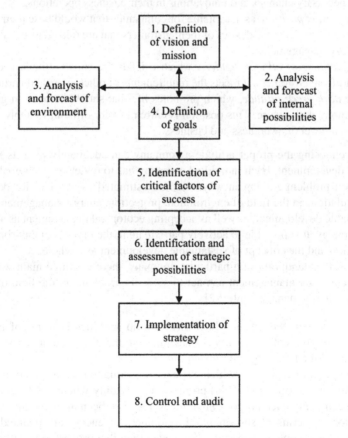

FIGURE 3.4 Elements of the process of strategic management as a starting point in defining the model of sustainable energy management

analysis of the environment this is of primary and probably the greatest importance, because it is very complex and multidimensional. It requires the ability to estimate and forecast in spheres that are not closely related to the sector of production and consumption of energy.

Each company should strive to fit, in accordance with their abilities and goals, into the above strategy of sustainable energy development and adjust their management according to it. However, all business entities are not and cannot at the same time fully implement the complete process of sustainable energy management. Modern management science recognizes this fact and offers four basic attitudes a company can adopt so as to implement the process of energy management, as follows: [67]

> *The legal position*, which means the company would voluntarily and without question apply all the laws, rules, and regulations in the field of sustainable energy management, but try to apply these rules to their advantage by making the necessary changes and innovations in their business operations;
> *The market position*, as acceptable for companies that would base their operations on meeting the desires of their customers that are related to sustainable energy management;
> *The attitude of all stakeholders*, which extends the above attitude because, apart from customers, it meets the requirements of other interested parties;
> *The dark green attitude,* which promotes life that should be lived in greater harmony with nature. This position is currently very difficult to apply at the current level of awareness and business.

By choosing the proper attitude a company can adequately plan its future energy development. Each individual company has to review its own goals and approach problem solving in ways that are estimated as optimal. Respecting the regulations in the field of environment protection, energy management, and sustainable development, as well as accepting ecological management as a business strategy, it is possible to actively participate in the process of environment protection and the concept of sustainable development as a whole.

From the standpoint of a particular company, there are three main ways in which company management can act to solve its problems in the field of sustainable energy management: [82]

- *Inactive management*, in which there are no activities in terms of understanding problems of energy management, so that there are no activities in terms of their solving and prevention;
- *Reactive management*, which reflects the situation in most companies today, and implies (possible) problem solving only when problems occur. In such companies no preventive measures have been applied, and certain actions in terms of sustainable development of energy are planned only when the company is forced to, and even then that process is usually slow and sporadic;

- *Pro-active management*, which is imperative of any modern energy-responsible business, where all activities are directed towards planning and prevention. The knowledge of the specifics of sustainable energy management and the considerable efforts towards its implementation are built into the business policy of the company and become its integral part.

Selection of the approach in the field of energy management is conditioned by the influence of a series of internal and external factors that define each company, as well as the social and economic circumstances in which the company operates. On the other hand, planning of energy management in terms of a particular state or region is connected with an even greater number of factors that determine it at a certain point and that are changing very frequently. Therefore, the development of an adaptive and yet clear model of sustainable energy management is crucial.

3.2 SUSTAINABLE APPROACH TO ENERGY MANAGEMENT

Previous experience, theory, and practice clearly indicate the need to develop special mechanisms aimed at proper management of energy in all stages and at all levels in all timeframes. The need for energy management has existed since ancient times in various forms, but only modern science and management practices, respecting the principles of sustainable development, have provided an adequate framework for the realization of sustainable energy management.

When designing sustainable energy management, it is necessary to start from the broadest framework that defines processes in this area. This relates primarily to two main factors that are extremely important: the size of the strategic issue that needs to be addressed and the estimated time required to implement the proposed strategy of energy management. A sustainable approach to energy management has been created because of the two characteristics that determine it more than numerous changes that mankind has encountered through history.

The *time dimension* of energy issues is particularly emphasized and greatly determines the ways of energy management, as well as consequences that may result from it. Meeting the demand for energy has existed throughout the history of modern mankind, and needs have increased as energy resources have been exploited faster and in a more efficient way, but the situation at the end of the twentieth century led to the need to reconsider the trend of uncontrolled energy consumption, to slow down and redefine it in the ways that would be in accordance with the concept of sustainable development.

The future of energy management is in a certain way defined by a clear determination to slow down the exploitation of non-renewable resources, to use energy efficiently, and to raise awareness of all energy users that they have to adjust to the new way of doing business and thinking.

These fundamental changes in energy management are extremely complex, comprehensive, and linked with a number of smaller and larger changes in all spheres of human activities. It is necessary to redefine the existing legislation, adopt appropriate strategies at global and national levels, and adjust business operations of each business entity and individual consumer.

Apart from the time required to use sustainable energy management strategies, in order to select the type of strategic implementation it is necessary to consider the size of the strategic problem. The energy problem is global and is characterized by a number of factors primarily related to the increase in global needs for energy, uneven distribution of energy resources, and high degrees of dependence of people on the energy they have at their disposal. Problems with energy lead to minor or major problems in economy, society, relations between countries, and relations in the global community. It is expected that these problems will increase and become more complex in the future. It is therefore quite correct to assume that the energy problem is one of the largest and most complex strategic problems of today.

Respecting the timeframe and estimating the scope of the energy problem, the following suggestion can be given for implementation of sustainable energy management, as shown in Figure 3.5.

When a company is faced with minor problems (small energy consumers) and there is adequate time available, the implementation of the strategy can include evolutionary or gradual (incremental) changes. If the problems a company encounters are minor, but the period of time is short in which it is necessary to solve the problems, which is often the case when new legal regulations are adopted that sharply define changes in energy consumption, activities are done by management (managerial intervention) directly where the problems have arisen.

If the company is a big producer or consumer of energy it is necessary to carefully select the method of implementation of strategy of the sustainable management of energy. Depending on how much time the organization has to solve the problem, there are two possible strategies. If the company has a long period of time, and the problems that emerged are big, sequential intervention, i.e., intervention in stages, is required. However, if the problems can not wait to be solved, complex intervention is applied, and the management of the company has to synchronize (adjust) the changes in all parts of the company.

| | | Timeframe for the change of implementation | |
		Long	Short
The scope of strategic problem	Big	Sequential interventions	Complex interventions
	Small	Evolutionary interventions	Managerial interventions

FIGURE 3.5   The types of possible implementations of the strategy of sustainable management of energy

Changes in national energy management largely depend on the resources the country possesses and the way in which energy is being spent. The strategy of sustainable development defines energy problems in general as the those that require too much time to be solved, i.e., longer than the lifetime of the current generation. The energy problem will be a part of mankind for an indefinite period of time in the future.

The selected strategy of sustainable energy management is a highly complex process influenced by many factors. On the other hand, regardless of the clear definition of the proposed models for energy management, it is often not possible to select and implement a certain strategy without intermediate changes and adjustments. In addition, changes in the energy sphere are numerous and diverse, and their impact on the sustainable management of energy is often associated with a series of economic and political changes. Because of these issues, there are difficulties in clearly defining what sustainable energy management actually is. Every possible definition can be redefined and changed depending on many factors.

Energy management can be defined as the process of planning, directing, implementing, and controlling the process of generation, transmission, and consumption of energy. Energy management is a certain synthesis of energy and the concept of modern management, i.e., the application of modern hypotheses of management in the energy sector. Apart from that, in stating the main hypotheses in terms of energy management, modern management starts with the assumptions that it is possible to preserve and keep energy stability for current and future generations. Therefore, modern energy management can be seen as a kind of synthesis of the three exact sciences: energy industry, sustainable development, and management, which are interconnected and mutually conditioned, as shown in Figure 3.6.

Sustainable energy management is a new concept, idea, and approach that requires many changes in the traditional way of understanding and interpreting energy management at all levels. Sustainable energy management integrates

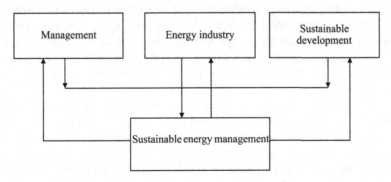

FIGURE 3.6 The main concepts of the sustainable management of energy

many features of the environment and cannot therefore be interpreted as a concept adopted and defined once and for all, but must be constantly modified and adjusted in accordance with changes in the three areas that define it, and in accordance with the specific country or region where it has been applied. Basically, the concept of sustainable energy management is defined by the following parameters:

- Management, i.e., accepted theories, existing experience, state regulations, company orientation, the requirements of all stakeholders, degree of education, awareness, and commitment as well as general orientation towards socially responsible business;
- Energy industry, legislation, existing energy resources, new energy resources, energy efficiency level, the level of technology development, production and consumption levels, system stability, connectivity with other systems, and the degree of self-sufficiency; and
- Sustainable development, i.e., acceptance of the concept at the state level, the ratification of international agreements, the parameters of the national strategy of sustainable development, and the list of priorities.

Accordingly, *sustainable energy management can be defined as the process of energy management that is based on fundamental principles of sustainable development.* Sustainable management of energy must therefore be regarded as a concept that greatly differs from traditional management, which has largely ignored the need for conservation, restoration, and saving of energy resources. Thus developed, the concept of sustainable energy management is a major global change whose effects are felt throughout the world and major changes in behavior and thinking of responsible institutions and individuals are required. Therefore, the concept of sustainable energy management can be seen as a challenge of modern mankind, which opens up possibilities for proposals and development of numerous solutions that will have long-term consequences for the development of human society as a whole.

The only possible way to approach the problem of sustainable energy management is by respecting the scientific achievements and practical experience as well as taking into consideration the particular characteristics of the subject that initiates, accepts, and implements the energy management in a sustainable way. In general, sustainable energy management involves the implementation of a number of different activities aimed at setting concrete goals of sustainable development into practice of production, transmission, and consumption of energy.

From the point of view of traditionally measured social and economic development, most important is the implementation of those measures and activities that bring about the most effective results in less time, but with respect to the principles of sustainability the imperative of time becomes less significant. Activities that bring quality results are acceptable, regardless of the time needed to detect and measure the positive effects.

Given the complexity and global scope of modern business, and taking into account a number of problems and disagreements in terms of global trends and priorities of development, sustainable energy management cannot be defined through a series of strictly determined activities that need to be undertaken. Modern science and contemporary practice show that the activities in the field of strategic and operational management, as well as the improvement of quality and socially responsible business, have become imperative in terms of implementation of the concept of sustainable energy management.

Theoretically, the problem represents the "gap between current and desired state." With this in mind, it is necessary to implement appropriate changes and undertake appropriate activities that would lead a company or community to an energy sustainable operation and therefore resolve the problems of that kind. The stages of more responsible operations are the following:

- Identification of problems (as a diagnostic activity) is a specific stage for each company because each company is characterized by a number of differences and particularities in relation to other companies. If the management of the company sets as the desired state the introduction of energy-sustainable business to a certain level (or completely), it should be in line with the ability to estimate its own strengths and define its own weaknesses. This is the stage of setting goals in the company that largely determines all other activities.
- Identification of development options includes prediction of activities that can be implemented to achieve the goals that relate to energy-efficient business operations. In predicting the ways in which it is possible to achieve the set goals, each option should be carefully elaborated and realistically estimated. At this stage, the experience of other companies in a similar situation can be useful (but not crucial).
- The selection of the most appropriate option leads to achieving the goal – energy-responsible business. This stage involves making appropriate decisions (among several options offered) that determine the future course of action.
- Implementation of the chosen method to achieve the goal includes a series of activities and requires the engagement of certain human resources, financial resources, and time.
- Control and correction of the identified deficiencies is a necessary activity that must be continuously carried out to correct irregularities and weaknesses that appear no matter how detailed the activities were planned.
- Achieving the goal, by which the company achieves a higher level of energy responsibility, about which all interested parties should be informed, mainly consumers, owners, and the community.
- Identification of (new) problems, thus returning the activities to the beginning, and the company strives to ensure that the activities on the improvement of ecologically suitable business are continuous.

In each compamy, the relevant service needs to monitor the situation in the environment, which is primarily related to new requirements for energy-responsible business (mostly coming from consumers and legislators) and to initiate the previously mentioned activities. In all stages leading to the achievement of environmentally responsible businesses the whole management team should be involved, and if necessary, professionals can be engaged.

Sustainable development itself is a kind of strategy of development of mankind that is not limited by time. It is subject to changes and adjustments. The effects of such development can be precisely measured at a certain point, but unlike traditional management, there is no strict orientation toward reaching goals. In fact, sustainable development is not defined in time and space and involves constant improvements so that it is not possible to reach a certain goal and then stop the implementation of the sustainable development management process. Sustainable development is not a goal in itself. It is a process whose goals are constantly redefined and corrected.

Thus defining the process of sustainable development imposes the need to apply certain traditional strategic management activities, which enable implementation and control of the process in an unlimited period of time. Sustainable energy management is not a simple application of traditional strategic management, but implies its modification, which has led to the development of a unique model of sustainable management that differs from the traditional one. At the same time it significantly questions the necessity of its implementation in the case of sustainable management in the field of energy.

To design a strategy for sustainable energy development it is necessary to set specific priorities for development. Five basic priority programs have been suggested, which are diverse in program content but complementary from the point of coordinating the operation and development of the whole energy system, i.e., energy production sectors and the sectors of energy consumption and a gradual but consistent achievement of the goals promoted during the implementation of the suggested strategy. The priorities are as follows:

- The first priority of continuity of the technological modernization of existing power plants/systems/sources, in the sectors of oil, natural gas, coal from surface and underground mining, electric power sector, with production plants and facilities: thermal, hydro and heating power plants and transmission system or distribution systems, and a sector of heating energy – city heating plants and industrial power plants.
- The second priority of rational use of quality energy and the increase of energy efficiency in production, distribution, and use of energy by the end users of energy services.
- The third-specific priority of using NRSE (new renewable energy sources) and new energy-efficient and environmentally friendly energy technologies and devices/equipment for energy use.

- The fourth-optional priority for the emergency/urgent investment in new power sources, with new gas technologies (combined gas-steam thermal power plants).
- The fifth long term development and regional strategic priority, building new energy infrastructure facilities and electric power and heat sources within the energy sector, as well as capital-intensive energy infrastructure.

In accordance with the stated priorities whose implementation enables achievement of sustainable energy systems in the world, it is necessary to undertake the implementation of local and national programs that will be defined in accordance with the new approach to strategic management of energy. However, the traditional approach to strategic management is not fully acceptable for the purposes of sustainable management.

The actual results of the planning process should be concrete actions (implementation of the strategy). Planning decisions are only an intermediate stage of the planning process. Through the implementation of the strategy, the company transfers ideas into concrete achievements. In the process of strategy development, entrepreneurship and visionary qualities of leaders are required, while in the process of implementation human resource skills are needed to achieve the set goals. Modern business is characterized by continuous change. The scope of the problem caused by change and the time required to solve the problem determine the complexity and the speed of implementation of the strategy. Depending on the size of the company and the style of management there are several approaches to implementing the strategy. Each of the approaches has advantages and disadvantages, and the task of strategic management is depending on the scope of changes caused by the implementation of the selected strategy and other conditions, to decide the way in which the strategy will be implemented.

The implementation of strategic orientation is not a single process, but a process that involves the provision and allocation of resources, creation of organizational, procedural, motivational, and other conditions, and development of a series of individual plans for undertaking and coordinating activities for the achievement of these goals. The implementation of strategy involves a complex process of creating conditions and integrating activities to achieve results. For the proper management of activities it is necessary to specify the responsibility and authority to identify tasks and budget and allocate them to the carriers through a system of coordination and overall management of the implementation process. In this case, emphasis is placed on global issues affecting the implementation of the strategy.

Sustainable energy management is an extremely complex process, both in planning and implementation. The very implementation of sustainable energy management in practice has the following main characteristics that define it and distinguish it from traditional management:

- The implementation is extremely demanding. The transition from traditional to sustainable energy management requires the adoption and implementation of a number of major changes, both in thinking and in the manner of implementation and measurement of financial effects. If all planned activities are performed in accordance with scientific and practical recommendations, there is still a high degree of probability that the realization itself will not be fully effective and will not achieve certain goals. Managerial practice implies that implementation that reaches at least two thirds of the defined goals can be considered very successful. When it comes to such complex changes, such as the transition to sustainable energy management, the degree of success is usually even lower.

- The implementation is a long-term process. Every large strategic change in any sphere of human activity requires enough time and its effects are most often visible after more than ten years. The transition to sustainable energy management in this regard is not an exception. It can even be said that according to the time criterion, such a strategy change is even more demanding. The planned activities usually last at least ten years, and the implementation itself, after which you can measure the effects, lasts at least the same period of time. Therefore, sustainable energy management is a process that requires extremely careful management, commitment, accuracy, continuity, and a clear commitment of all participants.

- The implementation is multidisciplinary. The problem of providing enough energy for the needs of modern mankind is one of the largest problems today. It will largely determine the direction and intensity of development of the planet in the future. The very acceptance of the concept of sustainable development takes a lot of time and is linked with a number of difficulties, and the transition to sustainable energy management is just one of the goals of sustainable development, which is also linked with a number of political, economic, and other characteristics. It is necessary to involve a number of stakeholders both on a local and national level. The problem of adequate energy supply that is not too expensive, will not cause too much pollution, and will be sufficient in both quantity and quality, is one of the most complex problems of modern development, because energy directly affects the possibility of realization of most human activities. Therefore, the solution to this problem must involve all stakeholders: industry, services, agriculture, legislation, executive authorities, and all participants in the process of solving economic, environmental, energy, development, social, and other issues.

- The implementation requires strong IT support. The traditional way of managing the work of the energy sector includes the mandatory application of modern IT tools and software support, but transition to sustainable energy management is an even more demanding process. It includes specific modeling of relationships between production and consumption of energy from different sources and monitoring of the pollution. The modern trend imposes the need for rational use of energy and energy conservation at all levels. The support provided by information technologies is the only tool of support.

- The implementation requires a specific engagement of human resources. So far energy management processes have not been particularly demanding in terms of engagement of human resources. The involvement of professionals in the energy sector is technically necessary and therefore enough. Essentially, sustainable development integrates primarily environmental, economic, and energy knowledge and requires the engagement of experts from these areas. In addition, it is necessary to engage people who are fully trained to monitor issues of sustainable development and, accordingly, modeling of the development of the energy sector.
- The implementation is subject to many influences. The development of mankind and activities of people have always involved the consumption of certain amounts of energy. Progress has never been possible without an adequate amount of energy available. On the other hand, the progress of mankind has led to a large number of economic, ecological, cultural, and many other problems. The energy crisis and a high dependence of man on the energy available emphasize the problems in this field and make them top priority. Future development should be able to provide enough energy without causing excessive pollution. Any change in the energy field affects all actors of social and economic life, so their claims and efforts are subject to special attention. Modern processes of creation of the development policy include consultation with all stakeholders, which further aggravates the already difficult transition to sustainable energy management.
- The implementation records negative financial effect at first. The need to move towards sustainable energy management was created primarily as a result of increased awareness of environmental pollution caused by production and consumption of energy. Necessary adaptation involves transition to new energy sources, which are mostly at the stage of examination, or their exploitation is insufficient so that a clear conclusion is not possible. The transition to a new way of energy management is necessary and occurs at the moment when there are not necessary prerequisites. Some technical solutions are not effective enough, exploitation is uncertain, distribution is difficult, legislation is often inadequate, and there are no financial investments planned in advance that are extremely high. Sustainable energy management was created as a result of the consciousness of mankind that the energy is a limited resource; the shift has already started, although there are no clear solutions to various problems encountered in the transition process. In many cases, the legislation imposes the transition to the production of certain amounts of energy from renewable sources without providing ways to realize that.

Therefore, the transition to sustainable energy management imposes the need for substantial financial investments, where there are no clear predictions of how long it will take for these investments to return the investment, although it is quite clear that all investments of this type are long-term. Therefore, the transition to sustainable energy management in the early years usually records more or less negative financial effects, because it requires significant investment, and production of energy from these sources is small and therefore generates low revenues.

3.3 STRATEGIC ANALYSIS OF ENERGY SECTOR

Strategic analysis is the initial phase in the field of energy management, both in the traditional and proposed model of sustainable management of energy. Various models are analyzed in various ways. Thus, analyses vary in the levels of detail, comprehensiveness, and the use of different temporal, geographical, political, and many other restrictions. Regardless of the scope and limitations, strategic analysis must be given special attention, because on the basis of this analysis further steps and management procedures are taken, and any error in the process of analysis significantly affects the final result of the management process, most often after a period of many years.

For the purpose of a comprehensive and realistic understanding of the situation and subsequent creation of the appropriate strategy of the sustainable energy development, three basic analyses are necessary:

- *Global analysis* – analysis of energy production at the global level is the first necessary condition for successful planning and implementation of the process of sustainable energy management. Energy is produced all around the world from different sources, in different quantities, and in different ways. Until the first half of the twentieth century energy was mostly spent in the place of production, i.e., it was transferred to shorter distances. With the increase in the number of people, globalization of the world economy, and exploitation of resources that are concentrated in certain parts of the world, the need emerged to produce increasing amounts of energy. Energy is becoming one of the key limiting factors of life on the planet, and it is needed to perform daily activities and complex human activities. Energy needs are constantly increasing, while the production of energy is trying to meet ever-increasing demands. Because of this, energy has become the subject of world trade and an important factor in business and political changes. In addition, awareness of the problem of limited energy resources raises the issue of energy production to a higher level, and energy management is becoming one of the key issues of sustainable development and development in general.

Global analysis includes the following:
  o Analysis of global consumption – consumption of energy resources in total, consumption in the regions and in the countries, consumption of energy according to energy types, consumption according to the purpose, average energy efficiency, energy efficiency according to the fields of industry, economic parameters of energy consumption; and

o Analysis of global production – global reserves of energy resources according to the types and distribution, availability of the renewable energy resources, production of energy from renewable resources, global trends, economic parameters of the world energy industry.

- *National analysis* – apart from the analysis of the global energy situation, availability of resources, production, and consumption, it is particularly important to make adequate analysis of the situation in the field of energy on the national level.

National analysis should including the following:

o Analysis of national consumption – quantity of energy spent in a certain country in a certain period of time, in total, according to the types of energy and field of activity, definition of the biggest consumers of energy in the regions and according to the field of activity, energy efficiency according to the field of activity;

o Analysis of national production – available renewable and non-renewable energy resources, legislation on exploitation, structure of ownership and capital, existing technologies, impact on the environment, regulation of energy waste;

o The degree of self-sufficiency – a real estimation of self-sufficiency is particularly important in creation of national strategy of energy development. Countries that can satisfy their needs for energy are few, and their monopoly provides them the opportunity to influence world politics in all domains. Most countries in the world are forced to import energy of different types and in different quantities and because of this are forced to adjust themselves to different influences; and

o Harmful emissions into the environment – consumption of energy resources necessarily leads to a variety of emissions into the environment. These emissions are the focus of observation of all stakeholders because of the long-term adverse impacts on the greenhouse effect and global warming. The analysis of the ecological cost of progress should be made, i.e., it is necessary to determine the load of the environment per GDP.

- *Analysis of the company's current* – a global strategy of sustainable development and global availability of energy resources certainly influences the framework in which a certain company operates or will operate, since the issue of energy industry is long-term and complex, but the situation inside the company is crucial to solving the problem, because companies are just agents who, along with individual consumption, largely affect the energy future of the world.

From the point of view of the company, the analysis should include:

o Internal analysis – energy consumption by type and quantity, in total, per unit, per unit of revenue, the advantages and disadvantages of using certain energy, technological solutions, possible improvements, harmful emission into the environment; and o External analysis – analysis of the energy situation in the immediate environment of the company, especially energy policy of suppliers, subcontractors, competitors, attitude and activities of the local community, support of financial institutions, tax incentives, special credit lines, subsidies, and the possibility of obtaining the necessary energy from renewable sources.

Global analysis offers general guidelines for creating environmental policies in each country. On the basis of general guidelines and specific analysis on a national level, each country develops its own strategy of sustainable development and, within it, a special strategy of energy development at the national level. Proper and realistic analysis is the basic precondition for the creation of adequate energy strategy, which recognizes the concept of sustainable development and will be the basis for all long-term activities in the sphere of energy production and consumption. Only on the basis of correct analysis of the situation is it possible to define the desired goals and activities needed to achieve these goals.

The analysis of the current situation in the company is of particular importance. The energy policy of any responsible company should be harmonized with the national strategy of sustainable development and the strategy of energy development, its basic principles, and priorities. Each company has to conduct its own analysis and come to certain conclusions as well as establish its particular objectives of sustainable energy development. Apart from this, only with a realistic assessment of the situation can a company define the resources required to achieve the defined goals with great precision (equipment, money, people, and time).

3.4 STRATEGIC OBJECTIVES OF SUSTAINABLE ENERGY MANAGEMENT

Energy management has always been a challenge for anyone involved in the design and implementation of the process, and in most cases, the consequences of energy management are far-reaching. The complexity of the global economic and environmental situation, and the full awareness that there is not enough energy, as well as that the processes of generation, transmission, and consumption of energy, which are the greatest polluters of the environment, have caused minor and major ecological disasters and led to major changes in traditional energy management.

The goals of traditional energy management were based primarily on the principles of economic growth and the demands of society, and the main objectives of the owners of energy resources were to find effective ways for successful and cheaper exploitation of energy resources. The next goal related to providing storage of certain resources until delivery, ensuring adequate safety standards, which were not related to the protection of the environment, but were primarily aimed at preserving stocks of resources. In addition, an important goal of the owners of energy resources was reflected in the fact that resources were considered the owner's property (company, state, private capital) and were sold as commodities in the world market at the highest possible price. Exploitation in order to gain economic profit and monopoly position was the only business objective of the owner of the energy resources.

Similarly, there were certain goals regarding energy transfer. Traditionally, the main goal of this segment of the energy management process was to transport the energy resource, no matter whether it was in a processed, semi-processed, or raw form, in a way that was most economical. Fast and efficient transport of energy in order to gain a competitive advantage was the only objective that guided the energy transfer processes.

Energy distribution in the traditional sense means the supply of a certain type of energy to end users. Accordingly, the main objective of traditionally based distribution of energy lied in the fact that the energy was bought at an adequate price and stored, transported, and sold to customers at a price satisfactory to the energy carrier.

The end users, as users of energy, set as the main objective the ability to buy certain energy resources at an adequate price and to make the supply stable. Production processes, as well as the products and services, are designed to use energy in a way that is conditioned by technology, without considering the possible savings. The value of energy that is consumed in the production process is compensated by the final price of products or services that are paid by the consumers.

As with the owners of energy resources, economic interest was the main goal of business. In accordance with that, all steps of management were defined, decisions were made, technologies were installed, and the whole system was defined by certain legislation that in a traditional way regulated the relationship among all participants in the process of energy supply. Thus the defined system of energy production and consumption survived proportionally long, although with the start of the first energy crisis (in the 1970s) all shortcomings of such energy management started to be identified.

The emergence of environmental problems, warnings about the limited supply of energy resources, pollution caused by the use of energy resources, and the increase in consumer awareness have led to the need to introduce significant changes in the process of treatment of energy from producers to end users. Modern sustainable energy management has emerged as a unique phenomenon with its own set of goals.

The main objective of sustainable energy management lies in the fact that energy is produced and consumed in ways that will not endanger the opportunities of future generations to meet their energy needs. Since economic growth is closely related to sufficient quantities of available energy, the main goal of energy management can be defined as energy management that will allow economic growth that will not burden the environment over the limits of sustainability. Other objectives derived from this objective are the following:

• To present the relationship between ecology, economy, and energy as a very important element of economic and political stability;

- To promote a public information system regarding the need to preserve the environment through sustainable management of energy resources (including renewable energy — small hydroplants, wind generators, geothermal sources, solar energy, and biogas), with a strong emphasis on the policy of national development and regional cooperation;
- To raise consumer awareness about the fact that energy use has a major impact on environmental issues and economic development;
- To encourage dialogue between various stakeholders in the field of energy and ecology, with special emphasis on rural areas where the production of energy from renewable sources can be used as an opportunity for job creation;
- To participate in creating a national energy strategy that will accept the contemporary experience of other countries in the region and European energy policy;
- To support the development of small and medium enterprises (SMEs), particularly in the production of energy from renewable sources;
- To promote sustainable development by appropriate policy and the use of renewable energy sources;
- To assist local initiatives in the process of public participation in decision making, support the implementation of appropriate procedures, and provide access to information as well as implementation of legislation;
- To establish strong communication with relevant government and other institutions in order to participate in the development of energy strategy. Joint working groups or committees should be established at the operational level to facilitate inter-sectorial cooperation;
- To provide access to information and education in the field of renewable energy sources to all stakeholders (different levels of management and decision-makers, professional institutions, NGOs, interest groups, and organizations) through the establishment of educational centers for renewable energy sources;
- To develop and implement projects at the local level in the function of education of all stakeholders in the process of planning and decision making and participation of the local business sector and local communities, in the broadest way;
- To initiate the establishment of a national network for management of renewable energy sources that will work on promoting the use of renewable resources and programs for conservation and efficient use of energy. The network should also support access to finance for small and medium-sized producers of energy;
- To introduce environmental/energy/economic topics, in the broadest sense, in the school system;
- To mobilize financial institutions to encourage investments in this sector;
- To work on promoting measures that will provide support through fiscal and financial policy (lower taxes, lower interest rates); and
- To support research and development of energy efficient products and technologies.

3.5 CASE STUDY – INDIA

Energy Poverty: A Special Focus on Energy Poverty in India and Renewable Energy Technologies

Abstract

As a large percentage of the world's poor come from India, development in India is a key issue. After the establishment of how access to energy enhances development and the achievement of the Millennium Development goals, energy poverty has become a major issue. In India there is a great interest in addressing the subject of energy poverty, in order to reach development goals set by the government. This implies an increase in India's energy needs. In a climate of change and environmental consciousness, sustainable alternatives must be considered to address these issues. Renewable energy technologies could provide a solution to this problem. The government of India has been focusing on implementing electricity policies as well as on promoting renewable energy technologies. The focus of this article is to bring to light the problems faced in India in terms of energy consumption as well as the hindrances faced by renewable-based electrification networks. Government policies aimed at addressing these issues, as well as the current state of renewable energy technologies in India are discussed, so as to analyze the possibility of a solution to the problems of finding a sustainable method to eradicate energy poverty in India. The research reveals that the government of India has been unable to meet some of its unrealistic development goals, and to achieve the remaining goals it will have to take drastic steps. The government will have to be more aggressive in the promotion of renewable energy technologies to achieve sustainable development in India.

1 INTRODUCTION

1.1 Energy and Poverty

As per the Human Development Report (1997), the definition of human poverty is seen more as the denial of choices and opportunities for living a tolerable life, than just the traditional definition of income poverty. Even though energy is not sufficient to bring about the required economic and social development to provide these choices and opportunities, it is essential.

The link between energy and poverty has become apparent over the past few decades. According to the IEA (International Energy Agency, Paris) analysis, around 1.5 billion people, i.e., a quarter of the world population, have no access to electricity. It predicts that in the absence of vigorous new policies, 1.3 billion people will still lack access to electricity in 2030.

The Millennium Development goals set by the UN provide concrete, time-bound objectives for dramatically reducing poverty in its many dimensions by 2015 — income poverty, hunger, disease, exclusion, and lack of infrastructure and shelter — while promoting gender equality, education, health, and environmental sustainability. Some of these goals include:

- Eradication of extreme poverty and hunger;
- Achievement of universal primary education;
- Promotion of gender equality and empowerment of women;
- Reduction of child mortality;
- Improvement of maternal health;
- Combating HIV/AIDS, malaria, and other major diseases; and
- Ensuring environmental sustainability.

Access to energy enables the achievement of the Millennium Development goals. Electricity provides lighting and permits usage of household appliances. This extends the number of working hours beyond daylight and enables studying and learning beyond daylight. Modern fuels or electricity can reduce the exposure to indoor pollution, the time spent in cooking inefficiently, and the collection of firewood. With refrigeration, food can be stored for longer periods of time. Opportunities for self-employment from home grow with access to machinery such as sewing machines. Electricity aids in drawing of water through pumps as well as its purification. With access to electricity, schools can have better hygiene and facilities to attract students. With electricity, clinics can be equipped with modern equipment, carry out proper sterilization of instruments, and store various medicines and syringes, thereby improving health services to the people.

It has been estimated that 2.4 billion people rely on traditional biomass, such as wood, agricultural residues, and dung, for cooking and heating. This number is estimated to increase to 2.6 billion by 2030. Biomass exposes women and children to indoor pollution every day. The World Health Organization estimates that 2.5 million women and young children in developing countries die prematurely each year from breathing in the fumes of biomass stoves.

Electricity does not directly replace biomass. The transition from traditional fuels to modern fuels is not straight. The three main determinants in the transition are availability, affordability, and cultural preferences. Even if modern fuels can be afforded, people continue to use biomass if the modern fuels are much more expensive than the biomass that is readily available and perceived as "free" (see Figure 3.7).

Lower-income households prefer to use biomass for cooking and heating. As income increases it is seen that electricity and modern fuels are used for lighting, modern appliances, pumps, and communication, but they do not substitute for cooking and heating. Only in higher income groups is biomass completely substituted in household consumption.

1.2 India and Energy Poverty

India has a population of 1.1 billion, and about 22% of this population lives in poverty. Around 70% of the poor live in rural areas. It has a fast growing

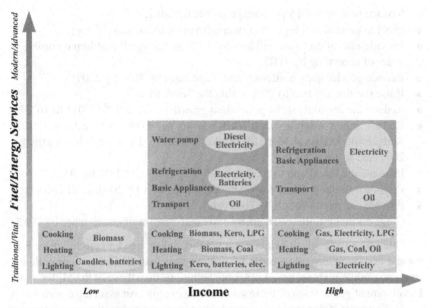

FIGURE 3.7 Household fuel transition. Source: IEA Analysis World Energy Outlook

population at an annual growth rate of 1.38%. The target set for poverty reduction in India is 19% population below the poverty line by 2015. The India Planning commission is expecting to meet this poverty target by 2015. The Indian government has announced plans to provide electricity to the entire population by 2012 to achieve social development.

According to the World Coal Institute, India is the sixth largest electricity-generating country as well as the sixth largest electricity consumer. Despite this, the electrification rate is only 44%. The population estimated to have no access to electricity is 582.6 million. Some 140,000 Indian villages out of 586,000 remain to be electrified and in many of the officially electrified ones, quality of service is such that it does not resemble true electrification. About 625 million people do not have access to modern cooking fuels and traditional fuels still provide 80-90% of the rural energy needs.

Detailed statistical research has revealed that in developing countries around 2.4 billion people rely on biomass for cooking and heating, and 1.6 billion people use no electricity at all. Most of these people are from South Asia and sub-Saharan Africa. The health burden as a result of the heavy dependence on biomass and the related indoor pollution is a loss of 1.6-2.0 billion days of work lost annually. The average time spent per month per household in collection of fuel wood is 40.8 h. These factors severely affect women and children, who are the ones who bear this burden the most. India's human development targets were outlined by the Tenth Five Year Plan and include the following goals:

- Reduce poverty by 15 percentage points by 2012;
- Provide gainful and high-quality employment to the labor force;
- Provide education to all children by 2003 and have all children complete 5 years of schooling by 2007;
- Reduce gender gaps in literacy and wage rates by 50% by 2007;
- Raise the literacy rate to 75% within the Tenth Plan;
- Reduce the decadal rate of population growth between 2001-2011 to 16.2%;
- Reduce infant mortality rate (IMR) to 45 per 1000 live births by 2007. Reduce the Maternal Mortality Ratio (MMR) to 2 per 1000 live births by 2007 and to 1 by 2012;
- Increase the forest and tree cover to 25% by 2007 and 33% by 2007;
- Provide sustained access to potable drinking water by 2007 to all villages;
- Provide electricity for all by 2012; and
- Clean all major polluted rivers by 2007 and other notified stretches by 2012.

Other than the goals to reduce poverty ratio, reduce gender gaps, increase forest and tree cover, provide electricity to all, and to clean all of the major polluted rivers, the other goals set have been more ambitious than those of the Millennium Development goals. Each of these goals has an energy cost associated with it.

In the course of this study we aim to look at the current energy poverty scenario in India and how the issues can be addressed by renewable energy technologies. We will then look at studies performed on the household consumption patterns in India in order to identify the consumption trends and establish how rural areas need more attention. As energy poverty is more of a problem in rural areas, this study will focus on rural energy poverty. We first look at the current energy scenario in India to get an idea of how electricity is currently being produced and then at renewable technologies in India and their current status. This will give us an idea as to how energy production is taking place and what role renewable energy technologies currently play. We will then consider issues with renewable-based rural electrification systems to see what hindrances exist and then we will focus on policies established by the government to understand how they plan to approach these problems. Existing facts on the current energy scenario in India and on the available renewable technologies in India will also be presented so as to get a picture of how renewable technologies can play a role in eradicating energy poverty in rural areas. By observing all these we aim to establish how development in India can be enhanced by the use of clean energy technologies.

2 HOUSEHOLD ENERGY CONSUMPTION IN INDIA

India is a country in energy transition. Through the study of the consumption patterns, a shift from inefficient solid fuels to more efficient liquid and gaseous fuels and electric power has been observed. According to the studies, currently in India, only approximately 8% of residential energy needs are met by the grid, while around 80% of the residential energy is met by the use of solid fuels, such as biomass and coal. The remaining fraction represents the use of liquid fuels in the residential energy mix.

Data collected by the National Sample Survey Office (NSSO) shows that, despite the decline in biomass consumption over the past few decades, even today biomass comprises about 32% of India's primary energy consumption. From 1980-1981 to the period of 2000-2001, the share of biomass consumed by households decreased from over 90% to a little over 80%. However, looking at the overall biomass consumption, it is observed that it has increased continuously.

From Figure 3.8, which compares rural and urban energy consumption, we can see how the rural consumption patterns have not undergone any significant transition, whereas the urban consumption has been marked by rapid substitution of traditional fuels. It can be seen that rural residential energy consumption still depends on biomass to a large extent. The urban household energy mix, on the other hand, is increasingly dominated by commercial fossil-based energy sources and consumption of electricity. Even with this change, however, the presence of biomass still continues even in urban households.

The overall energy trend is characterized by an increase in per capita electricity usage, but this is attributed mainly to the consumption in urban households, as the proportion of commercial energy used in rural households is still low. Also, even though gas, oil, and electricity consumption has risen in rural areas, it is still much lower in aggregate and per capita terms, as compared to urban households.

It is observed that the total residential energy consumption in rural households always exceeds that of the urban households. The cause of this is attributed to the increasing dependence on inefficient solid fuels in rural areas, while urban areas switch to more efficient fuel sources.

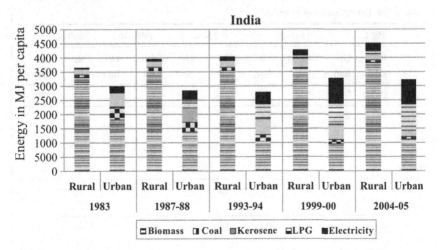

FIGURE 3.8 Per capita energy consumption patterns in urban and rural households.
Source: Household consumer expenditure surveys

The study of the use of different sources of household energy, in terms of percentage of population using them in rural and urban households, also gives insight as to what the trends are and how they differ between the two cases. From the table on percentage of population using different sources of household energy, it can be observed that the consumption of various fuel sources in rural or urban households does not add up to 100%. This reflects how in each household a mix of different energy sources is consumed (Tables 3.1 and 3.2).

In rural consumption, the most dominant fuel sources are kerosene, fuel, wood, and dung. The dependence on coal/coke, kerosene, fuel wood, and dung does not seem to vary significantly in the rural areas while we see the use of sources such as LPG and electricity rise. In the case of urban households, however, a significant halving of dependence on biomass sources is observed. Urban households have also shown a shift in transition towards greater dependence on LPG and electricity and less on biomass, coal/coke, and kerosene.

It is observed that as rural expenditure rises, the consumption of all types of energy sources also rises, without any bias. Only consumers in the top decile in rural areas show a preference for substitution of solid fuels by non-solid energy forms. In modern households, however, the transition to cleaner modern commercial fuels is more apparent with increases in household expenditure. The share of biomass used decreased from 65% among the poorest decile to 5% for the richest decile. The top decile exhibits a clear shift away from biomass towards more electricity and LPG use. Kerosene and coal are observed to be transition fuels as in the lower urban deciles the dependence on them increases, whereas in the upper urban decile the dependence decreases. In urban households the energy consumption in the urban middle-income decile has a mix of energy types, including more efficient modern fuel that is established to produce useful energy.

TABLE 3.1 Energy sources in India

	1983		1987-88		1993-94		1999-00		2004-05	
	Rural	Urban	Rural	Urban	Rural	Urban	Rural	Urban	Rural	Urban
LPG	0	9	1	20	2	33	6	47	12	61
Coal/coke	3	21	3	14	2	8	2	5	2	5
Electricity	15	58	24	67	36	77	47	84	54	91
Kerosene	95	92	96	88	95	83	96	75	91	55
Fuel wood	86	61	89	50	88	42	88	35	88	35
Dung	53	27	56	24	53	18	52	12	46	10

TABLE 3.2 Potential of energy sources in India

Source/system	Estimated potential	Cumulative installed capacity/number
Wind power	45,000 MW	3595 MW
Biomass power	16,000 MW	302.53 MW
Bagasse cogeneration	3500 MW	447.00 MW
Small hydro (up to 25 MW)	15,000 MW	1705.63 MW
Waste to energy		
Municipal solid waste	1700 MW	17 MW
Industrial waste	1000 MW	29.50 MW
Family-size biogas plants	12 million	3.71 million
Improved chulhas	120 million	35.20 million
Solar street lighting systems	-	54,795
Home lighting systems	-	342,607
Solar lanterns	-	560,295
Solar photovoltaic power plants	-	1566 kWp
Solar water heating systems	140 million m^2 of collector area	1 million m^2 of collector area
Box-type solar cookers	-	575,000
Solar photovoltaic pumps	-	6818
Wind pumps	-	1087
Biomass gasifiers	-	66.35MW

For the same levels of expenditure in rural and urban areas, the household energy consumption mix differs, with the urban household mix containing more modern fuels and services. As urban areas have higher population density and due to the restrictions in space for fuel storage and collection, a need for delivery of higher density fuel and electricity exists. Also, it is easier to provide services such as electricity and fuel supply in urban areas at a lower cost, as compared to rural areas that are remote and have a population of lower density and lower purchasing power. In fact, the quality of energy services, such as electricity, is poorer in rural areas as compared to urban areas. All these factors hamper the availability of different modern fuels in rural areas and decrease the possibility of enhancing rural access to modern energy services.

The fuel choices due to affordability and availability are also reflected in India on a state level. States in India such as Maharashtra, Delhi, and Gujarat, where urbanization and income levels are higher, use a larger portion of commercial fuels and electricity as compared to less developed and poorer states in India. Geography and resource endowments in a state determine the fuel choices in that state to a large extent as well. Availability of coal due to mining activities in a state, such as Bihar and West Bengal, leads to a greater dependence on coal as a fuel source, as does the availability of hydro-based electricity in the states with rivers and suitable terrain, such as the northeastern states of India.

As fuel choices are also determined by personal choices, demographic factors such as sex and education of the head of the household or decision maker, determine the fuel mix used in households. As women and children are usually the most exposed to indoor pollution, women tend to support the substitution of biomass fuel sources. However, the head of the family and decision maker is usually male in India, and hence this consideration of exposure to indoor pollution is usually overlooked in the purchase of fuel. A lack of education and knowledge of the benefits and adverse effects of different fuels also leads to an uninformed selection of fuel. It is observed that as the educational level of households improves a shift to the use of more efficient fuels is observed.

3 CURRENT ENERGY SCENARIO IN INDIA

According to the World Coal Institute, India is currently the world's eleventh largest energy producer and accounts for approximately 2.4% of the world's total annual energy production. It accounts for around 3.7% of the world's total annual energy consumption, which places it as the 6th largest energy consumer. India is also the 6th largest electricity generator, accounting for almost 4% of the global annual generation, as well as the 6th in terms of electricity consumption.

India's electricity production relies heavily on coal energy sources. A strong second, as can be seen from Figure 3.9, is hydropower, and the third is natural gas.

Including large-scale hydro projects, about one third of the total energy consumed is contributed by renewable energy technologies. Coal currently provides 69% of the electricity demand in India and will continue to be a major source in the future. India has around 10% of the world's coal reserves, but this coal is of low quality. This poor quality coal is an inefficient source and highly polluting. Growing concerns for the environment have also driven the need to find substitutes for this energy source.

To meet the growing demand, energy imports such as oil and natural gas have been increasing and energy security and less dependence on imports has become a critical factor to consider. As energy is crucial in terms of development, the Ministry of Power has targeted rural electrification of 100,000 villages by 2012. The Indian government has announced plans to provide power to the entire population by 2012 — this would require an additional 68,500 MW of base capacity.

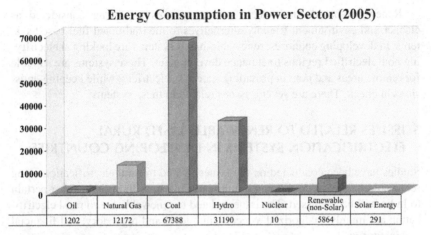

Energy Consumption in Power Sector (2005)

	Oil	Natural Gas	Coal	Hydro	Nuclear	Renewable (non-Solar)	Solar Energy
	1202	12172	67388	31190	10	5864	291

FIGURE 3.9 **Energy consumption in Indian power sector in 2005. Source: GENI (Global Energy Network Institute)**

It is observed that the effective price per unit of useful energy being paid by the rural population was significantly higher than that paid by urban residents, as rural households use a higher share of inefficient traditional fuels. By accounting for all fuels and electricity used, it can be seen that rural households in 1999-2000 spent a higher portion of their total household budget on energy, as compared to urban households.

India has highly subsidized energy for the household sector. It has undergone attempts at liberalization and price reforms in the past, but a large part of the supply still remains state controlled. The subsidies delivered in India do not vary across rural and urban households, despite the fact that on average, rural incomes and expenditures are significantly lower than those of urban households.

4 RENEWABLE ENERGY TECHNOLOGIES IN INDIA

India is rich in renewable energy resources whose potential has not been properly tapped. It has one of the largest programs in the world for deploying renewable energy products and systems. The government of India has, in fact, set up an exclusive ministry for renewable energy development, the MNRE, the first and only such ministry in the world.

The main sources of renewable energy in India are biomass, biogas, solar, wind, and hydropower. Renewable sources contribute about 5% of the total power production in India. India is also the world's highest biogas user and is fifth in the world in terms of both wind power and photovoltaic production. From the table on estimated potentials and cumulative installed capacity, it can be seen that none of the potentials of the various renewable sources have been reached.

Renewable-based rural electrification systems are being considered as cleaner and environment friendly alternatives to the traditional fuel-based systems. In developing countries, renewable-based systems are looking at electrifying non-electrified regions to enhance development. These systems are suitable for remote areas and help in providing access to electricity while keeping emissions in check. There are several issues related to these systems.

5 ISSUES RELATED TO RENEWABLE-BASED RURAL ELECTRIFICATION SYSTEMS IN DEVELOPING COUNTRIES

Studies have been conducted on the issues related to rural electrification using renewable energy in developing countries. Here, only those issues that pertain to India will be discussed. The issues related to renewable-based rural electrification are mainly categorized as economic, legal and regulatory, and financial and institutional.

The lack of subsidies, high initial capital costs, and high transaction costs for small decentralized systems are some of the identified economic barriers. Rural electrification programs involving renewable energy technologies usually involve consumers with low incomes who cannot afford to pay for technologies with high initial capital costs by themselves. A lack of subsidies, financing, and credit especially affects these users, rendering these technologies unaffordable and unfeasible to them [8]. In rural areas, consumers are also scattered over large areas and therefore the additional expense of distribution networks and high transaction costs for small decentralized systems add to the burden. The demand and use of electricity in rural areas is also low and this makes it less profitable for private companies to offer their services. Failure to incorporate fuel cost risks for fossil fuels and the lack of pricing policies that take into account the real economic costs of environmental damage also do not promote a shift to cleaner renewable energy-based electrification. The gradual extension of grid networks to rural areas also hampers the development of renewable systems in rural areas.

Identified legal and regulatory barriers to renewable-based rural electrification programs consist of inadequate legal and policy frameworks for renewable energy power sources and burdensome requirements of the small power producer that are set by their use. Without adequate legal frameworks in place, private companies are normally reluctant and unable to participate in the programs. A policy framework, on the other hand, enables the encouragement of the implementation of the technology and provides incentives for private companies, users, and other involved parties. The improper use of subsidies prevents the subsidies from effectively targeting the desired results and also slows down progress in the programs.

Donor dependency in such programs hampers the development since when there are restrictions in the funding from donors, the program comes to a standstill. Unrealistic political commitments are also identified as barriers. Unrealistic commitments by governments do not motivate the program implementers to achieve the goals as they know the set goals are too unrealistic, and in fact de-motivate them. Governments of developing countries are known to make commitments based on political interest, and these interfere with systematic development of such programs as the electrification plans can be altered by the politicians at any time according to their interests.

The identified financial and institutional barriers involve the lack of access to credit for both consumers and investors, and the lack of sufficient technical, geographical, and/or commercial information. A reliable and stable line of credit must be available to both consumers and investors to promote the setup and the use of these systems, despite their high initial capital cost. A lack of information makes the setup of these systems more onerous and decreases the attractiveness of investment in them as the market participants are unable to make sound calculated economic decisions. A lack of institutional capacity and technical knowledge prevents provision of guidance and support during the setup and promotion of such systems. Also, the problems related to such programs require decentralized problem solving and such institutional support is essential.

6 POLICIES RELATED TO ELECTRICITY IN INDIA

In this section the policies that are being put in place in India and how these policies are oriented towards renewable energy technologies are discussed. We will look at the National Electricity Policy of 2005 and the Electricity Act of 2003.

6.1 National Electricity Policy of 2005

Some of the relevant objectives of the National Electricity Policy (2005) include:

- Access to electricity: availability for all the households in the next 5 years (2010);
- Availability of power: demand to be fully met by 2012. Energy and peaking shortages to be overcome and spinning reserve to be available;
- Supply of reliable and quality power of specified standards in an efficient manner at reasonable rates;
- Per capita availability of electricity to be increased to over 1000 kWh/capita by 2012;
- Minimum lifeline consumption of 1 kWh/household/day by 2012;
- Financial turnaround and commercial viability of electricity sector; and
- Protection of consumers' interests.

6.2 The Electricity Act of 2003

Some of the salient features of the Electricity Act (2003) include:

- Regular preparation and publication of the National Electricity Policy and Tariff Policy;

- Preparation and notification of a national policy, permitting standalone systems (including renewable-based systems) for rural areas;
- Terms and conditions of determination of tariff to be guided by the promotion of cogeneration and generation of electricity from renewable sources and the National Electricity Policy and Tariff Policy; and
- State commissions will discharge the functions of promotion of cogeneration and generation of electricity from renewable sources by providing suitable measures for connectivity with the grid and sale of electricity to any person and specification of a percentage of the total consumption of electricity in the area of a distribution license, for purchase of electricity from such sources.

These policies address some of the issues with renewable-based rural electrification while also pointing out what the government is focusing upon. This will be analyzed later. Various renewable energy sources and the technologies associated with them as well as their roles in electrification will now be discussed. Only rural electrification as well as household applications that are useful for rural households and livelihoods will be discussed, as the energy poor are mainly concentrated in the rural parts of India.

6.3 Solar Energy

The current solar-based installed electrical capacity is approximately 1.4% of the total. Most parts of India receive 4-7 kWh of solar radiation per square meter per day with 250-300 sunny days in a year [69]. Solar energy intensity varies geographically, with western Rajasthan receiving the highest annual radiation energy and the northeastern regions receiving the least.

Solar energy can be used through the thermal route or the photovoltaic route. A few applications of the thermal route are water heating, cooking, drying, water purification, and power generation. Through the photovoltaic route it can be used for applications such as lighting, pumping, communications, and electrification of villages. There are several applications with respect to solar energy. Cooking is one such application of solar thermal energy. Solar cookers have no recurring fuel expense and are estimated to save three or four LPG cylinders per year on regular use. As they do not pollute the environment, exposure to indoor pollution by biomass is reduced with their use. Solar cookers cook slowly and hence produce better and more nutritious food, which is a concern for households with low income and scarce meals. Box solar cookers are usually the most suitable for small individual households. The cost of a box solar cooker varies from around $24 to about $51. Bigger and more expensive community solar cookers are available, which could be useful in schools to provide students with healthy and nutritious food and encourage attendance and prevent them from falling ill. The Indian Ministry of Non-Conventional Energy Sources (MNES) provides financial support for larger types of solar cookers but not for box solar cookers.

Agriculture involves various time-consuming stages in processing, and drying is one of them. Drying in the open sun is not only time consuming but also unhygienic. Solar dryers can be used to dry crops and other products. They come in a variety of types and sizes, and therefore they can be utilized for various domestic purposes as well as in agricultural processes. The disadvantages of these systems is that they are slower than the dryers using conventional fuels and they can be used for drying only at 40-50°C. The MNES is implementing a national program on solar thermal energy, which provides an interest subsidy in the form of soft loans, as these systems are quite capital intensive. The manufacture of specific purposes and sizes as per requirements is restricted in India and hence large systems have to be set up on a project-by-project basis.

Currently, there are 14 photovoltaic companies that manufacture photovoltaic modules and 45 companies that manufacture solar photovoltaic systems. The Bureau of Indian standards has established photovoltaic standards and the MNES has established facilities for testing solar cells, PV modules, and systems.

Solar photovoltaic lighting systems are becoming popular in rural areas of India. They are used in the form of portable lanterns, home-lighting systems with one or more fixed lamps, and street-lighting systems. Solar lanterns are light and portable, normally designed to provide light for 3-4 h of light daily. This aids in providing light in huts beyond daylight hours. They normally cost around $62-$68 depending on their capacities.

Solar home systems (SHS) provide comfortable levels of illumination in the rooms of a house. Various models of SHS feature one, two, or four compact fluorescent lamps. Small DC fans or 12-V televisions can also be run by these systems. These systems are designed to work for 3-4 h daily, with an autonomy of three days, which means they can function for three cloudy days. Different SHS models differ in cost depending on the number of compact fluorescent lamps, fans, or televisions they can run. They range from a model that can operate one 9 W compact fluorescent lamp at $124 to a model that can operate four 9 W compact fluorescent lamps at. Solar lighting systems are used to illuminate a street or open areas in villages. These are designed to operate from dusk to dawn automatically. The cost of these systems is about $39. The Ministry of New and Renewable Energy, India provides financial assistance for the promotion of these systems among eligible categories of users.

Solar photovoltaic (SPV) power plants generate electricity centrally and make electricity available to users through a local grid, standalone mode, or connect to the conventional power grid in a grid-interactive mode. Power plants are preferred over the individual SPV systems when users are in close proximity. Standalone SPV power plants are usually implemented where conventional grid supply is not available, or is erratic and irregular. They are used mostly to electrify remote villages. Their capacity varies from 1-25 kW_p and can even be higher depending on use. They provide constant, stable, and reliable supply to their customers. These systems can also operate where grid supply is available, and the power plant works as a hybrid power plant. The cost of such a power plant depends on the capacity it is built for. The approximate cost is between $6200-$6800/kW of photovoltaic capacity. The distribution costs are not included and would add to these estimates.

For an agricultural-based country and in a geographic region where sufficient drinking water is unavailable to a significant population, such as India, SPV pumping systems are another important application of photovoltaic energy. These pumps can be used to draw drinking water, as well as water for irrigation. The cost of these pumps depends on the capacity and the type of pump.

6.4 Wind Power

The most important application of wind energy is the generation of electricity. With 2,980 MW of installed wind power capacity, India currently ranks fifth in the world in terms of wind power capacity. India's current technical potential is estimated at about 13,000 MW, assuming 20% grid penetration. However, wind power potential has been assessed at 45,000 MW. A wind power program was initiated during the Sixth Plan, in 1983-1984, in order to survey and assess wind resources, to set up demonstration projects and provide incentives, and to make wind electricity competitive.

As wind power density is not uniform, only certain states have this resource, while others do not have sufficient wind. The windiest states are Gujarat, Andhra Pradesh, Karnataka, Madhya Pradesh, and Rajasthan. In fact, the Centre for Wind Energy Technology has set up a Wind Resource Assessment Program to tap suitable resources and has even published wind data it has generated in order to promote wind power. It has even set up master plans to provide information on the availability of wind, land, grid availability, and accessibility to the site for the benefit of project promoters.

Several promotional incentives have been made available for wind power projects. The incentives available are 80% accelerated depreciation in the first year and concessional import duty of 5% on five specified wind turbine components and their parts. Favorable tariffs and policies are also available in several states. These are aimed at promotion of wind power projects by private investors. India also produces wind turbines, mainly through joint ventures or under licensed production agreements.

For decentralized mode rural applications such as pumping and power requirements, wind energy can be harnessed through water-pumping windmills, aero-generators, and wind-solar hybrid systems Water pumping windmills can be used to pump drinking water and water for minor irrigation from wells, ponds, and bore wells. They do not need high wind speeds and can start lifting water when the wind speed approaches 8-10 km/h, and they have the capacity to pump around 1000-8000 l/h. Windmills, however, can operate only in locations with medium wind regimes and with no surrounding obstacles such as buildings or trees. The system costs are also very high and are therefore not affordable to many individual users. The cost of setting up a water-pumping windmill varies from about $1340 to about $3600. The Ministry of New and Renewable Energy, India provides for the previous investments of the cost of setting up the system, and even higher subsidies for unelectrified islands.

Small wind electric generators (aero-generators) can be installed as stand-alone systems or along with solar photovoltaic systems to form a solar hybrid system for decentralized power generation suitable for unelectrified areas. The optimum wind speed for aero-generators is at about 40-45 km/h. Batteries can be charged and used during non-windy periods. The cost of these systems is around $4000-$5000/kW, and they require maintenance work of about $40 per kW per annum.

Wind-solar-based hybrid systems produce mutually supplemented power, thus offering reliable and cost-effective decentralized electric supply. Batteries can be charged for use when required. The cost of the system varies from $5000-$7200/kW. Installation costs are about $205, while maintenance and repair costs are about $60 per kW per annum. The Ministry of New and Renewable Energy provides subsidies for community use as well as higher subsidies for non-electrified islands.

6.5 Hydropower

India is endowed with hydro resources that are both viable and economically exploitable. In fact, hydropower is the second highest contributor of energy consumed in the power sector. Large hydropower projects are being utilized for power production and account for most of the energy consumed that comes from renewable sources. The hydro potential has been assessed to be about 84,000 MW, at 60% load factor. In addition to this, 6780 MW of installed capacity from small, mini, and micro-hydel schemes have been assessed, and 94,000 MW of installed capacity have been assessed for pumped storage schemes. However, only 19.9% of the potential has been harnessed so far.

As hydropower has been tapped and used for grid purposes, small hydropower will be focused on in this discussion, as small hydropower has small-scale applications that would benefit the energy deprived. Small-scale hydropower (SHP) projects are hydropower projects with a station capacity of up to 25 MW each. These can be set up on rivers, canals, or at dams and are flexible in terms of installation and operation. The technologies and manufacturing base are indigenous in SHP projects. Small hydropower project equipment is being manufactured in India. According to the MNES, 11 small hydropower equipment manufacturers currently operate in India. Hydropower is also environment friendly as it causes little or no submergence, minimal deforestation, and minimal impact on flora, fauna, and biodiversity. These projects are in fact even compatible with use of water used for irrigation and even drinking water.

The cost of SHP projects depends on where they are set up, i.e., the location and the site's topography. The cost of the civil works and the equipment usually determine the cost of the project. They normally cost between $1 million to about $1.4 million per MW. The payback period of these projects is usually 5-7 years, and this depends on the capacity utilization factor.

The MNES has so far installed 523 SHP projects with an aggregate installed capacity of 1705 MW and has 205 SHP projects with an aggregate capacity of 479 MW under implementation. The SHP sector is increasing its competitiveness with other alternatives through increase in capacity by an average increase of 100 MW per year, and through reduction in the gestation period and capital cost. Databases containing potential SHP sites have been created and a program involving subsidies is being planned to encourage development of SHP plans in states across India. In order to be eligible for subsidies, standards for quality have been put in to place.

States in India have put policies such as low interest rate loans in place to attract private sector entrepreneurs to set up SHP projects. Tariffs are being determined by the state electricity regulatory commissions, keeping the interest of stakeholders, developers, and the MNRE in mind. Several leading financial institutions have started financing SHP projects, thereby providing stable and secure financing options.

Water mills are an example of converting the energy of water into useful mechanical energy. They have been used in north India for applications such as grain grinding and oil extraction. These normally have low conversion efficiency, and improved systems have been developed for electricity generation. The MNRE has a scheme for water mill development through associations, cooperative societies, NGOs, local bodies, and state nodal agencies. The scheme subsidizes up to 75% of the actual cost if lower than the ceilings. These have been effective in remote and rural areas with suitable topography in states such as Uttaranchal in North India.

6.6 Biogas

India is currently the fifth largest consumer of biogas. The biogas derived from animal waste in India is mainly from cow dung. The estimated potential of household biogas plants derived from animal waste in India is 12 million plants. Under the National Biogas Program over 3.7 million biogas plants of 1-6 m^3 were installed until December 2004. Since then, the MNRE has established larger units in several villages, farms, and cattle houses. An estimate of 3.5 million m^3 per day of biogas production is provided by the MNRE. This is equivalent to a daily supply of about 2.2 million m^3 of natural gas.

Biogas plants in India provide gas for cooking, lighting, and power generation. Biogas is used with specially designed burners for cooking and plants of 2 m^3 capacity are sufficient for families with four to five members. Gas lamps are fueled by biomass and for a 60 W lamp, 0.13 m^3 of biogas is required per hour. It can also be used to generate mechanical and electrical power by replacing 75% of the diesel used in dual-fuel engines. Efforts are being made to generate electricity from biogas in a decentralized mode by means of small-capacity engines. Thus, standardized models of biogas plants suitable even for households are available. For family-type biogas plants the fixed-dome and floating-drum plants of 1-4 m^3 capacity is usually used.

In India, not only is biogas plant-related technology and infrastructure being developed, but training and deployment of skilled manpower for plant construction and maintenance has also been taking place. The plants also take care of treatment of industrial and urban waste to a large extent, as the technology of using anaerobic digestion to treat these wastes and produce biogas has been successful. The biogas slurry is rich in a water-soluble form of nitrogen and hence can be used as organic manure. It is also rich in bacteria and can therefore be used for composting along with biomass. Hence, these plants not only provide a more efficient energy source, they also provide employment to a large number of people in the rural areas and provide solutions to environmental problems, such as waste and manure handling, water pollution, and carbon monoxide emission.

The lifespan of biogas plants in India can be estimated to be about 10 years [47]. Based on life cycle analysis, the estimated cost of generation of 1 m^3 of biogas is 6 cents, and this is said to be more cost-effective than the cost of diesel and kerosene when comparing energy values. The MNRE has also estimated that the construction of 1 million biogas plants generates about 30 million man-days of employment for skilled and unskilled workers and 1.2 million tons of organic manure per year. It also leads to abatement of greenhouse gas emissions through anaerobic digestion of cow dung and effective utilization of biogas.

Financial assistance for construction and maintenance of biogas plants, development of skilled manpower, training for use and maintenance, awareness creation, and support to implementing agencies and technical centers for implementation are provided by the MNRE. The National Biogas Program in India provides various central subsidies for setting up biogas plants and the amount depends on the users and the area, thus promoting the setup of these plants in remote and underdeveloped areas and for users who do not have access to electricity, as well as those who have greater development and energy needs. The government of India also provides repair and servicing facilities to support the operation of these plants. The state governments have also implemented programs targeted at implementation and training support. They regularly help organize training of workforce in the construction, maintenance, and use of the biogas plants. The training centers also provide technical training and publicity support for the National Biogas Program.

7 RESULTS

After observing the household energy trends it is obvious that there is still great dependence on solid fuels. The percentage of residential energy needs met by solid fuels was only 80% and as these fuels are highly inefficient, polluting, and create health burden, cleaner and more efficient alternatives must be provided.

Also, there was an increase in the overall biomass consumption even though dependence on it in households decreased. The drastic difference observed in the trends between urban and rural consumption reflects differences in preferences of fuels due to unequal development and differences in the availability and affordability of fuels. This is also supported by the fact that the population in rural areas that earns as much income as those in urban areas still consumes a greater percentage of solid fuels due to lack of accessibility to other alternatives.

Differences in consumption patterns varying geographically between states according to availability of resources as well as differences in the extent of development were also observed. All these factors imply that the government of India must play a significant role and put relevant policies in place to promote the use of grid electricity over solid fuels, as well as be more proactive by providing more energy source alternatives along with the pace of the growing population and its needs. Also, the policies must be targeted towards rural areas and the main consumers of the solid fuels. Subsidies should not only be determined by what resources are easily available, but also keep in mind the impacts and indirect costs, such as costs to the environment.

India encounters the problem of a negative balance in overall energy consumption and production, and the growing population adds to the problem. The majority of electricity produced comes from the use of coal, while only a small percentage comes from renewable energy sources, which does not include large-scale hydro projects. The commitment of the government is to provide the entire population with electricity by 2012, to increase per capita availability of electricity from its current figure of around 380-421 kWh/capita to over 1000 kWh/capita by 2012, and the achievement of the development goals. Each of these goals has an associated energy cost and entails an increase in the requirements of energy. In order to stick by the commitment made by the government through the ratification of the Kyoto Protocol, sustainable options must be focused on. The government of India has taken steps toward this by putting much needed electricity policies in place and by focusing on renewable energy technologies.

For remote villages that are not connected to a grid, decentralized application of renewable energy sources provide a suitable alternative. The Intergovernmental Panel on Climate Change (IPCC) report findings also show that it is the right time to look at alternative energies for rural energy supply. Successful rural electrification programs have shown that a sustained renewable energy market can develop quickly and efficiently through the right combination of institutional and financial regulations and if adequate energy policies are adopted. The government's electricity policies display their intention of overcoming hindrances to electrification, especially to rural electrification, but some issues are still not resolved and there is scope for further improvement.

The government's measures such as the setting up of the MNRE show its intention of having renewable energy sources play an important role in the further electrification of the country, and in the eradication of energy poverty. However, the potential of renewable energy sources has not been tapped in India.

This shows a great space for improvement, provided the right encouragement is provided by the government.

Solar thermal energy is a cost-effective and reliable form of energy and it has several household applications relevant to rural areas, where access to modern energy fuels and services is limited. The photovoltaic route in India has also been well established. Even the manufacture of photovoltaic cells takes place in India. The government's assistance in terms of setting standards and setting up testing facilities for solar cells, PV modules, and systems helps encourage high quality in the photovoltaic manufacturing sector in India. The targeting of categories of users for the solar home systems by the government is useful in providing benefits for those who need it.

For wind energy there are greater limitations, as the resources are not evenly distributed. For small-scale wind projects, there are obstacles that cannot be helped, such as the need for perfect isolated locations with the current wind speeds. These limitations do not make them a universally ideal solution to the need for rural electrification. However, India has not tapped its full wind energy potential and through the infrastructure and programs set up, wind energy sources will provide a lot more benefits in the future. By taking care of maintenance and providing installation subsidies, the government of India is trying to make wind energy attractive. State involvement assists in providing the required decentralized support to systems that provide rural electrification.

Hydro projects have excellent efficiency, low maintenance costs, and low costs of generation. Small-scale hydro projects have the advantage of using indigenous technology that is manufactured in India, and they cause minimal environmental damage. The technology for these systems is rapidly improving and becoming more and more competitive. The government's support, through subsidies and setting up of databases, is encouraging the development of these systems and making them more affordable. Financial assistance through loans and subsidies is also helping the setting up of these systems. The involvement of state nodal agencies thereby provides the decentralized support required.

The use of biogas in households also saves women and children from exposure to indoor pollution and the time as well as the effort they spend in collecting firewood and cooking with inefficient fuel sources. Thus this technology provides various environment-related advantages and social-related advantages, while providing energy solutions to energy deprived rural areas. Financial assistance for construction and maintenance of biogas plants, development of skilled manpower, training for use and maintenance, awareness creation, and support for implementing agencies and technical centers for implementation are provided by the MNRE. Therefore, the government is not just providing financial support and technical assistance for biogas plants, they are also enhancing development by providing job opportunities to those living in rural areas. They are providing training as well as employment for people from rural areas, thereby bringing about development in more than one way.

Although renewable energy technologies might require low-cost or free fuel, they incur high upfront capital costs. The cost of electricity produced (UNDP - United Nations Developed Programme, Annual report, 2004. 2004) by the solar photovoltaic route is about 25-160 cents/kWh and by the solar thermal route around 12-34 cents/kWh. In India, in fact, the cost of grid-interactive power-plant-produced electricity is around 31 cents/kWh and from standalone power plants it is more expensive. Cost of electricity production for wind energy is 4-8 cents/kWh and for small hydroenergy plants is 2-12 cents/kWh United Nations Developed Programme Annual report (UNDP), 2004. While based on life cycle analysis, the cost estimated for the generation of biogas from a family-type plant is 6 cents, and by conventional coal electricity generation is about 4 cents/kWh. In terms of costs of electricity generation, conventional coal technology is much cheaper in comparison to renewable energy source technologies. As the customers of electricity in rural areas are from lower income groups and private parties will not invest unless it is commercially viable, the burden of these costs must be carried by the government.

8 CONCLUSIONS

As a considerable percentage of the world's poor reside in India, the issue of sustainable development is a key concern. It has been established that access to clean energy enhances development by providing several opportunities and improving the quality of life. Most of India's poor live in rural areas and hence a greater focus must be placed on the access of energy sources to these regions. By announcing their development goals and by announcing the goals of providing electricity to the entire population by 2012, the government of India is committing itself to development and reduction of poverty [68].

By making electric power accessible to the whole population by 2012, the Indian government is ensuring the eradication of energy poverty. Increasing the reliability of quality of power and its efficiency at reasonable rates helps target the needs of people of lower income groups. This helps ensure proper electrification for the regions that, although declared electrified, have poor services and cannot reap the benefits of electricity. An increase in per capita consumption, which is stated as a policy, does not imply equitable distribution, but it does imply a greater level of development. Making electric power commercially viable will decrease dependence on sponsors and make it profitable for private parties to invest in, thereby reducing the dependence on sponsors. The involvement of private parties in fact will reduce maintenance costs, reduce overuse, and also maximize benefits from these systems.

The protection of consumers' interests ensures that the energy deprived are kept in mind and are not being exploited by private investors who enter the energy market. Regular preparation and publication of policy and tariffs will help establish fixed guidelines for the sector to avoid ambiguity and promote private investment. National policy for permitting standalone systems including renewable systems for rural areas implies making it easier for private investment to set up these projects and promotes such projects in general. The intended determination of tariffs based on generation and cogeneration from renewable sources shows how there will be an added advantage for renewable-based technologies, hence promoting investment in them. Establishing the responsibilities of the state government addresses the decentralized problem solving that is required to deal with issues related to rural electrification, which are decentralized in nature themselves. The Indian government has also established several programs and institutions to increase reliability in this sector that makes purchase and payment of these systems easier.

The government has also addressed the problems of maintenance and monitoring by taking up this responsibility through its various institutions. This shares the burden with the involved parties and also builds confidence in the systems. The policies of the government, however, do not explicitly address the current electricity subsidies. Subsidized electricity does not target the increase in energy access to energy-deprived rural households, and in fact it benefits richer urban households who have greater access to more modern energy technologies.

A large number of programs and initiatives has been carried out by the government of India. However, it has set unrealistic targets in terms of its development goals and reduction in energy poverty. To achieve these targets, the initiatives taken so far are insufficient. For the achievement of these targets through sustainable means, renewable energy technologies must be encouraged to a further extent. The Indian government has made an ambitious start. However, the different levels of local, state, and national government, need to increase their efforts to a much greater extent to achieve significant results.

Chapter 4

Energy Management – Implementation

4.1 DEVELOPMENT OF STRATEGY OF SUSTAINABLE MANAGEMENT OF ENERGY

The implementation of strategy is to the core step in the process of strategic management. No matter how difficult it is to develop the strategy for sustainable energy management, it is even more difficult to implement it. The strategy is the planning framework for taking actions that would use energy resources adequately. Therefore, the implementation of the sustainable management of energy is an important field in strategic management. The strategy is first developed and then executed, but the danger for the (lack of) success exists in both areas, in the process of developing a strategy and the process of implementing the strategy, as is illustrated in Figure 4.1.

In developing strategy, first it is necessary to consider the scope of the problem and the time available for its solution. The transition to sustainable management of energy is a significant change, which depends on the intensity of energy consumption in the company, but also on the method of management and current characteristic of organizational culture in the indicated company. The strategy can be planned and implemented in different ways, with different final outcomes:

- Success occurs when the strategy of sustainable energy management is developed well and implemented well, which means the internal and external opportunities have been identified on time and organizational, resource, motivational, and other conditions have been created for the implementation of the strategy. This situation is certainly the most desirable for the successful process of sustainable energy management.
- A roulette situation is characterized by a poorly developed but well-implemented strategy of sustainable energy management, where deficiencies have been identified during the implementation of the strategy and can be surpassed. This situation can be found relatively often in practice and shows that there are still major problems in the planning and implementation of sustainable energy management. This situation can be partly justified by the fact that the transition to sustainable energy management is a huge undertaking and includes a great number of uncertainties.

Sustainable Energy Management. http://dx.doi.org/10.1016/B978-0-12-415978-5.00004-7

Developing strategy

	Good	Bad
Good	Success	Roulette
Bad	Difficulties	Failure

Implementation of strategy

FIGURE 4.1   The relationship between the implementation and development of strategy

- Difficulties arise if the strategy is well formulated but poorly implemented. It often happens in practice, but in this case the organization would change the existing strategy rather than remove barriers that prevent implementation of the strategy. Taking into account the fact that the strategy of sustainable energy management is a very long process, this situation is particularly undesirable.

- Failure is a situation that occurs when a poorly formulated strategy is poorly implemented, leading to confusion and failure. This case can hardly be corrected. In an attempt to create and implement strategies for sustainable energy management, this is certainly the least desirable situation and involves a great number of financial, environmental, organizational, and other problems whose impact is great and long-term.

The actual result of the process of planning sustainable energy management should be concrete actions (implementation strategies), while planned decisions are only a byproduct of planning. Through the implementation of the strategy the company transforms ideas into concrete achievements. This process requires that the goals are transformed into concrete tasks, activities are identified, the necessary resources allocated, and the business culture integrated.

In the process of developing a strategy, entrepreneurship and visionary leadership qualities are emphasized, and in the process of implementation it is important to use human resources to achieve the goals. Therefore, the implementation of strategy means taking actions to bring into practice the intentions stipulated by the mission and goals. The implementation runs the activities of the company, i.e., people, technology, and resources, and the management of the company has to anticipate possible problems, conflicts, production bottlenecks, and other external and internal barriers.

Modern business is characterized by continuous change. Changes are quantitatively and qualitatively significant and far-reaching in the field of energy production and consumption. These changes can be different: a change in the size of company, mergers and acquisitions, new competitors entering the industry, the emergence of substitutes, changes in the organizational structure of enterprises, changing market conditions, and a number of other factors. The existence of several large global companies for the exploitation of energy resources, powerful monopolies, and the growing demand additionally influence the process of creating a sustainable energy management. It can be said that the twenty-first century is a century of change, and that change is the only certainty. The scope of the problem and the urgency to solve the problems influence the complexity and the speed of implementation of strategies, as shown in Figure 4.2.

When a company is faced with small problems and the period of time available is long enough, the implementation of the strategy would require a shorter period of time. At the same time, the implementation of strategy will include evolutionary or gradual (incremental) changes. This situation is characteristic for enterprises and sectors that are not highly energy dependent. In such cases, the company has enough time to design, plan, and carry out certain activities in the field of energy management. These companies are potentially in the best position, because they have enough time to understand the specifics of the transition from traditional to sustainable energy management, to quantify the necessary changes, and to predict results and timely direct the company in the right direction. They have time enough, given the low energy demand, to foresee and implement the use of renewable energy sources, to develop activities to improve energy efficiency, and to reduce the energy waste to a minimum. In other words, these companies can and should use all the available interventions and evolutionary mechanisms for systematic resolution of the problem of energy solving. In that way, companies are able to gradually transfer to a more acceptable way of using energy, without endangering their own stability.

Another situation is typical for companies that consume less energy but for some reason are forced to react quickly. If the problems the company faces are small, but should be resolved in a short period of time, the activities are performed by management (management intervention) directly at places where the problems emerged. As modern business is characterized by rapid changes that often result in serious problems, companies are often in a situation to respond quickly to certain influences. Factors that often force a company to launch its managerial mechanisms and adjust its own energy strategy in a relatively short period of time include sudden changes in energy prices, changes in legislation, the pressure of the local community, filed lawsuits, environmental pollution,

| | | Time needed for the change of implementation | |
		Long	Short
The scope of strategy problem	Big	Sequential interventions	Complete interventions
	Small	Evolutionary interventions	Managerial interventions

FIGURE 4.2 Types of strategy implementations

and the like. The response of management must be quick, but certainly in conformity with the long-term global orientation of companies on sustainability in the field of energy needs. Of particular importance is that modern companies, regardless of their size or energy intensity, integrate their strategy of energy management into the existing strategic plan for the development of the company and that any decision concerning changes in the business is compliant with the adopted view that sustainable energy is the only way of energy management in the future. Not a single goal of the company must be set, if, among other things, it does not comply with this requirement.

When there are major problems, the company must react in a different way. This situation usually happens in companies that consume large amounts of energy. Depending on how much time the organization has to solve the problem, there are two strategies. If the company has a long period of time at its disposal and the problems are huge, it is necessary to react sequentially, i.e., in stages. In fact, most modern companies are facing a gradual need to switch to sustainable energy management and thus have at their disposal a certain period of time to adjust its business operations to it. Most countries have legislation that stipulates quite a long period of time to solve this problem. In this case, it is optimal that the enterprise acts on the principle of sequential (gradual, phase) interventions because in that way they can optimize the use of its resources and gradually resolve the existing problems. Usually, the company first adjusts its own rules and regulations on business operations, which should be adjusted to positive legislation. Then, the company defines its own energy strategy within which it determines its own vision and mission and adjusts its goals to the principle of sustainability. After that, the company gradually moves on to solving concrete problems (introduction of energy consumption from renewable sources, improving energy efficiency, implementation of energy savings, etc.), depending on its capabilities, needs, and support provided by the environment.

However, if there are problems that cannot wait, the company has to react quickly. Most commonly, this situation happens in companies that are large consumers of energy and suddenly find themselves in a situation in which a change in the approach to energy management is required. The reason is usually pressure from the environment, where the changes of energy prices and lawsuits are the driving factors for the company to react urgently. In these cases, adequate complete intervention is applied, while the management of the company must fit (synchronize) changes in all parts of the company. Solving of the problem in this way is connected with increased risk and often represents a hazardous undertaking, requiring considerable experience of decision makers. The sudden switch to another type of energy, the termination of cooperation with existing suppliers, and the attempt to avoid paying fines or similar in most cases cannot be resolved successfully in this way. The only way to avoid this approach in problem solving is to avoid having to solve huge problems in the field of energy in a short period of time. This is possible only if the strategy is directed in a timely manner according to the principle of sustainability.

The following ways of reacting to problems can be practically realized only by introducing changes in management of the company. Depending on the size of the company and management style there are several approaches to implementing the strategy of sustainable energy management, as follows:

- The command approach is typical of smaller companies or of situations in which the implementation of the strategy does not involve a lot of changes. This approach is based on the fact that the formulated strategy is executed by lower levels of management. In fact, the tasks of strategic management should be separated into "creative thinking" and "implementing" operations. These cases are usually detected in companies that are small consumers of energy. Creating a plan of transition to sustainable energy management is simple, and realization of goals does not require the increased use of the resources of the company. A company can fully convert to the use of sustainable energy sources without specific changes that would threaten the stability of the business.
- The organizational change approach requires greater activity of the top management, thorough understanding, and designing of the necessary changes in organization, planning, rewards, and control. This approach takes into account the need for forecasting and applying various organizational and other changes in implementation of the strategy. The above case is specific to companies that consume large amounts of energy, and are aware of the need to change the energy strategy, but for some reason, such determination is not known to all employees. In these cases company management makes decisions and foresees necessary changes that are further implemented without any special effort to familiarize all employees with them. This approach is particularly effective when companies are in a position to have to quickly change their energy strategy due to particular pressure, while management has neither the time nor the mechanisms to inform all employees.
- The collaboration approach is a process where managers of business functions and units are involved in the process of implementation so as to obtain more realistic estimates of implementation tasks. This approach motivates managers and encourages them to take on greater responsibility for the implementation. This approach is possible in cases when companies are small or large consumers of energy, but what they have in common is that they have enough time to switch to sustainable energy management. This approach is, however, exclusively reserved for companies with strong management structure, where there is no comprehensive inclusion of all employees in planning and implementation of change.
- The cultural approach is a process of designing a strategy and its implementation that includes all those to which the strategy applies. This is a broader approach than the previous one and is based on good communication and encouraging employees to identify their culture with the mission and vision of the organization, which will result in better operationalization

of the strategy. This approach is suitable for organizations that insist on the business culture and have rapid growth. Usually, these are organizations that carry out activities in sectors with high technology. This approach is also an excellent solution in organizations whose long-term development strategies are aimed towards corporate social responsibility, where there is an appropriate organizational culture and practice of involvement of all employees in all aspects of business. The transition to sustainable energy management essentially considers these approaches an ideal solution.

- The bottom-up approach insists on more intensive engagement of the direct executives in designing and implementation of the strategy. In this way the executives are encouraged to have creative thinking and initiative. Top management has the role of auditors of the proposed strategy of sustainable energy management in the company.

Obviously, each of these approaches has advantages and disadvantages, and the task of management is to create a flow of implementing the strategy, depending on the size of the changes caused by the implementation. For the implementation of any approach, it is very important to clarify the solutions through communication, to operationalize the objectives, and to encourage further development and motivation of the executives. The transition from a traditional to sustainable approach to energy management is a complex and long-term process that greatly influences the stability of business operations and therefore requires special attention of all persons involved in the decision-making process.

4.2 TRADITIONAL APPROACHES TO IMPLEMENTATION OF SUSTAINABLE ENERGY MANAGEMENT

Implementation of the sustainable management of energy refers to running activities and resources of the company so as to achieve the set goals in accordance with the strategic orientation. The transition from traditional to sustainable energy management is a major strategic change and involves huge utilization of all resources. Implementation of strategic orientation is not a single process, but a process that involves provision and allocation of resources, creation of organizational, procedural, motivational, and other conditions, and making a series of individual plans for undertaking and coordinating activities for the realization of the objectives, taking into account the sustainability of the environment.

The implementation of the strategy of sustainable energy management is a never-ending process since the external and internal environment is continuously changing and completely new and unique situations occur that require understanding the relationship between the functions of the company. It is necessary to establish priorities, activities, and holders. All predicted activities should be properly connected and synchronized. The management of implementation involves several elements or areas of activity, as shown in Table 4.1, and discussed in the following sections.

TABLE 4.1 Stages and elements of implementation of sustainable energy management

Creation of organizational structure for implementation of the strategy	Allocation and focusing of resources towards strategic goals	Galvanization of implementation of the chosen strategy	Establishment of an internal administrative system for support	Providing strategic leadership
1. Development of organizational structure for the support of strategy	1. Provision of budget and programs for each organizational unit	1. Motivating organizational units and individuals	1. Adjustment of policies and procedures to support the strategy	1. Creation of new business climate and new culture
2. Development of skills and distinctive competencies on which the strategy is based	2. Directing and encouraging efforts of individuals and organizational units toward achieving goals	2. Creation of working conditions and business culture to support the strategy	2. Ensuring adequate information system	2. Innovations and dynamics of the company
3. Selection of people for key positions		3. Orientation towards results and achievement of supreme results	3. Institutionalization of the internal system of control	3. Managing conflicts and negotiations
		4. Harmonization of the system of rewarding with strategic performances	4. Good relation between requirements of the strategy and current way of doing business	4. System of control and auditing for management of implementation

Creation of Organizational Structure for Implementation of the Strategy

This is an essential element for the implementation of the strategy that is used to create a new or adapt the existing organizational structure, which is required for the smooth implementation of the strategic directions and ways of doing business in the organization. At this stage the focus is on establishing adequate internal organization and management skills that match the strategic determination and vision. The transition to and implementation of sustainable energy management involves major changes in the entire organization and acceptance of change by all employees. There are a number of strategic decisions that should be made at various levels of management, and these decisions need to be communicated to employees. Sustainable management of energy requires the full commitment of management, but employees should be familiar with the changes in order to change their way of thinking and take an active part in the new strategy.

Allocation and Focusing of Resources Toward Strategic Goals

Since operationalization of the strategy requires engagement and using a number of resources, it is important to take into account the anticipated changes, the scope of activities, and provide a budget for each group of tasks and their appropriate use. It is necessary to understand the nature, importance, and priorities of different parts of the strategy and methodology of calculating costs and expenses. Determining the required resources and their allocation is the most crucial prerequisite for the implementation of the strategy of sustainable energy management. A particular problem arises in companies that are major producers and/or large consumers of energy. Their traditional way of work involves the use of energy without limits, in accordance with needs, and the payment of energy is through the price of the final product. Studies have shown that in the cost of a product, the value of energy is often unrealistically high. In general, poor indicators of energy efficiency give much room for savings. Besides the rationalization of energy consumption, companies are often faced with the objective costs of the transition to using a certain amount of energy from renewable sources, which requires investments that are not paid back quickly or easily. For these reasons, companies transferring to sustainable energy management must provide certain financial resources for these needs, from their own funds or by using other financial sources, including the use of financial sources and subsidies predicted in most countries which adopted new legislations and legal mechanisms for promotion of sustainable energy development.

Galvanization of Implementation of the Chosen Strategy

This stage includes understanding, maintaining, and coordinating overall activities of the company in order to implement the strategy and create the infrastructure needed to ensure efficient and effective implementation. The commitment of management and all employees is crucial to the success of implementation of the strategy for sustainable energy management. It is, therefore, necessary to design and develop specific mechanisms for monitoring and directing behavior of employees in the direction of the company goals. Most often, tools such as training programs, development of a system of rewards and punishments, and other mechanisms that are considered acceptable and effective in terms of employee motivation are used.

Establishment of an Internal Administrative System of Support

This stage includes establishing the activities that institutionalize the implementation of the strategy by providing appropriate procedures, policies, and other internal regulations. Under the influence of new circumstances, new business policies are incorporated in revised procedures and efforts, including:

- Adapting policies and procedures for the support of the strategy, which clearly define the rules and procedures for the implementation of sustainable strategy. A clearly defined strategy of general sustainable development, as well as clearly defined procedures for carrying out the activities and delegated responsibilities, creates a great positive precondition for successful implementation of the strategy. Energy management in companies is a complex process and requires involvement of employees of different profiles, and clear definition of individual responsibilities and duties is of special importance.
- Ensuring an adequate information system, without which the transition to and implementation of sustainable energy management cannot be realized. Contemporary information systems have to be designed so they can accept all necessary input parameters, to process them, and to develop a specific portfolio of energy consumption that is assumed to be the most effective, i.e., that will be compatible with the objectives as much as possible. The information system must follow all the parameters of production and consumption by type and quantity, the places of consumption, and the amount of energy waste. Also it has to propose the ways of energy saving and to predict effects of energy saving. In addition, it must show all financial and non-financial effects of the suggested changes.
- Institutionalizing the internal system of control implies creating and constantly improving the control system. Modern business imposes strengthening of the system of control, but it is an absolute imperative when it comes to the implementation of the strategy of sustainable energy management since it strongly affects all aspects and effects of business. Energy is a resource that considerably limits the business of a great number of companies, so the system of control should be properly structured and properly implemented. Strategic objectives must not remain objectives only. The success of implementation of a certain strategy should not be determined by the application of general measures and qualitative evaluations. Achieving goals and effects of the whole strategy of sustainable energy management can be monitored

only if the goals are measurable, and all parameters of control (the essence, indicators for monitoring, frequency) must be harmonized with the specific features of the existing and new system of control in the field of energy.

- Ensuring good relations between the strategy requirements and the current way of doing business. Development and implementation of the strategy of sustainable management of energy has to be integrated in all other development planes of the company. It should not be carried out suddenly, nor should the company set unrealistically high goals. Their implementation is usually impossible, bringing about great difficulties and the expenditure of time and resources, and failure often leads to the loss of motivation and temporary or permanent loss of vision for the company. The strategy of sustainable energy management has to be created in accordance with the current operations of the company, and all planned changes must be gradual and clearly oriented towards the goals.

All these basic managerial measures and mechanisms are the factors that define possibilities for successful implementation of the strategy of sustainable energy management in all its stages. The mistakes that often occur in this field include vaguely defined vision and mission, a lack of integration of the strategy of sustainable energy management into the existing strategic planes, unrealistic goals, and an inefficient system of control.

Providing Strategic Leadership

For the implementation of the sustainable management of energy, leadership is very important. Managers need to have entrepreneurial and leadership skills, because they are expected to inspire, create policies, resolve crises, allocate resources, represent the company well, etc. The leader is a person who, without being forced, has the ability to motivate and gather together employees based on the company's vision, ideas, and charisma, while managers, through planning, organizing, and control, ensure the running of the business.

Implementation of the strategy of sustainable energy management is a complex process of creating conditions and integration of activities to achieve expected results. To manage these activities it is necessary to specify the responsibility and authority and to identify the tasks and budget and allocate them to the carriers through a system of coordination and overall management of the implementation process. In this regard, special emphasis is placed on global issues affecting the implementation of the strategy, as follows:

- Creation of adequate organizational structure and corporate culture of the company;
- Transformation of long-term goals into the current levels of performances as well as allocation of tasks and budget to functional areas;
- Management of activities and management of the process of implementation; and
- Control and audit of strategy.

These factors represent only a few preconditions for successful implementation of the strategy of sustainable energy management, but theory and practice has shown that these factors have the strongest individual and synergy effects.

4.3 CREATION OF ORGANIZATIONAL STRUCTURE AND CORPORATE CULTURE

As the implementation of the strategy for sustainable management of energy involves mobilizing people and resources in a rational way, one of the key issues to be addressed is the establishment of an appropriate organizational infrastructure that can ensure harmonious operations of the company as a whole. If you understand the structure of an organization as a way of identifying the top and parts of the enterprise, as well as the relations between them, then its structure is a key element in the implementation of the strategy. The implementation of the strategy of sustainable management of energy is by rule integrated into all organizational units of the company and as such creates additional organizational difficulties. Organizational structure coordinates the company philosophy and anatomy in an effort to ensure optimal working abilities and vitality in the specific economic conditions. Objectives, responsibilities, abilities, and tasks of employees should be coordinated, so as to create significant competitive advantage by implementing the strategy. In that way the system of support is provided for implementation of the activities in a value chain of the company. Only the optimal organizational structure can ensure the realization of strategic goals of energy management. If employees are not adequately organized and motivated, implementation of the best-designed strategy of energy management should be questioned.

The organizational structure is not only an instrument for implementing the strategy, but also the basis for creating strategic performances in the company. The organizational structure is the infrastructure that provides a rational relation between objectives and those who perform the activities in the company. This is even more important when it comes to creating energy performances, which are long-term and often associated with high costs.

Today businesses can be organizationally regulated in a relatively small number of ways. There are certain organizational structures, with minor or major modifications, that can be applied in business, regardless of the size of the company, the form of ownership, energy consumption, or the activity that the company deals with. Small businesses that are minor consumers of energy or large companies that consume huge amounts of energy can have the same organizational structure and be equally successful or unsuccessful in business. The impossibility of creating a large number of varieties of organizational structure is one of the limiting factors in trying to adapt the organizational structure to the new way of energy management. Building organizational structure is based on differentiation, coordination, and integration.

Differentiation is a way of identifying parts of the company that will, according to similarity of businesses, provide opportunities to achieve the maximum effects of specialization and concentration. Differentiation of individual company units must be coordinated with all the functions of the company, not only with the needs of sustainable energy management. Additional problems arise from the stipulated fact regarding organizational differentiation that can be considered acceptable for sustainable management of energy. Very often the organizational structure of a company is adapted to the form of ownership or is determined by the activities the company does, but at the same time it can be completely inefficient for the needs of energy management.

Therefore, in most cases, at the beginning of the process of planning the strategy of sustainable management of energy it is necessary to create a detailed review of the existing organizational structure and adjust it to new needs. The most effective technique for differentiation is often separation of parts of the company by activities or work processes performed in that part of the company. In order to consider some unit acceptable in terms of sustainable energy management, that unit needs to be described with certain energy parameters, independently of other parts of the enterprise.

If the selected organizational unit cannot be accurately described in this way, the stipulated unit may not be well differentiated. If the differentiated organizational unit, compared to other units or the company as a whole, can be described by certain indicators of energy consumption based on the quantity and/or the type of energy used, or based on energy efficiency or energy price that was spent in that unit, it can be considered that the stipulated unit is well differentiated and that being such, it can constitute a separate unit of the organizational structure.

Coordination involves establishing relationships between organizational units in order to provide synergy in business. Very often the differentiation of organizational units can be implemented successfully, but the organizational structure is not successful. In addition, it often happens that the same organizational structure in companies with the same scope of activities may show different effects. Therefore, it is clear that the best differentiated organizational structure cannot show its positive effects if the differentiated parts are not linked in an efficient way.

In theory and in practice there are certain suggestions and solutions to connect the differentiated parts of the organizational structure. The success of the links between the elements of organizational structure therefore largely depends solely on the respect of these links. In certain operating systems, the organizational structure that is observed only in one way, as a bond that is directed from higher to lower levels, may prove to be most appropriate. However, planning and implementing the strategy for sustainable energy management requires a strong one-way communication in the initial stage, but during the time, it is necessary to redefine the organizational structure and make it more flexible and two-way. Since each part of the organizational structure must be described with certain energy parameters, efficient connection between the parts is almost essential.

Integration refers to the detection of the number of hierarchical levels, their responsibilities and powers and grounded on that, provision of effective management, and control. The companies that are large consumers of energy, and therefore particularly motivated to implement a strategy for sustainable management of energy, often have complex organizational structures and operations at more hierarchical levels. Very often these are the companies that operate in many locations in several states or even on different continents. All this affects the need to properly and timely determine the levels of decision making.

The strategy of sustainable energy management is primarily in the domain of activities of top management of companies, or the highest hierarchical level, but the practical realization and particularly the control are in the domain of lower hierarchical levels, and activities are conducted in the immediate realization of activities, at the lowest level that usually is not able to make decisions. The long-term impact, sensitivity, and economic value of changes in the energy sector give additional importance to the need to precisely determine the hierarchical levels of decision making.

As the implementation of the strategy often leads to changes in size, structure, and activities of the company, and its bonds with the environment, it is necessary to anticipate the new organizational structure in the business. Internalization of business and geographical spread of activities change the existing relations between the top and the parts, so it is necessary to establish the centers of responsibility with new competencies. Implementation of the new strategy requires adjustment of the organizational structure; otherwise, it leads to dysfunction and reduction of the expected results. Apart from that, a good organizational structure that measures responsibility for results and the system of motivation and rewarding can encourage entrepreneurship and generation of new strategic ideas. The transition from traditional to sustainable management of energy is a big strategic change and inevitably leads to the need to review the existing organizational structure and redefine its parts and hierarchical levels.

Development of Organizational Structure of the Company

Growth and development of a company do not follow a straight rising line, but undergo stages of evolution and revolution, characterized by various problems and managerial efforts. All companies spend more or less energy during a certain period of time, and in many companies the cost of energy is quite high. Besides, globally accepted legislation imposes the need to implement the sustainable management of energy as soon as possible and in a more efficient way. However, the company has to justify its business to fulfill its mission and vision and survive in the market. Quantitative growth in the size of the company causes qualitative changes so as to make the effects of size possible (economies of scale) and diversification of activities. In the periods of evolution and revolution (Figure 4.3.) that characterize the interventions in the field of organizational structure, crises occur, or stages come in succession, which require changes in organizational structure and management style. Izsák Adizes described the process of growth and aging of organization through different stages of the life

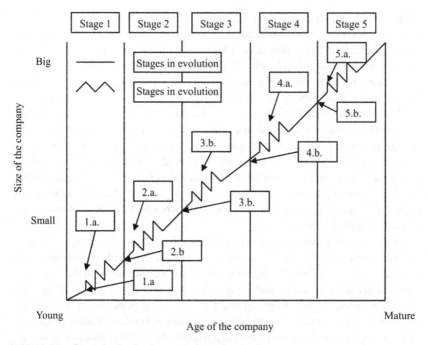

FIGURE 4.3 Stages in the development of an organization

cycle, as follows: courtship, baby, go-go, adolescence, top-form, stability of the aristocracy, early bureaucracy, bureaucracy, and death [1].

This interpretation can serve as a starting point for companies that plan to switch to sustainable energy management and to keep such a way of doing business, where potential changes are anticipated in terms of the size of the company over time. During this time, each company works within a smaller or larger capacity, and therefore consumes lower or greater amounts of energy. It is therefore necessary that management of the company be aware of the changes that can inevitably be expected.

Many factors in the external and internal environment affect the form of organizational structure that will be accepted by the company as a basis for implementing the strategy. As mass production causes excessive greenhouse gas emissions, the gases that cause the "greenhouse" effect, it is of paramount importance to take into consideration economic and ecological implications of the strategy implementation when choosing the appropriate organizational structure.

In this regard it is important to determine the range of management and control. There are two possibilities: a high (deep) organizational structure with many hierarchical levels between top management and direct supervisors, or a flat (shallow) organizational structure with few hierarchical levels between the highest and lowest levels of management. Since a good organizational structure should provide effective management and control, the choice between shallow and deep organizational structure must be geared toward gaining competitive advantage, organizational units, and business functions. The selection of the depth of the organizational structure is mostly determined by the energy intensity of the activity that the company deals with. For companies that spend large amounts of energy, a deeper organizational structure is preferred, although one has to be careful here, because too many hierarchical levels often impose burdens and slow down the system of decision making.

Stages in Creating Organizational Structure

Because of the relationship between strategy and organizational culture, the following stages can be defined in the process of creating an organizational structure that will effectively support the implementation of the sustainable management of energy:

- Identification of the key functions and activities for implementation of the strategy of sustainable management of energy, by which it is necessary first of all to define the activities and responsibilities, rights and responsibilities, and the chain of responsibilities;
- Understanding the relationship between activities in the company's value chain, by which all activities in the process of sustainable energy management are integrated into the existing system of values of the company;
- Grouping activities in organizational units, by which special attention should be paid to defining every organizational unit as an entity with its own peculiarities of production and/or usage of energy, which can be described and set independently from other organizational units;
- Determining the degree of autonomy and independence of each organizational unit, which further enhances the possibility to determine energy characteristics for each organizational unit; and
- Ensuring a system of coordination of activities of organizational units as a natural consequence of interconnectivity of all organizational units, i.e., natural and technologically conditioned cycle of changing the matter and energy.

A particular issue relates to the organizational structure in cases of internationalization of business, because in that way a company significantly expands the scope of its activities and its influence, thus accepting greater social responsibility in the field of sustainable energy management. In these circumstances it is important to determine the relationships between the peak (the center) and organizational parts that are located in different countries. Apart from the production and geographical criteria, in order to identify the organizational structure it is necessary to determine the areas of communication between the center and the affiliations in countries and the degree of their independence. In this sense, the company chooses one of the following forms:

- Division structure, which is based on creating divisions as units for coordinating and managing international operations in a given country;
- Global production structure, which is based on the division of the company's organizational units according to groups of products that operate as international units. This form is used when a company is diversified and has significant market share; and
- Matrix structure, which implies that the central unit of the company is responsible for production planning, while geographical divisions are responsible for local sales, distribution, and administration.

Defining Corporate Culture

When it comes to corporate culture, organizational structure is viewed as a framework and corporate culture is an invisible hand, i.e., a catalyst that provides harmonious functioning of the company. It is usually expressed through a system of values, decision making, and the dominant model of organizational behavior.

Environmentally friendly and socially responsible business is an imperative of contemporary business. A responsible attitude toward energy, energy resources, and energy consumption must be integrated into a business.

Some authors describe corporate culture as the style of doing business, organizational climate, the ego of the company, or beliefs and ideas shared by the employees. It is compared with the personal identity of individuals, as an accepted system of values, norms, and expectations that determine the way in which companies resolve concrete business situations. Corporate culture gives direction and sense of action; it helps employees identify priorities and properly implement formal activities. In practice, it determines the way in which employees and managers of companies treat customers and suppliers, how they respond to competition, and perform all other activities. Corporate culture is vital if it is characterized by the following: unity, innovation, a clear understanding of company goals, a high level of loyalty, and responsibility to the company or customer orientation.

Since corporate culture consists of value attitudes and norms of behavior according to which employees behave, it is hardly visible, unlike other components on which the implementation of the strategy is grounded. Thus, implementation of the strategy involves harmonization of the structure, competence, allocation of resources, methods of selecting and rewarding human resources, activities of business functions, policies, procedures, and so on.

In the implementation of sustainable energy management, building appropriate business culture is a matter of great importance. In this way, the opinions and attitudes of both management and employees are changed, and the economic performance becomes the only measure of the value of the company. Accepting the need for a responsible attitude towards energy is a slow process that requires an exceptional commitment from management, as well as education and motivation of all employees, along with the participation of external stakeholders.

From this point of view, the concept developed by McKinsey Consultants Comp [62]. is useful. This concept is based on the mutual interaction of seven key categories, known as the concept of 7-S. The 7-S framework includes seven categories: the strategy, structure, systems, human resources (staff), shared values, style, and skills. This concept emphasizes the necessity of putting corporate culture in the function of variables such as strategy, system of business flows and information, organizational structure, management style, personnel structure, and the adopted system of values and skills. As shown in Figure 4.4, an accepted system of values puts corporate culture in the middle, where it permeates all other components and drives the behavior of the company.

The strategy of sustainable energy management is directly or indirectly connected to all other categories. The strongest interaction is with the category of staff and with the category of shared values, which is the result of the need to implement the strategy by people who share the same values. Sustainable management of energy requires huge changes in understanding and accepting new values of doing business by employees. Doing business just to gain economic benefits is no longer an imperative in business.

In addition, there is a strong interaction with the categories of skills and structure. Energy management requires certain skills and engagement of people at different structural levels. Companies that are major producers or consumers of energy are also the companies with complex organizational structures.

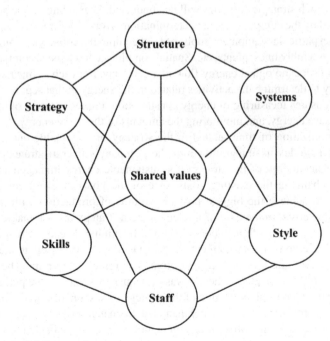

FIGURE 4.4   7-S framework

Integration of sustainable energy management into all levels of an organizational structure is often difficult but essential.

4.4 TRANSFORMATION OF GOALS AND ALLOCATION OF RESOURCES

As the strategy defines the ways of realization of the mission and goals of the company, an important issue for the implementation of the strategy is the transformation of the new goals into current tasks performed by certain business functions and organizational units of the company. In order to implement the strategy of sustainable energy management in practice, resource solutions related to space, time, and carriers are required. In fact, this part of implementing the strategy is about transforming strategic plans into tactical operations, which refer to shorter time intervals, with fewer alternatives and concrete structured problems. For this reason, the implementation of the strategy requires spatial, temporal, and subject-compliant action plans for one accounting period.

Energy management is inherently a long process and is connected with numerous changes, which significantly extends the time needed for implementation and assessment of positive or negative effects.

A spatially harmonized plan of activities implies that all activities related to sustainable energy management must be precisely determined in terms of space. First of all, it is necessary to determine the parameters of the physical environment in which strategic activities will be conducted. Depending on the activities conducted by the company, spatial determination involves defining activities performed on particular equipment, tools, parts of equipment, connections, important devices, assembly lines, plants, etc. Spatial compliance has to be accompanied by a natural or technological energy flow in a company. For each defined area it is necessary to determine the activities related to the energy in that segment, which actually ensures measuring of energy consumption, implementation of the measures to save energy, and monitoring the amounts of the waste energy.

Harmonization of time related to the strategy of sustainable energy management includes its integration into the company's overall strategic plans. The implementation of the strategy for sustainable energy management often requires changing the existing goals. One of the goals of every company is certainly to achieve the biggest possible volume of production with the lowest possible costs. In the case of implementing a strategy of sustainable energy management, this is often impossible in the beginning, because the transition to a new method of management is associated with increased initial costs. Therefore, the time for resource allocation is extremely important. The capital has to move from the place where it was previously assigned to the place where it will be used for implementation of the strategy of sustainable energy management. Often this means that the company has to change existing goals and give up some specific goals. Most often companies give up on goals that require the use of large amounts of energy and/or production of energy-inefficient products.

Operationalization of the strategy may relate to the operationalization of new business operations, or improving or maintaining the current level of operations. Thus, project plans related to the implementation of complex and unique activities may be needed, and the standard plans are used for regular and recurring business situations. The introduction of the strategy of sustainable energy management is usually a major strategic change and requires significant changes. A company will usually create a new product and business portfolio and keep the products and operations that are acceptable in terms of the new strategy but will gradually reduce and eventually abolish energy inappropriate aspects of business. All this requires good planning and significant investments and, certainly, enough time. Changes should be planned and not jeopardize the business ability of the company. The transition to a fully sustainable business in terms of energy can be a significant change and as such it should be carefully approached.

Implementation of the strategy refers not only to the identification and implementation of projects but also to the planning of their performance through the network action plan with clear allocation of resources. Investment programs are developed and selected, as well as the selection of those who would implement the program and models for carrying out programs and projects. In practice, it is not just about standards and familiar and repetitive projects, but it is also about the implementation of previously started but radically new business activities. For these reasons, apart from the *know-how* in the field of project management it is necessary to build in the *project culture*. The project culture requires teamwork, precise responsibility, and parameters for the evaluation of workers involved, their charisma, and other skills.

As implementation of the strategy requires dimensioning the role and involvement of all business functions, it is important to establish priorities in selecting individual activities and creating resource allocation. Although the company sets the issue of energy management as a priority, it is necessary to implement it in line with other business activities. These issues can refer to the company as a whole or be treated as issues relating to a certain level of organization. The implementation of the chosen strategy can assume the following:

- The need to increase total resources, where the company can realize the investment in purchasing new energy efficient equipment or its parts or to provide energy from alternative sources;
- The change in the existing allocation of resources, which happens if a company does not have enough money but still wants to do certain adjustments. In this case, the company transfers the existing equity from one place to another. It ignores the previously defined objectives and priority is given to solving the problem of unsustainable waste of energy. In this case, the parts of the company are being threatened or even deliberately destroyed in order to create new, energy-friendly parts; and
- The reduction of the required resources and thereby their redirection to an area where resources are needed, as the case which is theoretically possible

but not recorded in strategies of energy management because they essentially require increased investments.

It is evident that a certain allocation of resources is usually applied in the company for the needs of sustainable energy management. Allocation of resources can be implemented in various degrees, depending on the degree of centralization and intensity of change, as shown in Figure 4.5.

When the implementation of the strategy does not require a significant change in resources, the allocation of resources will be done as it was done before. This is often the case in companies that need to implement low-level changes, and most often these are companies that are engaged in activities with low energy consumption. In this case there is no need for the allocation of resources. It is usually enough to monitor them precisely and control them.

If the management style is based on the partial involvement of individual units in the allocation of resources, it is possible to change the allocation of resources instead of the current practice ("formula"), and the decision will be based on cost analysis and alternative use of resources. This situation is called *conditional bargaining*, and it is specific to companies in which no significant correction of the existing energy flows is necessary, but there are some difficulties in implementation because the system of decision making is centralized.

When the decision making is decentralized, changes in the volume and allocation of resources can occur through *free negotiation*, especially if implementation of the strategy does not require the excessive use of resources. In this case changes in the field of energy are small but employees are free to implement them in their own organizational units on their own. This model has its advantages, because employees in certain business units often best know the specific characteristics of that unit, are informed about energy consumption and the reasons for inefficient spending, and can see opportunities for saving energy.

In the case when the implementation of the strategy involves changing the size and priorities of the engagement of resources (in the case of the development of intensive strategy), and centralization is strong, top management *imposes priorities,* and based on this the redistribution of resources is done. This is very

		Low	High
	High	"Formula"	Imposed priority
Degree of centralization of the direction	Medium	Conditioned negotiation	Conditioned bidding
	Low	Free negotiation	Open competition

FIGURE 4.5 **Allocation of resources at the company level**

often the case with companies that transfer to sustainable management of energy. In this case, energy management becomes a priority selected by the management of the company. This imperative becomes a priority of business and is transferred to employees and the requirements they have to meet. Such an autocratic approach can show good results in a short time, but should later be abandoned.

In order to stimulate organizational units and consider the environment realistically, these values should be a part of the budgeting process, which is often called *conditioned bidding* (and in practice is very similar to conditioned negotiation), where organizational units of the company apply for funds by bidding. In practice this means that the transition to sustainable energy management is not an absolute priority and its implementation is conditioned by the ability of the organizational unit to fight for the necessary funds. Concretely the organizational part of the company that plans to introduce changes in energy management is seeking to mobilize funds from the existing budget by providing the proof that the change is needed and sustainable in the long term. This model is especially acceptable for big companies that operate in many locations.

If an *open competition* between organizational units is chosen, organizational units are encouraged to do better. Such an approach respects the autonomy of organizational units and the need for self-responsibility and achievement of specific organizational units. It is therefore important to estimate the importance of the activities in the company's value chain, respecting the diversity of the focus of some business functions. In this way, conflicts or bias in the distribution (allocation) of resources can be avoided. The above model is acceptable for companies with a high level of corporate culture and is not recommended in the first stages of transition to sustainable energy management.

4.5 MANAGING THE PROCESS OF IMPLEMENTING SUSTAINABLE ENERGY MANAGEMENT

The essence of the implementation of the strategy of sustainable energy management is the realization of all predicted activities. In order to make the implementation of the strategy of sustainable management of energy efficient, it is necessary to design the organizational structure, allocate resources and related procedures, create an effective system of coordination and information system, make the selection of human resources for key positions, and create a motivating system of rewards.

The implementation of the strategy of sustainable energy management mainly depends on people, and for their motivation and encouragement for action it is necessary that there is *a leader* who has charisma, knowledge, information, and the authority to reward and punish. However, at the same time, there should be *a manager* who has conceptual and technical knowledge, and who through the processes of communication and coordination with average people achieves above average results. The management process generally requires the coexistence of the leader as the creator of the business idea; the business portfolio of the company; and the manager who through planning, organizing, managing, and controlling manages the work processes. The leader deals with challenges and encourages employees to work efficiently, or find the right ideas. He is long-term oriented and motivated to innovate and ask questions. On the other hand, the manager is focused on knowledge that refers to effective management of work processes in order to realize the vision of the company, i.e., he is oriented toward the realization of ideas in the right way.

By vertical differentiation a division is made in the number of hierarchical levels and by horizontal differentiation related activities are grouped into productive units and business functions to achieve efficiency. Therefore, coordination is necessary (vertical and horizontal) for different levels of activity as one of the central issues on which the success in implementing the strategy depends.

Coordination of the Strategy

Coordination can be defined as the process of integrating the goals and activities of differentiated business units to effectively implement the strategy. If differentiation is more prominent, coordination is more complicated, because it is difficult to find a link and the ability to achieve synergy among the parts of the company. This situation is most common in companies that do business with several major dislocated organizational units, and the need to transfer to sustainable management of energy is not considered equally important in all parts of the company.

At the same time, it is difficult to preserve the coherence of activities and flexibility that often do not go together. A significant and relevant source for acquiring the competitive advantage often lies in the possibility of the transfer of knowledge from one unit to another or combining activities that are used in several business units in a diversified company. It is necessary to transfer expert knowledge in the field of sustainable energy management from one part of the company to another. In practice it is often justified to consider the engagement of external experts and consultants, who will monitor the implementation of the strategy of sustainable energy management.

The danger of differentiation may occur due to possible conflicts and bias of management in relation to the organizational unit. The task of coordination is to detect potential conflicts and misunderstandings and manage them in a timely manner and, if possible, avoid such situations. Conflicts often arise because different organizational units understand the basic strategic orientation differently, use the same resources, and have different views on their role and importance in the realization of strategic orientation. Very often, the company adopts the implementation of sustainable energy management, but it is implemented differently in terms of its effectiveness in different parts. In order to identify the conflict areas and situations in a timely manner, top management of the company can:

- Rely on basic management techniques;
- Increase cooperation potentials; and
- Reduce coordination needs.

Rely on basic management techniques. In this type of management, coordination is realized through a single chain of command with clearly defined hierarchical relationships between the units and the flow of information adjusted accordingly. Procedures and stipulating rules can be defined so they provide independent and rapid implementation of the tasks of organizational units. To a certain level unity can be provided through insisting on general objectives and corporate culture that will provide links for coordination.

Increase cooperation potentials. This type of management is used when due to the diversification of activities and shallow organizational structure it is not adequate to rely on prescribing rules or a hierarchy of relationships. In this case there is a need to increase the potentials of coordination by using different measures and mechanisms by:

- Encouraging direct contacts between leaders to address common issues, in order to familiarize all with the particular features of the sustainable energy management;
- Appointing a liaison officer between the various entities so that each unit has a person to coordinate its work with other organizational units;
- Formulating an "ad hoc" group, comprised of representatives of concerned units that will coordinate joint activities, which is especially acceptable for introducing the method of sustainable management, as this represents a big change and requires a special attitude of the involved;
- Establishing a permanent committee responsible for the coordination of joint work, resulting in the possibility that the strategy of sustainable energy management will be implemented in the whole company simultaneously;
- Appointing a deputy general manager or a special manager for the integration of cooperation among organizational units as a necessity in large companies that spend significant amounts of energy;
- Shaping matrix organization in which activities of business functions and organizational units are cross-integrated; and
- Establishing a complete division at the top of hierarchy of the company for coordination of the activities of organizational units.

Reduce coordination needs. This approach insists on the reduction of the need for coordination, by allowing some organizational units greater flexibility in creating their programs and transforming them into independent entities. This approach is acceptable in companies with a high degree of organizational culture, because it means freedom for each organizational unit to select and implement changes in the field of energy independently.

Resistance to Change

The introduction of sustainable energy management can often be a big strategic change that requires intensive changes in the company's processes, but also in the attitudes of all employees. This major strategic change always encounters smaller or greater resistance that may last for either a short or long period of time, but it can also completely prevent the required changes. Therefore, for effective coordination in implementing the strategy it is very important to anticipate the active and passive resistance to change. The reasons for resistance can be numerous, including jeopardizing individual interests, misunderstandings, differences in understanding the directions of change, inertia, time period when the change is made, etc. Hence, in order to achieve efficient coordination and effectively manage the process of implementation it is necessary to make an extra effort in the field of communication, clarification, education, involvement in proposing changes, setting up systems of rewards, etc.

As a method of effective implementation of sustainable strategy of energy management that reduces the potential resistance, the concept of management by objectives can be used. It is a concept in which the involvement of executives in the process of defining goals provides greater motivation and responsibility, defines strategies and objectives in a more realistic way, provides a stronger cooperation in implementing the strategy, etc. The goals and tasks are set in a separate agreement that deals with relations between junior and senior executives and includes:

- Sphere of activity for each supervisor;
- Desired goals of activities (short-term and long-term);
- Ways of measuring implementation of goals;
- Limitations and freedoms that each executive has in achieving stipulated goals; and
- Assistance executives can expect from authorities and other organizational units.

Top management of the company has the dominant role in managing the process of implementation of the strategy of sustainable energy management. These bodies interpret and symbolize the strategy, initiate change, and are responsible for effectiveness and efficiency in the execution of the defined strategy. Centralized management is the management in which the hierarchy top defines the strategy for each organizational unit. Participatory management is applicable to cases where the process includes the heads of organizational units. In practice there are combinations of methods and management styles that depend on effort in terms of process planning and implementation of the strategy.

Selection of Managers

The strategy of sustainable management of energy requires engagement of appropriate managers who will deal with all phases of management of the process. The first question concerns the selection of managers. The selection of managers depends on the abilities and needs of the company. Also, the choice of the staff is important, i.e., whether to choose from the staff already employed with the company or to recruit new employees. The advantages of selecting the existing workforce is that they are already familiar with other employees, business practices, work environment, their characteristics are well known to other staff members, they have established contacts with customers, suppliers, and colleagues, and thus show their loyalty to the company. Disadvantages may arise due to the inability to accept change, workload, personal attitude, and the like. Using people who have gained experience in implementing similar changes is an advantage, since they desire to prove themselves in a new environment and are not overburdened by inherited relations towards associates. The disadvantages of this solution include the high cost of attracting such experts, the difficulty in finding the right person, the negative impact on loyalty of other employees, etc. The implementation of sustainable energy management is a major strategic change and involvement of experts outside the company is almost inevitable, but it is certainly desirable to create mixed teams of external experts, full-time employees, and experienced managers and engineers.

When selecting key managers two important aspects should be taken into account: 1) changes necessary for implementation of the new strategy and 2) past performance of companies. Taking into account these two coordinates, the following matrix shown in Figure 4.6 can be obtained.

Selective mixture describes the situation when the implementation of strategy requires a lot of changes, but those changes are not initiated by the previous performances. Since the previous performance was good, it is a good choice to rely on the strength of existing management, and selective outsourcing of external managers can ensure changes implied by the new strategy. This situation arises when the company has effectively run the process of energy management and gained valuable experience.

| | | Past performance | |
		Effective	Ineffective
Neccesary changes for the implementation of new strategy	Many	*"Selective mixture"* In-company managers are combined with the new managers (outsiders) through promotion and transfer and provision of knowledge and experience	*"The turn"* Hiring new managers (outsiders) so as to provide new ideas, skills, enthusiasm, and motives for success
	Few	*"Stability"* Orientation towards internal promotion of in-company executives so as to develop, keep, and reward their managerial flair	*"Reorientation"* Hiring managers (outsiders) so as to surpass weaknesses along with promotion and reallocation of the existing executives

FIGURE 4.6   Approaches to selection of managers

Stability is a situation that arises when the implementation of the strategy does not require a radical change, i.e., it is intended to preserve current performance and relies on the promotion of existing managers in order to build customer loyalty and further develop management talent. This situation is typical for companies that are usually small consumers of energy and sustainable energy management is not a big change.

The turn is characterized by a situation where it is necessary to intensify efforts to implement defined strategies due to ineffective operations. In this case, bringing in new managers becomes more important in order to provide fresh ideas, new skills, motives, and enthusiasm. In most companies sustainable management of energy is well-planned but it is not implemented, mainly because of the absence of commitment at top management level. In such a situation, bringing in new managers is quite justified, because the old ones usually are big potential source of resistance to change.

Reorientation characterizes the situation where the implementation of the strategy does not require a lot of changes and previous performance was bad. In this situation it is best to bring in external managers (outsiders) to help with critical issues in the field of business functions or reallocate existing managers.

Hiring of new managers and their participation in the activities of the company and the implementation of strategies for sustainable energy management largely depends on the company's ability to pay for their engagement and other benefits offered to new managers. Therefore, the structure of the reward system is very important for the process of implementation of the strategy. These are sensitive issues of great importance for the success of the implementation of the strategy. Within these considerations the form of giving incentives should be defined (profit sharing, stocks, cash, etc.), as well as techniques and the basis for calculation, bonuses, rights of the managers based on hierarchical position, the frequency of payments, consequences of risk, and other related issues.

4.6 CONTROL AND AUDIT OF THE STRATEGY

The final stage of the process of strategic management is the control and audit of the implementation of the strategy of sustainable management of energy. This phase should provide a more realistic basis for implementation of the strategy and should enable the timely implementation of strategic options, as well as help to promptly take any corrective action.

Timing of the Control

According to Herny Fayol, control of an undertaking consists of seeing that everything is being carried out in accordance with the plan which has been adopted, the orders which have been given, and the principles which have been laid down. Its object is to point out mistakes in order that they may be rectified and prevented from recurring [35]. Control of the implementation of the strategy of sustainable energy management should be conducted according to the plan and in the time determined. In this regard, and from the point of view of time, control may be conducted during the following periods:

- After formulating a global strategy but before its implementation;
- During the implementation of the strategy; and
- After implementation of the strategy.

The first control, which refers to the formulation of the strategy of sustainable management of energy, is aimed at checking the compatibility of the strategy with the mission, resources, weaknesses, and strengths of the company, external conditions, corporate culture, and time interval. Control during the implementation has the task of timely pointing to possible deviations during the execution in relation to the plan. Finally, the control after the implementation of the strategy is the basis for a change and a new plan.

The basic purpose of control and review of the strategy is to timely identify and evaluate possible deviation from the plan and take corrective actions. If the process of planning is the first in the chronology of strategic management, when decisions are made about the objectives, policies, strategies, plans and programs, then the mechanism of control that timely qualifies the flow of these activities is completely logical. In this sense, control should meet two purposes:

- To discipline those who plan and execute planned activities; and
- To contribute to the realization of the process of planning in the next period.

It is easy to see the correlation between planning and control. Planning and control are linked through the implementation of the strategy, and it makes no sense to plan if that plan will not be controlled. On the other hand it is not possible to control activities that were not planned previously. These two processes are mutually encouraged and derive one from another. It is necessary to determine the basic parameters and the scope of control as follows:

- Subject of control.
- System of Control.
- Control and Audit of the Strategy.
- Control of the Strategy in the Environment.

Subjects of Control

From the point of view of the dimension and the subject of control, as shown in Figure 4.7, it is possible to differentiate strategic, managerial, and operational control. The implementation of the strategy involves the activities of all hierarchical levels in the company: from the top of the company through strategic business units and business functions. This significantly determines the system of control.

On the other hand, the types and standards of control depend on the structure of activity, degree of decentralization, management style, and reward. Typically, strategic control refers to control of the implementation of basic strategic orientation within the competence of top management, which is of great importance to the company as a whole. This control is based on long-term, broad, and quantitative standards. The shareholders' assembly, supervisory board, and the external audit (auditing performed by certified organizations) assess the skills of top management. The rating of management refers to their ability to adequately evaluate the general and branch environment, identify critical factors of business success (hot spots), and choose the appropriate overall strategy. Figure 4.7 shows that management control is related to the control of the implementation of strategies at the level of individual business or programs, or whether it is in the domain of top management and heads of business units. In general, the control of the operation involves the control of the implementation of strategy of business functions and control of individuals and groups in accordance with the prescribed standards and procedures.

The type of system of control and efficiency of implementation depend on management style and the system of reward, or set of procedures and standards for monitoring. This is necessary because it is hard to distinguish when strategic, managerial, and operational control start and when they

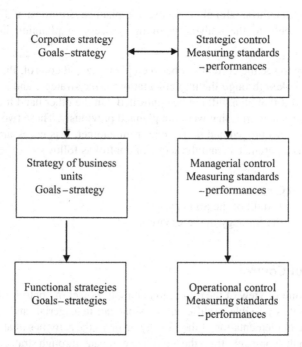

FIGURE 4.7   The relationship between strategy and control

finish, due to the interconnection of hierarchical levels in the implementation of the strategy. To understand this sensitive area in a realistic way, it is important to determine the standards and procedures more accurately. As synthetic indicators related to strategic control it is possible to measure the level of productivity, product quality, work morale of employees, earnings per share, the dynamics of introducing new products, market share, the costs of obtaining resources, etc.

The degree of success at the strategic level depends on the success of organizational units and business functions that are integrated and differentiated at the same time in the overall process of implementation of the strategy. For this reason, in addition to strategic control, operational control is elaborated which directly identifies performances toward carriers and the places of strategy implementation. Operational control is performed more frequently than strategic control and performance indicators are more specific. The very process of defining standards, measuring achievement, comparison with standards, qualifying, and if necessary taking corrective actions, is adapted to the planned implementation of the strategy. Figure 4.8 shows the importance of the system of budgeting, which is considered the forerunner of strategic planning.

The system of budgeting sets the network for planned allocation of resources and coordination and control of their use for their intended purposes by the carriers of certain activities. The budgeting system establishes standards that bind together human, physical, and financial resources for the accounting (planned) period. This system is of great importance because it determines the investment dynamics and priorities. By monitoring the indicators of the fulfillment of the plan (budget) of sales, it is possible to estimate the effectiveness in implementing the strategy and to timely take corrective action if the practice requires.

The plan of expenditures, if structured according to functional areas, makes it possible to monitor the implementation of each activity according to the prescribed standards, for example, sales per employee, ratio of sales and investments in marketing activities, increase in productivity, etc. A capital budget has similar characteristics and provides the ability to control types, levels, and areas of investing the capital.

System of Control

The necessity of reliable and comprehensive control, along with the existence of different standards, has influenced the development of a number of different control systems. These control systems are shown in Table 4.2.

Objective criteria, such as stock prices, return on invested capital, and the level of transfer prices, are the characteristics of *market control*. *Output control* is a control that takes into account the fulfillment of the division goals, business functions, and an individual's goals and requires a careful approach

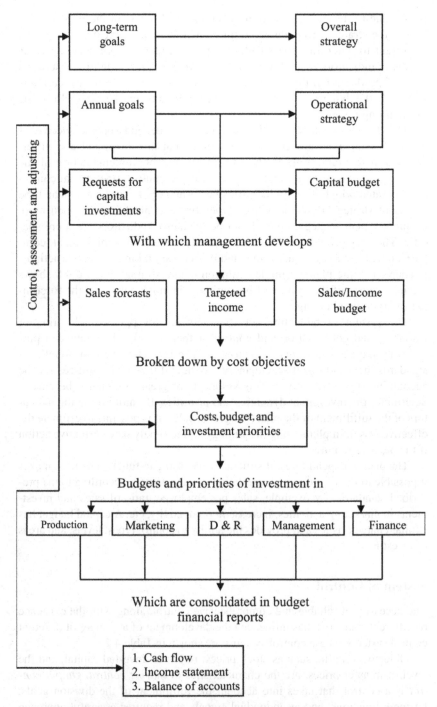

FIGURE 4.8   Budget system for control of implementation of the strategy of sustainable management of energy

TABLE 4.2 Types of control systems

Market control	Output control	Bureaucratic control	Group control
Stock price	Division goals	Terms and procedures	Norms
Return on investment	Function goals	Budget	Values
Transfer prices	Individual's goals	Standardization	Socialization

because the compliance of the quality and quantity of goals must also be taken into account. *Bureaucratic control* is more difficult to implement in practice, because it requires the definition of standards, procedures, and budgets for carrying out certain activities. *Group control* should identify the system of values, i.e., corporate culture, which directly depends on toleration and interpretation of certain activities as well as the ability of the employees to develop and maintain good communication.

Control and Audit of the Strategy

There is a great similarity between control and audit, and these two procedures should not be separated if we speak about the strategic component of control. While control determines the relationship between the achievement and the plan, the audit finds out whether the plan has been achieved and fulfilled in accordance with the possibilities. The control has a framework that entails the reality of the plan and the intention to keep business within the domain of the plan. On the other hand, apart from the plan, the audit takes into account the new challenges and opportunities that result from changes in the external and internal environment and suggests a possible change of plan. The audit, along with the control, is the element of planning, but it is based on understanding the trends in the environment and new management ideas.

The audit of the strategy of sustainable management of energy is based on two important assumptions: 1) predicting as a basis of input of planning is limited and 2) business planning depends on the style of management. In the first case the audit will be based on multivariety (contingent) planning, i.e., the planning will include multiple scenarios and plans. In the second case, the emphasis is on the change of management style as a direct consequence of new leaders and managers, leading to the need to review business philosophy, content, and the planning process itself. The market environment is very unpredictable: the longer the time horizon of the future, the less reliable the forecasting. This indicates that the audit becomes increasingly important when it comes to strategic control.

An audit can be defined as a comprehensive, independent, systematic, and periodic review of environment, strategy, objectives, and activities of companies or business units with the aim of identifying problems and opportunities, in order to propose a plan of action that will improve the business performance of companies. The relationship between audit and control is tight, because an adequate audit requires, apart from making predictions, taking into consideration the results given by the system of control of the previous activities.

The process of strategic management involves iteratively linked steps that are aimed at ensuring efficient and effective business. On the other hand, the process of making strategic decisions usually requires long-term engagement of the company and during that time circumstances in which the company makes changes. Accordingly, assessment, control, and audit are needed in order to determine whether the strategy as a result of strategic management is in accordance, above or below expectations or planning. For these reasons, the audit itself involves several areas, as follows:

- Evaluation of internal harmonization of strategic goals with other goals of the company, because the implementation of the strategy of sustainable management of energy often has direct or indirect effects on achieving other goals;
- Evaluation of the quality of the process of analysis which was the basis for the development of the strategy, as a way to determine if the previous assessment of the condition in terms of energy in the company was correct;
- Evaluation of the very contents of the strategy (direction, pace, and method of growth), because the strategy of sustainable management of energy usually takes longer and usually goes at an uneven pace;
- Evaluation of the opportunity of the company to achieve a formulated strategy, since regardless of orientation, it can happen that a company is unable to move to sustainable energy management in a way and to the extent envisioned by that company; and
- Evaluation of results brought about by the implementation of the strategy, which is achieved by measuring concrete outcome results—the most important are savings in energy and increased energy efficiency.

Control of the Strategy in the Environment

It is necessary to distinguish between an unstable and stable environment and adjust techniques on which the control of strategy is based. If the environment is stable, it is relatively easy to predict, and the role of the control is to observe the progress of the company. In an unstable environment, when predictions are hard to make, the task of the control is to assess the assumptions, goals, and components of the strategy. Two concepts of control derive from this attitude: 1) control of the current trend in the company and 2) control of changes undertaken by the company.

In case of the control of the current trend, which is applied in a relatively stable environment, the aim is to analyze where the company is in relation to the trend. Strategic control is intended to determine whether, despite the changing environment, assumptions on which the strategy is formulated are still valid. The strategy of sustainable energy management is carried out primarily on the basis of the strong influence of the environment, and in that sense it can be expected that the requirements would be the same and even emphasized. This largely guarantees justification of the change brought about by the new strategy.

Corrective actions are taken based on the control if the implementation deviates from the plan. In this case, previously defined general strategies expand. This form of control is based on creating special centers of responsibility, identifying success factors, and identifying performance standards for general strategies.

These are units for which the domain of responsibility and authority is determined, as well as standards against which achievement of the formulated strategy of the company will be monitored. In this sense, it is usually the creation of *revenue centers, cost centers, profit centers,* and *investment centers.* In the case of *cost centers,* the costs of production and supply are controlled. *Revenue centers* control realization of the planned performance in terms of marketing and sales. *Profit centers* control management of revenues, costs, and profit, thus they are considered a fairly comprehensive form of control. The most comprehensive form of control is the control by *investment centers,* which involves, apart from revenues, costs, and profit, investment management and assets management.

Control by the creation of centers of responsibility is a subsequent control of costs, revenues, profits, and investments in the field of energy management. In contrast to this concept, *the control based on factors of business success* is proactively oriented, because it examines the factors that contributed to (non-) success of the strategy and assesses whether the formulated strategy is still valid.

The domain of the previous forms of control that starts from the premise that there are no radical changes in the environment is not enough to propose the strategy through assessment and audit, if radical changes occur. In this case, it is necessary to apply *control based on the changes undertaken by the company.* In such cases it is necessary to rely on control of innovative undertakings that aim to audit the entire process of strategic management. This concept of control should help in volatile business conditions in the environment, which are characterized by irreversible changes. This form of control is predominantly proactively oriented and involves a much wider scope of testing and is also based on more sophisticated techniques: the subtle techniques of evaluating the components of the company, the possibility of combining for the needs of synergy, the concept of simulation, the concept of managing important strategic issues, methods, and scenarios, i.e., contingency planning, etc.

Such an audit has the characteristics of business restructuring, as it aims at removing the gap between the current and potential state and requires a diagnosis, a prognosis, and proposed changes. However, the audit differs from restructuring because the restructuring involves intervention to remove the already evident crisis, and revision is the result of the proactive attitude that it is possible to achieve more and better business operations. Restructuring is a subsequent, often belated response to changes in the environment, and the audit is the expression of proactive action, thanks to periodic and timely review. If the audit is well implemented, restructuring will not be necessary.

As the basic characteristic of a strategy is to adequately regulate relations with the environment, evaluations and audits are identified with the strategic control. Therefore, this stage of the process of strategic management is the responsibility of top management of the company.

4.7 CASE STUDY – CANADA

Energy Efficiency in the Production of Corrugated Cardboard Packaging – Case Study: Kruger LaSalle, Canada

1 BACKGROUND

Through the Canadian Industry Program for Energy Conservation, Natural Resources Canada's (NRCan's) Office of Energy Efficiency offers a financial incentive to help industrial companies increase energy efficiency, improve production processes, and cut costs. Program funding is available for up to 50% of the cost of an energy audit performed by a professional energy auditor, to a maximum of $5,000.

With funding from NRCan's Industrial Energy Audit Incentive program, Kruger Inc., a pulp and paper manufacturer specializing in the fabrication and multicolour printing of corrugated cardboard packaging, had an energy audit conducted at its plant in LaSalle, Quebec, in 2003. By implementing measures identified in the audit, which included plugging leaks in the compressed air distribution lines and replacing air compressors, Kruger was able to reduce the annual operating costs of its compressed air system by 42%. It was also able to reduce its space heating costs by recovering heat from the system and eliminate its use of municipal water for system cooling. Other highlights included:

- Electricity consumption for compressed air system reduced by 25%;
- Cost of operating the compressed air system reduced by 42%;
- Water consumption reduced by 38,000 m^3 annually; and
- Production increased due to better quality of compressed air.

2 PLANT PROFILE

Kruger Inc. is a major pulp and paper manufacturer with operations in British Columbia, Alberta, Ontario, Quebec, Newfoundland, and Labrador, the United States, and the United Kingdom. It employs more than 10,500 people worldwide.

Kruger's LaSalle plant produces 100% recyclable corrugated cardboard packaging for companies in the food and beverage, chemical, textiles, clothing, and agricultural sectors. It has an annual production capacity of 111.63 million square meters (m^2) of corrugated cardboard.

The corrugated cardboard is manufactured from rolls of paper produced in other factories and shipped to the LaSalle plant. The packaging is produced by printing, cutting, folding, and gluing the cardboard. The LaSalle plant specializes in four-color flexographic printing, a printing process that uses a flexible, relief-type printing plate.

The manufacturing process uses electricity and natural gas. Natural gas is also used for space heating. In 2001-2002, the plant used more than 7 GW hours of electricity, at an approximate cost of $425,000, and approximately 1.4 million m^3 of natural gas, at a cost of about $580,000.

3 ENERGY AUDIT

The LaSalle plant audit, conducted by R.O. Poirier Inc., examined the air compressors and the compressed air distribution lines and identified energy efficiency measures that would reduce the amount of electricity used by the compressed air system. Kruger received financial support from NRCan's Office of Energy Efficiency equal to 50% of the professional fees for the audit.

Kruger's LaSalle plant uses compressed air to operate its production equipment, including the printing presses, conveyors, and splicers on the corrugated cardboard machine. Compressed air is also used for cleaning some equipment.

At the time of the energy audit, the compressed air system comprised two single-speed, water-cooled compressors, rated at 112 kilowatts (kW) and 149.2 kW (150 horsepower [hp] and 200 hp), respectively, and a compressed air reservoir to which both compressors were connected. In 2001-2002, it cost the company about $131,000 to operate the compressed air system, including the cost of electricity and water used for cooling the compressors.

The audit found a very high rate of leaks and 45% of the compressed air was lost. The leaks meant that the compressors had to operate for longer periods with consequent increases in electricity consumption. The annual cost of producing compressed air was calculated to be $126.50 per m^3/h under standard conditions of temperature and pressure (15°C [59°F] and atmospheric pressure). This is considerably higher than the $64.70 per m^3/h (standard) cost usually associated with screw compressors. As well, the leaks reduced the system's capacity to provide compressed air at the volume and pressure required by the plant's equipment.

The proposed energy efficiency measures included plugging the leaks and improving the current system by replacing the two compressors with a new variable-speed, air-cooled unit. The ability to control the air pressure was also improved by adding a new air reservoir.

The amount of the leaks was determined by measuring the flow of compressed air produced by the system on a weekend when there was no activity at the plant. The audit found that there were leaks throughout the plant, with the biggest ones in the roller conveyers. The plant's staff worked to correct the loss, and the demand for compressed air dropped by over 176.4 m³/h, or about 40% of the compressed air produced by the system.

To improve the performance of the compressed air system, the auditors proposed measures applicable to both the individual components of the system and the system as a whole. However, because the compressors had been in service for many years and because their performance had dropped to below 70% of their rated volume, the plant decided to replace the entire compressed air system. This option also had the advantage of a shorter payback period.

The plant's compressed air system was replaced with a single variable-speed, air-cooled compressor rated at 112 kW (150 hp). Unlike a single-speed compressor that runs at the same speed whenever it is on, a variable-speed compressor adjusts the speed of its drive motor so that the production of compressed air matches the demand placed on the system, thus reducing energy costs. Given its increased efficiency, the new compressor is sufficiently powerful to meet the plant's reduced demand for compressed air, which resulted from eliminating leaks.

Since the new system is air-cooled, water is no longer used for cooling as it was with the old system. Air compressors give off a lot of heat and generally 80-90% of the electrical energy supplied to the compressor is converted to heat. At Kruger's LaSalle plant, the heat given off by the air compressor is now recovered and used as a supplementary source of space heating during the winter. It is vented outside during the summer, when space heating is not required.

A new compressed air reservoir, larger in volume than the existing reservoir, was also installed. This reservoir receives the compressed air as it is produced and acts as a buffer downstream from the compressors, cushioning fluctuations in the airflow and allowing for better control of system pressure.

Installation of the new compressor, which was done by a team of subcontractors working under the direction of the plant's engineers, did not affect production, since the old compressed air system was kept in operation until the new one was up and running. Purchase and installation costs of $170,000 were partially offset by $67,000 in financial assistance from Hydro-Québec, through its Energy Wise Industrial Initiatives Program.

4 RESULTS

By implementing the measures proposed in the energy audit, Kruger's LaSalle plant was able to reduce its production of compressed air by more than 40% and the annual operating costs of the compressed air system by 42%. Blocking the leaks alone generated savings of over $60,000 annually.

Reducing the demand for compressed air and installing a new variable-speed, air-cooled compressor helped the plant to reduce the annual electricity consumption of its compressed air system by approximately 25% and reduce the plant's water consumption by 38,000 m^3, generating savings of $55,000 annually. By recovering energy from the air used to cool the system, the plant has reduced its natural gas consumption by 21,000 m^3 annually, resulting in savings of $5,000 on space heating costs.

Considering the net cost of $103,000, savings of $60,000, and a payback period of approximately 20 months, the installation of a new compressed air system represented a financially attractive investment. In addition to the significant energy savings, the new compressor ensured better regulation of the air pressure in the network. This in turn eliminated breaks in production at the plant due to the drops in network pressure that occurred with the old compressed air system.

Kruger's commitment to fostering the environmentally friendly use of natural resources and its LaSalle plant's determination to increase the energy efficiency of its operations by improving the performance of its compressed air system ensured the successful outcome of its energy audit.

Chapter 5

Methods and Techniques for Implementation of Sustainable Energy Management

5.1 BASIC APPROACHES TO IMPLEMENTING STRATEGY OF SUSTAINABLE ENERGY MANAGEMENT

The implementation of the process of sustainable energy management can be undertaken by using appropriate techniques that are applicable no matter whether the management process will be applied in a particular company, region, or a country. Application of these methods of energy management is not recommended on a level broader than national, since the process requires analysis of the environment and the control involves measuring performance in each state, so the accuracy is acceptable up to this level. A particularly unclear situation will result from the analysis of an environment that is very complicated in an area wider than one country, as among countries there are significant differences in terms of social, political, economic, and environmental situations.

The main role of the methods discussed in this chapter is to implement the previously selected strategy of sustainable management of energy. The strategy of sustainable management is the only acceptable strategy for energy management and its implementation can be carried out by selecting one of the proposed models. All of the following models are designed to allow segmentation of the subject on which the control is performed. The company, region, or country in question is characterized by its specific characteristics, determined by many parameters, and in the given time it undergoes various stages of development.

Traditional management approached energy solely on the basis of its cost effectiveness. Modern sustainable management imposes the need to analyze various aspects of energy efficiency, consumption of energy, and harmful emissions that result from it. Cost effectiveness is no longer the only factor that influences the decision-making process. The environmental component, and energy sustainability within it, is becoming a very important factor that to some extent changes the traditional concepts of management and absolutely disrupts the traditional primacy of cost effectiveness.

The economic aspect of management is still the subject of study, but by no means is it the only and the most important aspect. The purpose of the new

Sustainable Energy Management. http://dx.doi.org/10.1016/B978-0-12-415978-5.00005-9

analysis of management techniques serves to find a method that harmonizes economic growth and environmental concerns – energy responsible business operations – without substantial harm to the development of a country or company as a whole. In current practice, in order to implement sustainable management of energy, six basic approaches have been defined:

Traditional approach;
System approach;
Eco-management approach;
Total quality management approach;
Life cycle analysis approach; and
Gap analysis approach.

It cannot be said which of these approaches is the best, because each has its advantages and disadvantages and specific features that make it more or less acceptable under certain conditions. It is often desirable to combine these approaches to get the best results.

5.2 TRADITIONAL APPROACH

The traditional approach is a logical way to implement the sustainable management of energy; its basic model is shown in Figure 5.1.

Development of Program

The traditional approach is a "top-down" approach. This approach begins with obtaining management support and determining program goals before collecting necessary information. The aim of the first step is to provide an executive-level management decision to start developing the strategy of sustainable management of energy that must be incorporated into the business policy of the company.

At this stage a team is created to collect necessary data, inspect the areas of application, and set priorities for implementation of the program. A preliminary evaluation is necessary to obtain the information needed to create the initial

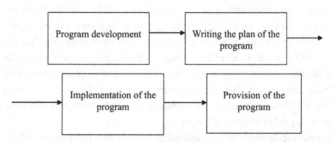

FIGURE 5.1   The traditional approach to implementing the strategy of sustainable management of energy

strategic plan. This part of the strategy is carried out by a team of employees or a group of external experts.

At this stage, the team collects information that might be potentially important: the amount of energy consumed, energy sources, owned energy resources, energy consumption per workplace, per process, and per product, and the level of energy efficiency and energy waste, as well as economic aspects of energy consumption. This information should provide a realistic picture of the current state of energy management, and therefore, identify the shortcomings and problems that will enable clear insight into the priorities of the energy policy.

Writing a Plan of the Program

The second step of the traditional approach is to write a program plan for the sustainable management of energy. A good plan should indicate what all stakeholders want from the program, including consumers, suppliers, employees, legislation, environmental organizations, local communities, shareholders, and others who have any interest in the program implementation.

The plan should indicate the goals clearly and the ways to reach them. It is not enough to decide to increase energy efficiency. It is necessary to clearly define the level of energy efficiency that is to be achieved and indicate the means and resources necessary to achieve that goal. Possible obstacles need to be anticipated in terms of implementation, and the plan should be made so as to overcome the obstacles. Finally, the plan should determine the schedule of activities that will, among other things, enable management to monitor the process and progress during the year.

Implementation of the Program

Implementation of the program of sustainable management of energy by the traditional approach can be summarized, as shown in Figure 5.2.

The aim of the detailed evaluation is to help the team determine alternatives, and for this, *brainstorming* is used. The traditional approach does not include a formal analysis of the root of the cause before the alternatives have been determined. This phase involves appointing the team for evaluation, data review and organization, and documenting of information. It is necessary to precisely determine the existing gaps in energy policy, which includes estimation of the type, intensity, and durability of problems. Detected shortages are documented and ranked according to their urgency.

Based on the detailed assessments of the possibilities for resolution of the identified problems, the evaluation team proposes a certain number of alternatives and screens to help determine the focus of the implementation that follows. Generally, a limited number of alternatives is recommended. The traditional approach for the assessment of alternatives uses a matrix of criteria. For defining the problem to be solved the following criteria are often used: degree of

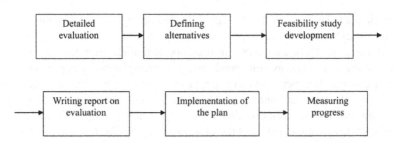

FIGURE 5.2   Traditional implementation of a program of sustainable management of energy

urgency, technical and technological limitations, availability, capacities, and economic aspects.

Upon completion of the review, the analysis of the feasibility of some alternatives is started. Of course, it is not necessary to conduct a detailed analysis for each alternative respectively, but on the other hand, there are those that are particularly demanding. Feasibility analysis should indicate the technical, environmental, and economic feasibility of each alternative. In most cases, it is first necessary to consider the technical feasibility of the proposed solution model. In fact, some energy problems could be resolved in a certain way, but usually there are technical limitations that hinder the application in some way. It may be that it is necessary to use a more expensive solution, because the cheaper and more accessible ones may not be technically feasible.

After evaluation of the feasibility of the alternatives, a report should be written indicating the details of the analysis and pointing to information that should be presented to management. In addition to technical feasibility of the chosen solution, it is necessary to point out the specific benefits of the proposed improvements, such as the amount of energy that can be saved, the degree of increasing energy efficiency, reducing waste, the amount of energy gained from renewable sources, etc. In this way, management can estimate the positive financial and non-financial effects of the proposed change.

Upon approval by management, the next stage entails implementation of the plan, which includes project selection, engaging necessary resources, and starting implementation of alternatives. The sustainable management of energy is approached through prior achievement of the strategic goals that were selected based on the previously defined criteria. These criteria create the traditional method of management, which is closely problem-oriented because it responds only to the perceived problems and thus has a predominantly reactive nature.

During the stages, the whole process is monitored and adjusted in order to achieve the set goals. The traditional approach does not require preparation of a specific action plan. The last step of this phase is to measure progress. The traditional approach requires the collection and analysis of the data, as well as measurement of economic results.

Provision of Program

In this stage five activities are performed, as shown in Figure 5.3.

Provision of the program begins with the integration of the program of sustainable management of energy in formal corporate initiatives and plans. These plans may include safety, quality, preventive measures, and the like.

Education of people involved in the program must be specified. Participants must be familiar with the process, especially if there are new employees. Employees must undergo an annual testing of knowledge about the program and constantly follow up on innovations.

Communication is important for the implementation of each activity. The traditional approach seeks to develop all types of communication. Communication should be two-way, and it is necessary to consider the suggestions of employees. The traditional approach recommends particular consideration of the program of rewarding employees.

5.3 SYSTEM APPROACH

The organization is a system if it operates as a whole and if its parts develop certain interactions. The system approach to sustainable energy management observes the whole organization, as well as its parts and connections among these parts. The functionality of the organizational units largely depends on how they are connected, more than on the parts themselves. [88] This model envisions that the effectiveness of the sustainable energy management system is more affected by the way and efficiency of the connections that exist among organizational units than by the specific features of the parts. However, the system approach is based on the fact that energy is not lost but is transformed from one form to another. According to this model, each system on which sustainable management of energy should be implemented is made of a set of components and connections between them, which are characterized by certain flows of energy. Parts of the system are connected directly or indirectly, and the change in one part affects all other parts. The system approach is unique for several important reasons.

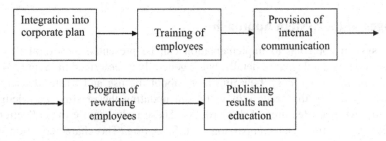

FIGURE 5.3   Provision of program

Processes that consume energy do not always provide information that is easy to present. Instead, these processes are sometimes defined so that the amount of energy consumed per process, per product, or per work task cannot be determined. For this reason, too much time and energy are spent in trying to understand the situation before moving on to the next stage of solving the problem.

The system approach cuts across this very situation. System tools indicate how the situation can change and how the resources can be saved. This is the basic difference between the application of the system tools and engagement of an external expert for energy management while the employees are not involved in the process. The system approach involves decision making by teams within the organization, rather than individual experts. The team members are required to analyze the problem of energy management to determine the cause of the problem and offer alternatives for solutions. That does not mean that teams should not use the services of experts. However, the responsibility for making decisions should be given to the team members who will go further in the implementation and evaluation.

Since the system approach is interactive, team members often believe they are responsible for making all decisions. Of course, other employees have their own opinions. Another problem is the very process of management in the organization. It is important to identify the reasons for the use of resources and for losses before arguments have been provided so managers can make decisions about the program. In this context, the system approach provides the management a clear view of processes and problems. Only in this way can management effectively plan strategies and objectives.

Many organizations believe that adjustment is the basis for survival in the global economy. Managers learn new ways to run their organizations and employees learn how to contribute to the improvement of business by using their knowledge and skills. By learning how to supervise, control, and continuously improve production, organizations are able to meet the demands of consumers and other stakeholders. [67] The principle of qualitative improvement can be a useful tool for achieving environmental excellence. Application of tools for improving quality is a powerful means to eliminating environmental inefficiency and preventing pollution.

Stages of the System Approach

The system approach to implementation can be presented in several ways, depending on the degree of details, but is generally as described in Figure 5.4.

Mapping the process. Learning more about the use and waste of energy resources during the process of work and updating the existing knowledge are the two main features of this process. These tasks require the collection, reviewing, sorting, and comparison of information. The characterization of the process is a step during which most knowledge about energy cycle in the

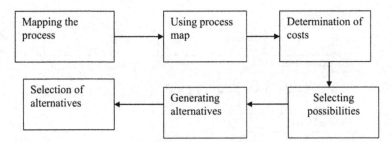

FIGURE 5.4 Stages of the system approach to implementation of the strategy of sustainable management of energy

company is acquired. At this point, the existing knowledge of the process is reviewed and reshaped, and new knowledge is easily acquired and complements the existing information, because the procedure of process mapping allows the process to become "visible." It is now possible to observe the process in a new way and look at all operations for support, in order to identify how they affect the progress of the main process.

Characterization of the process of flows and consumption of energy is more efficient than preparation of the classic worksheets. Diagrams are always a better and clearer way to represent information. Links that are identified between the process steps help to identify more clearly the causes and places of the use of resources and waste generation. Every place resources are used is an opportunity to save those resources. Each loss of energy resources is an opportunity to avoid that loss.

Characterization of the process includes collection of all information which describe the activity which is under observation. An effective way to create the characterization of the process is the use of the so-called hierarchical process maps. In most organizations, process documentation is created under categories such as company, equipment, production line, or department. The information is given in the form of diagrams, maps, machine configuration, plans, or in some other way. Each of these methods of characterization of the process has a fundamental flaw due to its great complexity because there are too many objects on a single sheet of paper.

Research has shown that most people can simultaneously observe six objects and express what the objects mean and the connections between them. Hierarchical process characterization shows a maximum of three to six objects on a single sheet of paper, because almost every process can be shown in such a way. Subprocesses are also shown at the next level, but again, displaying only three to six objects. In this way, a sort of "tree" is constructed which shows the relationship between process steps. The system approach indicates that each step is connected with the other steps and that they all create the system as a whole. The hierarchical process structure can be summarized as shown in Figure 5.5.

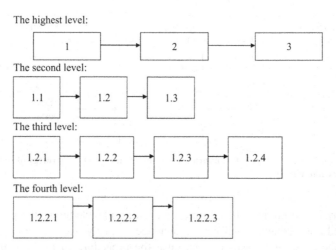

The highest level:

The second level:

The third level:

The fourth level:

FIGURE 5.5 **Hierarchical structure of the process map**

During the mapping process two basic rules need to be observed. First, the process map should help the program implementation team understand the process better than using other methods. Second, process maps should help the team communicate better with management and other stakeholders.

Using process maps as patterns. Some organizations think about the process as a box with inputs and outputs. The use of this method is appropriate for introducing only limited changes and enable the application of the concept of sustainable energy management. Using process maps as patterns, process documentation can be organized into separate steps in the process at the least number of levels. All procedures for the implementation of standards, the best management practices, implementation of recommendations, and others can be presented through the steps in the process map. What can be noted when using the process maps is that many problems are associated with a single step, and this makes it easier to focus activities on that particular step.

To present the process maps many designers use simple pencil and paper, although they can be presented using a computer and special software. If the organization decides to computerize process information, everyone involved in a single working step should have access to all information. The use of process maps as patterns helps organizations track the use of resources and waste in each process step in the main or side processes. All resources (energy, water, raw materials) can also be followed up through some process map. A simple review is shown in Figure 5.6.

The term "non-production use" means that the observed energy resource does not become an integral part of a semi-product or the final product. The term "non-production loss" means that the energy resource is lost in that step as waste, discharge, or emission into the environment. Process losses can be

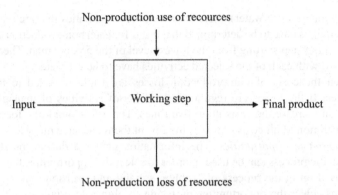

Non-production use of recources

Input ⟶ Working step ⟶ Final product

Non-production loss of resources

FIGURE 5.6   The use of a process map as a pattern for following energy resources

classified on the basis of the medium (air, water, solid waste). Costs can also be monitored on the basis of the process steps. Monitoring of this information is useful to rank the changes according to costs. All this can be provided by application of software, as well.

Determining the cost of losses. Collecting accurate data on costs is important to justify the investment in alternative solutions. These costs are collected by the operating steps because these are the places where the program of sustainable energy management will be implemented. There are three types of costs to be monitored: total costs, costs incurred by the loss of energy resources, and costs based on the cost of management and other intangible costs.

For each loss identified in the process map, the program should examine the so-called calculation map in order to check whether the cost is tracked by accounting services. The costs map shows a number of costs and/or a different code for this category of payment. It is important to remember that usually only the money that comes into or goes out of the organization is tracked. Internal transactions are not tracked. The internal transactions are the costs of activities. All data on costs obtained in this way are accurate in most cases. The existing procedures must be applied to energy costs, too.

The second category of costs is associated with the costs of resources that are non-production output or process losses. For example, painting of new cars in the car industry is connected with the use of sprays. Some cars will not be completely covered in paint, because a certain amount of spray is retained on the filter or ventilation system. It can happen that 60% of the parts are well painted, while 40% are not. In this case, the whole process of painting should be conducted again. The 40% is the loss of that working step, and that is a non-production loss. The costs of wasted color and energy can be added to the total cost for color filters, which increases energy non-production costs.

The third category of costs is associated with the losses regarding the activities of management. If the costs are covered (e.g., hazardous air pollution, hazardous waste, priority waste water), there are a number of activities that are regulated. First, activities have to be determined that occur for each non-production cost for each working step starting from the lowest level of the process map. Then, costs associated with each of the selected activities have to be estimated.

Often the costs of non-production losses are triple if added to the lost resources. If the losses are regulated, the costs of activities of managing the losses can increase the costs up to five times. The effort should be focused on the registration of all costs of losses to prevent them and save money.

Selection of opportunities. The information gathered during the stage of mapping the process can be used in order to select the opportunities that should be focused on in the process of problem solving and decision making. Some programs select the opportunities by trying to eliminate problems associated with the use or loss of energy resources. Other programs seek to identify opportunities among those that were previously selected by managers or specialists in the field of energy. Each organization has its own method for selecting opportunities, and there is a special tool (which dates back to 1897) that can help to easily perform this operation, discussed in the following.

If all opportunities identified in the process map are ranked according to the sequence of their costs, it can be noted that 20% of them carry about 80% of all costs, and vice versa. In most cases, this 80/20 rule (the so-called Pareto principle) should be used for selecting opportunities. Most organizations apply this rule in their work.

Quality improvement experts recommend focusing on several vital sources of the problem, thus avoiding their interference with the issues of minor importance. The term for this process is called ranking. Pareto analysis is a tool for implementation of the ranking. Potential opportunities should not be rejected easily, although in the early stage of implementation of sustainable energy management a careful selection has to be made so as to choose the most likely chance of success. Most often, teams have to work with projects that are either very large or very difficult to handle. We must focus attention on the 20% of causes because they bring about long-term benefits. In the process of sustainable energy management it always happens that in 20% of the places in the organization, 80% of problems have been identified regarding management of energy. Thus, attention should be directed towards the possibility of their elimination.

Cause analysis. The cause is the main reason why the energy resource is used or the loss occurs. If this cause cannot be eliminated, the use or the loss of the energy resource should be prevented.

Analysis of the root of the cause refers to identification of causal factors. Most people involved in the program of sustainable management of energy are committed to resolving the problem, but in their attempt to find a solution they may quickly overlook some very important activities in the process of problem solving. A team that skips important steps will often take the obvious actions instead of those that provide the best possible solution.

For example, when faced with environmental problems caused by excessive use of fossil fuels, the members of the problem solving team may assume from the start that the best solution to the problem is to find an adequate and safe substitute – energy from alternative sources. In fact, the problem may be caused by the way in which the fossil fuels are used in the organization, and not by the features of the fuels themselves. The change in procedure or equipment or staff training is often the best solution for reducing costs of this kind. Root cause analysis teaches organizations to consider all potential causes: material, technology, work processes, and employees.

Root cause analysis can be an effective management tool to determine the real cause of the problem, the ways to modify equipment, etc., and the preventive measures needed. It also provides opportunities for improvement.

A cause and effect diagram, also known as a fishbone diagram, is the most commonly used tool for solving these types of problems. To be used effectively, it requires some training and experience and should be used by the whole team. This diagram clearly indicates to management and other stakeholders what may be the cause of the problem in energy management. Once a diagram has been made, the team members need to accurately count the causes that have been identified. The 80/20 rule can help to focus attention on the most likely cause by drawing circles around the 20% of causes that result in 80% of problems. The team will be more effective if they focus on the process of identifying problems before attempting to propose solutions.

Generating alternative solutions. Each approach has a certain method for generating alternatives to solving problems in energy management. Some experts believe that practical applications should be limited to a small number of alternatives for solving a specific problem, either because they have not performed the analysis of the root of the problem, or because the team members are not adequately involved in the process of generating alternatives. References and practical examples provide just a few ways to specify each problem. The team members in each organization should be satisfied if they can develop even a few alternatives.

Tools for Improving Implementation of the System Approach

Tools have been developed in order to make the system approach easier to use and more efficient. The two tools that are most commonly used are action plans and checklists.

An action plan should be prepared for each activity that is planned for each year of implementation of the program of sustainable energy management. Regardless of how the organization plans to implement certain new solutions, the stage of formalizing its own action plan should not be neglected. Each action plan should show all the alternatives that will be applied and indicate substages and steps that will take place. The action plan should indicate the name of the person responsible and accountable for the performance of each step.

Each step should lead to achieving a recognizable goal that should also be clearly indicated. It is necessary to mark the measurements necessary for monitoring the progress during implementation in order to reach the defined goal and determine the time frame for completing each step. Finally, it is necessary to determine the resources for realization of the whole process. An example action plan is given in Figure 5.7.

Checklists are another helpful tool for the sustainable management of energy. Checklists help conduct organizational activities and progress and provide important information on the steps and methods for measuring organizational performance and effectiveness. They also help organizations collect and organize data necessary for evaluation of the current status and for monitoring the process. It is useful to compile a list of questions and answers for everything associated with each tool for problem solving and decision making.

Checklists help the team properly conduct future activities and suggest which things should be done, the people with whom to contact, and what questions to ask. Checklists also help the team organize tasks and gain general insight into the situation, recommendations, characteristics, alternatives, and consequences. The following steps should be taken to create a quality checklist:

1. Determine the purpose of using the checklist.
2. Conduct research to ensure the checklist covers all recommendations and answers to questions that should be collected.
3. Provide space for checking steps that are finished.
4. Ask experts in the concerned area to evaluate the final draft of the checklist to ensure all important details are included.
5. Provide audit and pilot testing of the checklist prior to its application in practice.

Teams should create a checklist that follows the process envisioned by their program. Some components of typical efforts use system tools, so teams can use a checklist to:

Determine the real costs for each working step in the process;
Collect information necessary for ranking alternatives;
Select alternatives that will be analyzed;

Selected alternative:			Date:		
Activity	Person responsible	Performance	Monitoring technique	Deadline	Required resources
1.					
2.					
3.					
4.					

FIGURE 5.7 Example action plan

Ensure that all causes of the problem are taken into consideration;
Document the search for potential solutions and alternatives;
Collect information on each alternative in order to determine priorities;
Test complete action plans;
Track information during the process;
Test the use of each element in the process; and
Test the total efficiency of the program.

Periodic use of checklists creates the impression of continuous assessment of progress. Checklists should be designed to provide managers and teams a starting point for the assessment of important characteristics of each step in the system approach, giving answers to questions of what to do, and analyzing whether the tools are properly used.

Combining the Traditional and System Approaches

Most managers use the traditional approach for the implementation of sustainable management of energy. The recommendation is to start by considering changes offered by the system approach that will improve business. The area that can be particularly improved in this way is the characterization of the process. The system approach requires development of process maps that allow people to easily identify and get insight into areas on which they should focus. The system approach is particularly effective for monitoring the flow of energy.

Mapping of the process should be done before implementation of the program is proposed because it will be easier to point out the reason for the change. Another potential improvement offered by the system approach is using problem cause analysis in order to select options. Experience has shown that teams that do not use cause analysis are able to identify no more than two or three alternatives. Teams that use the root of the cause analysis indicate that there as many as 40 alternative solutions.

The traditional approach emphasizes setting the objectives first in any business. The system approach determines objectives in the action plan but only after the information has been collected and analyzed. The goals are changed according to the needs and vary from one project to another, from one year to another. Some experts believe that in terms of quality the biggest problem of the traditional approach is inadequate setting of goals.

5.4 ECO-MANAGEMENT APPROACH

The latest series of international voluntary standards known as the ISO 14001 series is an effective tool for improving organizational environmental performance and implementation of a sustainable approach to energy management. The purpose of the standards is to establish and implement a systematic management plan that is designed to continually identify and reduce environmental impacts arising as a result of organizational activities, products, and services.

The eco-management approach may be particularly suitable because its implementation is a possible way to replace the widely used environmental control system based on legislation and controlling of the application of environmental and energy regulations. Eco-management systems can help organizations integrate environmental efforts into everyday efforts, decision making, and business practices.

Application of an eco-management system improves compliance with environmental regulations and prevents organizations from making progress without taking into consideration environmental regulations so that in both cases it has positive effects on reducing negative environmental impacts. Today, the number of companies in the world that integrate the eco-management system in their business strategies is rapidly increasing. They accept advanced technologies and achieve some savings. Systems of eco-management encourage companies throughout the world to consider environmental consequences of their operations and define strategies to help them reduce waste, risks, and costs.

As with other management systems, eco-management is a formal approach to setting goals, decision making, information gathering, measuring progress, and improving business performance. Eco-management promotes important elements of planning and improvement required to make an impact on all forms of environmental pollution. In addition, certain elements of eco-management provide an opportunity for implementation of assessment and continuous evaluation by management. A simplified description of the implementation of a sustainable energy management system using eco-management is given in Figure 5.8.

The eco-management approach is based on a documented policy that includes three key principles: compliance with changing environmental regulations, pollution prevention, and continuous improvement of environmental performance. In some cases, corporate policy can be complex and general, and as such unclear to employees and the public. Therefore, the eco-management policy should focus only on the three basic principles given above.

The eco-management system identifies and adapts environmental, legal, and voluntary standards related to the work of employees and business partners. Voluntary recommendations include a pollution prevention program, health and safety prevention, and sustainable development. In addition, eco-management

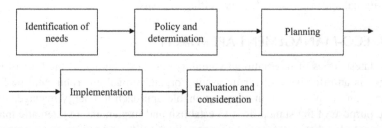

FIGURE 5.8 Implementation of a program of sustainable management of energy using the eco-management approach

approach for implementation of energy management elaborates procedures on how to achieve objectives, and also determines environmental responsibility and ways to respond to requests of individuals, organizations, and other interested parties.

Identification of all aspects and determination of their significance is usually the biggest problem in most companies. Because of that, the eco-management system determines procedures necessary for identification of environmental impacts of business activities, products, and services. Thus, most organizations are focused almost exclusively on their negative environmental impacts. Positive environmental impacts are also important, such as recycling and hazardous waste collection sponsored by the company.

The eco-management system establishes specific goals to implement energy management initiatives and ensures the participation of employees, as well as technical and financial support. In setting goals for environmental improvement of each business function in the company, the eco-management system recognizes existing environmental laws, recommendations, and requirements of all stakeholders. The eco-management system is based on documented procedures for the prevention, detection, investigation, correction, and reporting (internal and external) of active or potential environmental impacts and emergency situations. If the environmental incident occurs, the eco-management system should be reviewed immediately and adjusted so as not to allow such an incident to repeat. The eco-management system also provides for adequate training of employees, suppliers, and other business partners. Organizations should review the competencies and capabilities of its employees to meet the requirements of implementation of the eco-management system.

A particularly important factor in the successful implementation of the eco-management system is the determination of top management, who should accept that the eco-management system is more adequate than other management systems. With this regard, top management should provide maximum efforts to implement eco-management as well as review it periodically to improve it. Over time, investments in the implementation of the eco-management system prove justified, due to the improvement in environmental performance of the business which satisfies all the parties involved.

Identification of Needs

The aim of the ISO 14001 standard is to establish a common approach to eco-management that is internationally recognized. The ISO 14001 standard is a standard for management and not a standard for performance. Eco-management provides the system approach for the integration of environmental protection in business functions and management strategies, primarily because it requires developing appropriate business policies by top management. The eco-management approach involves several initial activities that take place in the sequence of events given in Figure 5.9.

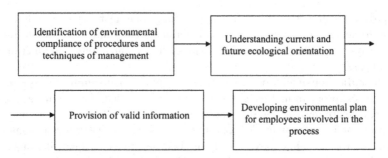

FIGURE 5.9   Identification of needs

Many organizations find it useful to use gap analysis of the existing energy system as an initial step in the development of an eco-management system. In this way the organization is able to review the current status of its own energy performance and identify areas that require attention when applying the eco-management system. [41] Gap analysis results provide a kind of benchmark regarding compliance of organizational performance with ISO 14001. The gap analysis should include areas of business related to ecological systems as well as links between specific equipment.

The eco-management approach to implementation of energy management encourages organizations to coordinate their activities with relevant environmental legislation and regulations along with other recommendations. The eco-management system together with the ISO 14001 recommendations offers good potential for achieving environmental efficiency and cost reduction. In this way, top management in the company observes the eco-management system as a way to achieve competitive advantage, and not as a place for incurring costs. Inclusion of top management in the procedures of policy development, considering the existing plan, and ensuring the application of eco-management is a positive step.

Energy Policy and Determination

Management is responsible for developing the vision and direction of development of the energy system in accordance with the environmental assessment of performance. When defining the energy policy, management should define the scope of observation and ensure consistency with organizational vision, core values, goals, and other efforts. The energy policy of the company should include a plan to implement the pollution prevention program, harmonization of business operations with existing recommendations, and continuous improvements of the system. The process of defining an energy policy can be summarized as given in Figure 5.10.

An energy policy is used as a specific guide for developing and understanding organizational energy goals. The eco-management approach to implementation

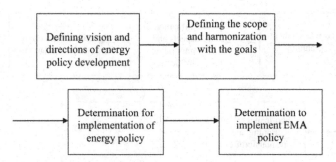

FIGURE 5.10   Process of defining energy policy

of energy management does not require defining specific energy goals. Instead, it provides a general network for organization and implementation of goals that are necessary for improving energy performance.

After defining the energy policy, it must be documented, updated, and delivered to employees. Many organizations have already developed procedures for internal and external communication of their own policies. The eco-management approach to implementation of energy management recommends announcing the energy policy of the company publicly. Many organizations already provide certain environmental and energy information in annual reports and plans for dealing with emergencies.

The eco-management approach to implementation of energy management defines the procedure for responding to external demands for environmental and energy information by which it also determines who is responsible for contacts, who notes the date and the nature of the request and response, and whether and what kind of written material is sent in reply.

The eco-management approach to implementation of energy management also recommends that organizations develop and implement procedures for the purpose of internal communication. The eco-management system and environmental aspects should be associated with all levels of organization and business functions that may have an impact on the environment. The procedure of internal communication outlines the responsibilities in the chain of communication developed for eco-management and environmental issues. Changes may include environmental information, such as review of the objectives, changes in procedures, environmental incidents, etc.

Planning

Planning is needed as the organization grows and production lines change. Planning should go along with other business planning efforts. The stage of planning within sustainable energy management can be summarized as shown in Figure 5.11.

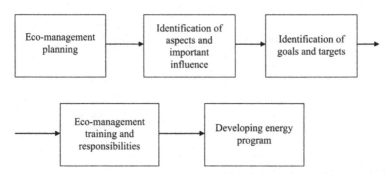

FIGURE 5.11   Eco-management planning

Insufficiently focused or badly managed efforts will inevitably lead to poor implementation of energy performance and higher costs. On the other hand, well-developed and effectively implemented programs lead to the achievement of high performance and cost reduction. Improved energy performance is an important benefit for most organizations that use the eco-management system for development. Also, some organizations operate under an eco-management system that systematically follows up on environmentally relevant activities. This eco-management system covers the organizational structure, procedures, processes, and resources needed to implement effective environmental management.

Top management appoints the company managers who will monitor the achievement of the objectives set out in the eco-management system. Managers in charge monitor and evaluate the system and report to top management about eco-management activities. Coordinators work with lower organizational levels and try to generate new ideas that may lead to modification of the eco-management system in terms of its improvement.

Teams are used to multiply the force of the organization. The team approach allows discussion and comparison of different viewpoints and opinions. It may be useful to form independent teams whose members will be selected from all levels of the organization. Involvement of people from different levels of business can only improve the overall efficiency of these issues. Clearly identified teams are of particular importance in the organization of activities and are opposed to individual business functions or independent work areas. These teams can be used for identification, evaluation, and implementation of opportunities. Teams are empowered to take direct action, make decisions, and initiate changes that will result in continuous improvement of the energy system in order to comply with policies and achieve organizational goals. Once the team has been established, it is necessary to clearly define the role of each individual in the team.

Identification of significant environmental impacts. The eco-management system is essentially derived from environmental problems and encourages organizations to systematically monitor the environmental impacts that result from carrying out their activities, products, and services. This approach can be very effective and can encourage organizations to take a proactive and sustainable approach to managing their environmental impacts and programs. Aspects of impacts are in fact elements of organizational activities, products, or services that may affect the environment. The organizational aspects may include generation of waste and pollution, use of resources, use and losses of energy, and other environmental impacts. The standard proposes some basic planning activities that are used in most organizations in order to implement sustainable energy management, as follows:

Identification of activities, operations, processes, products, and services that have environmental impacts;
Identification of legal recommendations in the field of energy that influence organizational activities, products, and services;
Evaluation of ecological impacts of energy consumption according to the degree of importance;
Setting goals whose achievement reduces negative environmental impacts generated by the use of energy; and
Selection and implementation of activities during the eco-management program that are necessary for achieving the desired goals.

During the procedures of identification and ranking of environmental impacts, planning can serve as a key element of the eco-management plan. Planned assessment is a systematic, periodic research of organizational operations designed to identify potential areas for change of energy performance. A properly designed eco-management system can have a far greater range than a traditional approach to energy management. In order to understand all environmental impacts, the organization should consider the use of energy and water, conservation of landscape, the problem of noise, and other impacts. In addition, an organization may request information from its information suppliers about the content of certain materials, the method and type of packaging, and methods of delivery. The procedure to identify aspects can be conducted by using the following tools

1. Process mapping
2. Interviews
3. Questionnaires
4. Checklists
5. Benchmarking
6. Cost-benefit analysis
7. Life cycle analysis
8. Analysis of input-output balance

A particularly important aspect of determining environmental impact includes consideration of the state of the control systems in energy management. The organization should develop appropriate criteria for determining the degree of significance of these effects. Criteria may include regulatory activities and management costs and risks associated with the use of energy resources. The most important feature of these criteria lies in the fact that they reflect the organizational values stipulated in the business policy.

Identification of operations, as well as monitoring and measuring of activities that are associated with significant environmental impacts, leads to the creation of procedures for minimizing risks from environmental influences. A systematic approach encourages the organization to find new solutions, to identify different opportunities, and to surpass the traditional way of considering energy issues, according to which only people in charge of health and safety of employees deal with environmental problems.

Typically, organizations separate their own environmental impacts to the impacts on soil, air, and water, and compare them with current regulations. This approach leads to individual problems and application of the so-called *end-of-pipe* strategies that consider only the consequences and essentially lack efficiency and are linked with higher costs. Corrections should increase efficiency and reduce costs.

The organization can have benefits if it includes its suppliers and partners into procedures of eco-management when it comes to important environmental impacts they can make on the environment. It can be necessary and useful to organize training for external colleagues if they directly influence environmental impacts. In this way, a forum of the two organizations is created to reconsider goods and services to improve them.

The organization may choose to modify the existing assessment tools or develop its own procedures to identify all organizational environmental impacts and the level of their significance. Using teamwork during the planning process, as well as sticking to the list of activities and time frame, is of particular importance. Possible improvements should be made first for the problems of the utmost importance. Finally, the importance of considering these issues beyond the organization and the existing regulations should be emphasized.

Identification of objectives and targets. The eco-management goals represent the most important part of the process of achieving the planned efforts to implement sustainable energy management (objectives and goals). *Objectives* are general environmental efforts that the organization wants to accomplish, while *goals* are detailed performances that are recommended in order to achieve the previously defined environmental efforts. Regardless of the specific nature of some goals, they have to be consistent with the existing organizational environmental policy.

Program techniques are most successful if achievable and measurable goals are set. In the stage of determining the goals, the organization should consider the significant environmental impacts, such as energy consumption, legal and other regulations, public attitudes, technical abilities, and flexibility of equipment. In this stage, the possibility of return on investment and business opportunities to correct its position in the market and gain profit should be considered. The eco-management approach encourages innovative solutions to the problems of using energy and materials at all levels of the organization. The objectives must be documented for all levels and business functions of the company that may affect the environment. The goals may be different at different levels such as the level of management, engineers, supervisors, and operatives. In short, the eco-management standard recommends that organizations set their own goals to reduce harmful environmental impact, choose the activities necessary to achieve their goals, and then carry out continuous monitoring in order to correct and improve the system.

Training and accountability. After defining the goals it is necessary to conduct appropriate training. Employees must be familiar with their own environmental responsibility and therefore be trained to behave in an environmentally responsible manner. Depending on the situation it is necessary to develop appropriate training that is required in order to achieve the desired objectives and integrate them into the existing environmental, health, and safety programs. Training should be provided to all employees in the implementation of sustainable energy management. Experience has shown that quality training of the participants is one of the most important factors for successful implementation.

Employees should be trained on the relevant procedures, their individual tasks and responsibilities, and be familiar with the expected results. After conducting training for all employees, the company will soon receive numerous proposals from employees about how to improve certain activities. Depending on the nature of the program, each company will predict the level of knowledge and training required for each individual to be able to achieve the environmental and energy requirements of his or her job.

Developing an energy program. The last element of planning within the eco-management system is the formation of an eco-management program. The program should contain an action plan, specified responsibilities at all organizational levels, the execution plan, timeline, and the resources required to meet objectives. Upon formation of the program, each individual who has been assigned certain responsibilities develops his or her own plan. Although the stage of achieving goals is separated from the stage of planning, they are closely related. It is necessary to have an idea in advance as to how the objectives can be achieved before the program is integrated into the company eco-management system.

An organization may have a number of eco-management programs. Sometimes a company develops eco-management programs for all determined objectives. In other cases, a company establishes specific eco-management programs for regulatory aspects, solid waste, energy use, etc. Finally, each company should follow the implementation of eco-management programs and make adjustments as needed.

Implementation

There is a clear need to conduct a careful evaluation of the ways in which the implementation of an eco-management system will affect organizational energy performance. At this point it is necessary to decide how to incorporate the eco-management approach into the existing environmental regulations. The stage of implementation of eco-management system can be represented as shown in Figure 5.12.

The first substage of implementation involves the development and control of documents. Today, most organizations have already developed procedures for monitoring energy processes, which include work instructions, training plans, the results of testing and monitoring, and calibration instructions. It is important that organizations create their own documentation whenever appropriate. The documentation provides an answer to the question of how to perform certain actions and provides step-by-step instructions for task execution. System requirements necessary for the operation have to be documented. The documentation should be as clear and as simple as possible.

If the instructions and the documentation do not add some value to operational control, are they necessary at all? Detailed documentation is not required for all departments in a company. The main factor affecting the need for documentation and its scope is the risk and complexity of activities as well as the frequency of control necessary to perform a particular activity. The organization team needs to identify differences in relation to existing documentation and, if necessary, initiate the creation of new procedures.

Operational control develops procedures for performing certain activities and defines the operational criteria as well as preventive and corrective measures. Implementation of operational control is in the true sense the working part of the eco-management system. Procedures are instructions that organizations use to implement environmental activities. Procedures determine who, what, when, where, and why for some of the planned activities. At this point, most organizations face a great number of problems. Since the developed procedures are extremely important, the organization will benefit if it properly determines what procedures will be documented and, if it documents them, the form of guidelines for training and implementation. Written procedures are an essential element of operational control and if they are lacking, this often leads to large deviations and discrepancies.

An important step in the phase of operational control is the identification of individual business activities and functions that may have active or potential impact on the environment. Operational control that is specified for the most important environmental impacts helps the company define the roles,

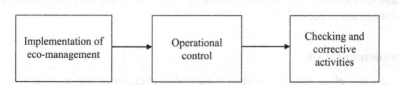

FIGURE 5.12 Implementation of eco-management system

responsibilities, and authority necessary in order to achieve the desired performance. Quality operational control can be briefly defined as a procedure or a process within an organization that reduces the generation of waste and preserves natural resources.

If, for example, there is a change in the working process in which the waste is generated – for example, air pollution due to using fossil fuels – it is necessary to adequately modify operational control. For example, by appropriately marking materials and waste it is possible to reduce the risk of incorrect disposal of substances. If it properly separates and disposes the particular elements of waste, the organization can access the recycling or reuse of materials. Operational control should also ensure the proper use of equipment. In order to avoid possible failures and inefficiencies, preventive measures should be developed and implemented.

After the substage of operational control, there is the substage of checking and implementation of corrective actions. It is necessary to control each active system in order to see whether the system is working well, to identify current or potential problems, and to determine ways in which identified problems can be solved. Certain measures of environmental performance have been specified and, according to the results, corrective actions are taken if needed.

Evaluation and Observation

The eco-management system requires certain procedures in terms of monitoring and measuring energy performance, collecting information needed to monitor and assess the potential for achieving the goals, and the assessment of compliance with environmental regulations. A particularly critical moment is the determination of what should be monitored and measured and which information needs to be collected. As previously set goals are measurable, the whole process leading to their implementation must be monitored. The evaluation phase and observation of the functioning of the eco-management system is shown in Figure 5.13.

Eco-management measurements are used as indicators of energy performance and must be in accordance with legal and other regulations. The number

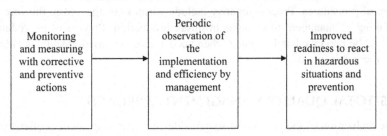

FIGURE 5.13 Evaluation and observation of the eco-management system of sustainable management of energy

of indicators that will be established and monitored must be carefully determined. A large number of indicators may produce a huge amount of information that surpasses the abilities of the information processing system, while an insufficient number of indicators does not provide enough information needed for effective decision making. In order to conduct high-quality measuring, it is necessary to determine who is responsible for monitoring, analyzing, collecting, and processing data, the frequency of measurement required to obtain quality information, the way the data will be analyzed, and the way in which the data will be presented.

A company may do monitoring every day or every hour, depending on needs. Monitoring will be used to determine the existence and nature of the trend so that deviation can be quickly and easily identified. Monitoring of eco-management performance is specific in that it requires monitoring at different levels of work and business functions, i.e., it is focused on employees and their effects.

The review and improvement by management is conducted in order to provide continuous improvement, which is one of the basic characteristics of this approach. Top management should periodically examine the implementation of the eco-management system and its efficiency. Experience has shown that the efficiency of management directly affects the chances of successful implementation of the eco-management system. The eco-management system is a business system that directs the organization to manage its environmental issues in a way that is based on continuous improvement. The eco-management system is focused primarily on the support of top management and involvement of all employees.

If discrepancies are identified during observation, the person responsible should determine the way in which adjustments will be implemented and how the system of prevention will be developed so as not to allow similar situations to repeat in the future. Readiness to respond in emergency situations is a necessary step because these situations can happen even in the best-prepared organizations. In order to reduce the probability of such events to a minimum, the eco-management system develops a system of procedures for the implementation of preventive measures.

Studies on the application of preventive and corrective measures and the ways of reacting in emergency situations must be integrated into operational control procedures. Procedures may include various changes, from the way of thinking of managers to changes in direct production. Preventive procedures should be perceived and updated when needed after an environmental accident or emergency situation.

5.5 TOTAL QUALITY MANAGEMENT APPROACH

The realization of the process of sustainable energy management is often considered a job to be dealt with by the staff responsible for ecology. It is considered much more effective if the program is integrated into overall organizational business practices. This approach allows the engagement of both top management and employees at all levels. In this section the most commonly used model of quality management will be described for the purpose of implementation of the sustainable management of energy, which is based on the *Baldrige Quality Program*.

The *Malcolm Baldrige National Award* was established by the U.S. President and Congress in 1987 in an effort to improve the quality of business in American companies. However, the criteria for the award soon surpassed American borders and became an example for quality improvement in the global market. Currently, the *Baldrige Award* is given in almost 50 countries around the world.

The first well-documented example of use of quality management was created by the integration of the quality model in order to monitor the environmental performance dates back in 1998, and was applied as the *Green Zia* program in New Mexico. It was then that the criterion of environmental excellence was established for the first time (with the analogy to the existing criterion of performance excellence, defined by the quality model). Environmental excellence can be defined as the ultimate goal that can be achieved by applying the quality model in the management of EHS (Environment, Health, and Safety) management.

An environmental excellence program establishes an appropriate range of objectives that can be achieved by reaching the best status in those areas that best support prevention-oriented approaches to EHS management. The results that can be achieved by implementation of goal-oriented activities are the output of EHS performance but not the measuring of the performance itself, and thus organizations cannot indicate whether their own results are better than the results achieved by their competition. By focusing on performance, an organization can contribute to the development of the program and provide a means of measuring their efficiency in terms of the functioning of the process of sustainable energy management.

Using the quality model, the range of goals can be defined as *zero – zero defects* (zero rejects), *zero equipment breakdowns* (zero equipment failures), and *zero waste* (zero waste). Many organizations today have expanded the zero concept of their EHS program to zero incidents, zero accidents, zero emissions, and zero drain – extinction of the world's resources which is the main precognition of sustainability).

The *Green Zia* model has all the elements necessary to realize the vision of zero waste. First, the criteria were given to define what is best in the class, allowing each organization to be able to measure its own progress towards achieving excellence. Second, the model offers a guide that can be used for integration of the criteria into the program. Third, it includes a rigorous summative system that measures the results of environmental excellence on the scale of 1000 units. Fourth, it provides a feedback report on the strengths and weaknesses when applying the program for the achievement of environmental excellence in comparison to the criteria given in the basic principles (guidelines). The summary report and the feedback report represent important tools for the organization trying to move towards achieving environmental excellence. Organizations achieve excellence in their own way by applying this model. They choose the performance they want to improve and determine how to incorporate the achieved results in the total efforts of the company.

In practice, the two models are commonly used – one contains seven criteria and the other contains eleven criteria.

The Seven Criteria of the Quality Model

Since the implementation of the program of sustainable energy management is a complex process, the most important factors must never be ignored. Based on many years of experience in applying quality management, the seven basic criteria have been defined, as follows:

Leadership;
Strategic planning;
Involvement of stakeholders;
Involvement of employees;
Process management;
Information analysis; and
Results.

These criteria are incorporated into the basic principles of the *Green Zia* program and are used for the purpose of application of the quality model in the implementation of the energy management program. Based on these criteria, the organization will be able to determine how to integrate the energy management program into its own business. In the past, organizations mainly received information about why something needed to be done, but not how to achieve it.

Leadership. Intensive use of *top-down* direction in the application of the quality model increases the chances of success of the program and helps to integrate it into the organization as a whole. A sustainable energy management program will be considered to the extent that the leadership of the organization considers it necessary and important. At this point it is necessary to raise two basic questions:

How can management of the company transfer its commitment to its employees in terms of implementation of sustainable energy management?
How can management demonstrate its determination and commitment?

The answers to these questions can only be obtained through discussions with managers, which can often take too much time.

Strategic planning. Management of a company usually uses a specific form of strategic planning in order to lead the organization. Sometimes it uses a certain strategic plan as well. In other cases, strategic planning can have a less formal form. The basic questions that arise during strategic planning include:

How do we identify long-term and short-term aims and intentions?
How do we develop these aims and intentions?
How do we apply achieved aims and intentions?
How do certain aims and intentions influence general business goals and intentions of the organization as a whole?

Ideally, the sustainable energy management program will be considered important in the minds of managers and an important part of strategic planning. There is a strong link between strategic planning and leadership.

Involvement of stakeholders. Organizations do not operate alone. There are many other organizations that have an impact on the implementation of sustainable energy management. Interested parties include a wide range of stakeholders such as consumers, suppliers, contractors, regulatory bodies, NGOs, environmental associations, social associations, and the general public. This involvement can be understood by raising the following questions:

How does the organization involve the stakeholders in the development and implementation of the program of sustainable energy management?
Is the organization involved in the program of sustainable management of energy in other organizations, and if so, how?

The greatest attention should be paid to the analysis of the influence of the employees as stakeholders.

Involvement of employees. This criterion analyzes the bottom-up orientation of the program of sustainable energy management, which is as important as the top-down orientation, included in the criterion that analyzes the attitudes of management. Employees are a very important element of the program because they know best the details of the business operations of the organization. The following questions should be answered:

How can the organization prepare and involve employees in the development and implementation of the program of sustainable management of energy?
In what ways are the values and well-being of employees involved in the program of sustainable management of energy?

Process management. This is the criterion by which the ISO 14001 standard or other eco-management standards help the organization implement sustainable energy management. Process management considers how to manage work processes in a manner established by the program. Issues to be considered include:

How does the organization identify the basic and auxiliary work processes that influence the program of sustainable energy management?

How does the organization analyze these work processes in order to understand their causes and effects?

How does the organization manage work processes to achieve program excellence?

The criteria stipulated above are closely linked with the criteria of information analysis.

Information analysis. The *Green Zia* program is, overall, based on the fact that all criteria are mutually linked. Each time one criterion is taken into account in order to make a change it is necessary to consider how this will affect the other criteria. It is important to emphasize that it is not necessary to be perfect in all seven criteria of the Quality Model, but to understand that some criteria will perform better or worse than others.

Results. The above seven criteria of Quality Model will enable the creation of images of the situation in the organization and, grounded on that, certain results of the analysis will be obtained. They will reveal the weaknesses, strengths, and improvement opportunities.

The Eleven Criteria of the Quality Model

Principles that are often referred to as core values are used in the context of all activities of an organization. They are designed to provide a guide for decision making at all levels in an organization. Each organization should choose the way in which each of these principles is going to be integrated into relevant criteria in the quality model to implement the program of sustainable energy management in its business. If the organization already has the basic principles in place, it is necessary to consider the application of sustainable energy management program in all these areas.

If the organization does not have the previously defined basic principles, it should first become familiar with the way in which they can first become a part of the organizational culture. In this case, the initial activities may take longer because it is necessary to implement the changes that have to become an integral part of the organization's culture. Experience has shown that it takes at least two years to engage in certain activities, including incorporating environmentally responsible activities in the core business values. When this is achieved, the integration of sustainable energy management can begin. Within the quality model it is necessary to consider the following 11 criteria:

Stakeholders-oriented program;
Leadership;
Continuous learning and improvement;
Evaluation of employees;
Fast response;
Effective design of products, services, and processes;
Long-term planning;
Facts-based management;
Partnership development;

Public responsibility; and

Focus on results.

Stakeholders-oriented program. A sustainable energy management program evaluates all interested parties, so it must be expressed through all products or services whose characteristics meet the values of stakeholders.

A sustainable energy management program oriented toward the stakeholders is thus a strategic concept. It focuses on customer retention, conquering a new part of the market, growth, and ensuring long-term communication with the environment. It requires a constant willingness to change the requirements of stakeholders and the market as a whole in terms of facts affecting their satisfaction.

A program oriented toward all interested parties is much more than a simple reduction in the amount of waste and reduction of emissions into the environment, meeting the statutory provisions, or reducing the number of complaints. On the other hand, minimization of waste or elimination of other causes of dissatisfaction of interested parties is a particularly important part of the program. The success of an organization in terms of recovery from possible EHS issues is key in establishing better and more stable position of organization in its environment.

Leadership. Top management of an organization is a team of people who create and direct the focus of interested parties and create clear and visible values and high expectations. The above directions, values, and expectations should be directed to all stakeholders. Leaders can ensure the creation of strategies, systems, and methods for achieving environmental and energy excellence. They should stimulate innovation and promote knowledge and skills. The strategy and values will help manage the activities and decision making in the organization. Managers who are focused on the development will encourage the participation of all staff by teaching all participants to be innovative and creative.

Based on the behavior and the role in the process of planning and communication and review of performance, leadership serves as a kind of model of behavior and actions, which gives a stimulus to activities in the whole organization.

Continuous learning and improvement. Achieving the highest levels of energy performance requires high quality design of the executive approach to continuous improvement and learning. The term continuous improvement refers to the improvement of existing activities and the adoption of completely new solutions and major changes. The term implies willingness to change, moving towards new goals and approaches. Improvement and learning need to be a regular part of daily activities. They should be practiced at individual and all other organizational levels and should be oriented towards creating opportunities for innovation and better functioning of sustainable energy management. Improvement and learning include increasing the value expected by stakeholders through new and improved products and services; developing new business opportunities; reducing waste, emissions, and costs; improving the level of responsibility as an activity without creating a direct revenue; increasing efficiency in productivity by using all resources (energy, water, and raw materials); and increasing the value of organizational performance in order to comply with the full social and civilizational responsibility. Therefore, improvement and learning are directed not only to better products and services, but also to increase the level of environmental responsibility.

Evaluation of employees. Success in the implementation of sustainable energy management in an organization depends largely on the knowledge, skills, innovation, creativity, and motivation of employees. The success in the work of employees largely depends on the ability to get a chance to learn and apply new knowledge and skills. An organization can gain considerable advantage by using proper work potential through investments in learning, training, and creating opportunities for continuous improvement. Adequate training provides a very effective way to improve work processes and training on new technologies. Education and training should be adapted to the requirements of each job. The main challenges of evaluating the performance of employees are integration of human resource management activities (selection, performance, recognition, climbing the career ladder); development, education, and transfer of knowledge of employees; and harmonization of the human resources management with strategic processes. The answer to these challenges requires monitoring of the knowledge, skills, satisfaction, and motivation of employees in terms of sustainable energy management. Data on the above can serve as a useful indicator of the conditions in the organization or in some of its parts.

Fast response. Obtaining permissions and compliance with the regulations may require additional time in the decision-making process. Success in the global market requires shorter cycles for introduction of any improvements and innovations. Also, faster and more flexible response to requests from stakeholders becomes increasingly important. Great progress is made in terms of time needed for responding to the requirements towards simplification of work units and processes together with the incorporation of the program in the design stage (the so-called *design for environment*). Other time improvements are related to the fact that the time required for improvement leads to simultaneous changes in organizational behavior, quality, productivity, and costs.

Effective design of products, services, and processes. Organizations need to integrate the program of sustainable energy management into the design stage itself. Experience has proven that the program that is integrated at this stage is much more effective than if integrated into the processes that occur in the stage of production or even that stage. The design includes the creation of such products, services, and processes that will be environmental and energy-friendly by their nature. Costs arising from preventive problem solving are much lower than the costs that arise after the problem has been caused. The design stage is critical and demanding, especially in terms of social responsibility.

Long-term planning for the future. Survival in the market implies a strong orientation toward the future and the willingness to take on long-term obligations toward all stakeholders. Organizations should assume and predict a number of factors in their own strategic plans, such as expectations of stakeholders, new business opportunities, changes in the global marketplace, technological developments, new market segments and new consumers, changes in legal regulations, social expectations, and strategic efforts of competition. Short-term and long-term plans of sustainable energy management, strategic objectives, and allocation of resources should reflect these influences. The most important factor that takes into account the long-term nature of obligation is the development of the awareness of employees and suppliers that they are the key participants in the process.

Facts-based management. Organization is largely dependent on the measurement and analysis of energy performance. Each measurement is determined by an organizational strategy and provides important information about key processes, outputs, and outcomes. To measure energy performance different types of data and information are needed. The areas of observation include satisfaction of all stakeholders, operations, market position, and financial indicators. Only on the basis of the collected facts should management make its decisions.

The analysis involves extraction of the basic meaning of the energy data and information that can help evaluation, decision making, and improvements in an organization on the basis of established trends and projections, which are impossible to reach without adequate analysis. Identified trends and projections are of particular importance for understanding the progress of the process and can serve as a starting point for benchmarking the results of the competitors.

Partnership development. Experience has shown that organizations can quickly and efficiently achieve their desired goals by building appropriate internal and external networks and partnerships. Partnerships are more desirable in the modern world, where energy responsibility is one of the pillars of corporate social responsibility.

Internal partnerships might include cooperation of employees in the same area, including mutual training, sharing experience, and creating teams. In addition, internal partnerships encourage the creation of internal links between business units, improving flexibility and acquiring new knowledge.

External partnerships can be built with customers, suppliers, NGOs, environmental associations, and educational institutions for various reasons, among which the main one is training. A particularly important form of external partnership is a strategic partnership (strategic alliances). A high-quality external partnership can help with entering new markets. Strategic partnerships allow combining of organizational core competencies and leadership skills with complementary strengths and capabilities of partners, thereby resulting in the increase in total capacity of all partners. Internal and external partners can develop their own long-term energy-related goals, and based on them plan and implement joint investments.

Public responsibility. Today, organizations gain a special advantage if they prove their environmental responsibility according to the demands of society and behave as civilized citizens. Responsibility is primarily related to an organization's core values, which should include health care, safety, and the environment. In this way organizations create a positive image and gain the loyalty of consumers and society as a whole. Particularly important aspects that should be considered at this point are the promotion of education, promotion of health care in the community, improvement of environmental quality in the region, and participation in local activities that improve quality of life.

Focus on results. By measuring energy performance it is possible to get the proper results and meet the requirements of stakeholders. Results that do not meet the stipulated requirements suggest the need to make corrections without delay. With this approach, the organization is constantly in need of balancing between its own abilities and requirements of interested parties, without jeopardizing the requirements of any interested party. This is possible only if the organization focuses on the energy results, which will be the only measure of success of implementation of sustainable energy management.

Application of the Quality Model in Realization of the Pollution Prevention Program

For the purpose of the integration of sustainable energy management into organizational practices, the application of the quality model, which contains the five basic steps, is given in Figure 5.14.

Planning and Development of the Strategy of Sustainable Management of Energy

This step can be summarized as shown in Figure 5.15.

Planning the strategy of sustainable energy management. The first step in application of the quality model involves determination of the elements of the quality model, which refer to planning of the strategy of sustainable energy management. This primarily refers to strategic planning, focusing on the demands of stakeholders, leadership, and involvement of employees. Furthermore, it is necessary to identify inconsistencies and deviations (gap analysis) that may arise between the detailed issues that are anticipated by *Green Zia* program for each of these criteria and what is currently happening in the organization. Gap analysis

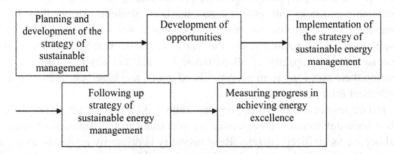

FIGURE 5.14 **Basic steps in application of the quality model in the implementation of the strategy of sustainable management**

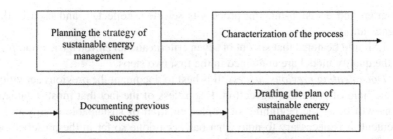

FIGURE 5.15 Planning and development of the strategy of sustainable energy management

should also consider how the strategy of sustainable energy management can be integrated into the business practice by using certain combinations of the presented eleven principles of the quality model.

Furthermore, it is necessary to consider the element of leadership, which helps in preparation of the policy of sustainable energy management. Examination of leadership will point out how top managers demonstrate their own commitment to the need for continuous environmental improvement to employees and other stakeholders.

Including employees helps in preparation of the training program for employees because its consideration brings insight into the condition of interest and training of employees for the implementation of the program of environmental improvement. In addition, an insight into their expectations will be made in terms of improving health and safety.

Strategic planning is a criterion that indicates how the organization identifies, develops, and implements long- and short-term goals and efforts in the field of continuous improvement of the environment and how the goals are related to general business aims, mission, and vision of the organization.

Finally, the criterion of stakeholders determines whether and how the organization involves stakeholders in the development and implementation of the program of environmental improvement.

Characterization of the process. At this point it is necessary to characterize the energy processes that take place in the organization and for this purpose the most useful is the application of the process mapping that is envisioned in the system approach to implementation of pollution prevention. The determined process maps will be used as a template for information in the process, which will consequently identify production units that require further analysis.

Besides the mapping process, it is necessary to review the criteria of the process management and information analysis. The analysis of the process management criteria will provide information on how the organization identifies, analyzes, and manages processes that potentially affect the environment or may cause injury and illness of employees. Analysis of information provides insight into how information that affects the efficiency of the realization of the program and decision-making process is selected, collected, and used in the organization.

It should be noted that six out of seven criteria and all eleven basic principles of the quality model are contained in the first two steps.

Documenting previous success. It is best to document the previous activities in the field of pollution prevention. Regardless of the fact that most organizations will be at the beginning of implementation of sustainable energy management, it is necessary to note what has been done so far in the preservation and improvement of the quality of the living environment. The time frame that should be examined entails the previous period of 2-5 years. Employees and management should be familiar with the current activities and try to build in all positive experiences and steps into their future program of prevention of pollution.

Drafting the plan of the sustainable energy management. Having implemented activities, it is necessary to prepare a draft plan in terms of pollution prevention, which will later be discussed by the internal and external stakeholders.

Development of Organizational Possibilities

The planning stage of sustainable energy management is followed by the determination of the opportunities for improvement within the program, which can be summarized as shown in Figure 5.16.

Collection and analysis of information. A hierarchical process map, prepared in the previous stage, is used to collect information in defined production units. After that, the team will be able to identify and develop opportunities. If the analysis, for example, identifies the use of toxic materials, it is an opportunity for improvement to avoid the use of toxic materials. The analysis can provide a huge number of potential opportunities for improvement. It is necessary to rank them by application of so-called Pareto analysis. It is optimal to get eight out of eleven opportunities for improvement at this stage. These options should be selected in accordance with the objectives to be achieved in one year. The program team should develop their own criteria for a fast analysis. If necessary, the

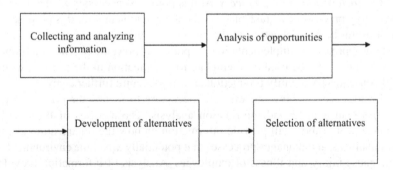

FIGURE 5.16 Development of opportunities of sustainable management of energy

team can undertake further analysis of one or two opportunities, which may take more time but should not last longer than two years.

Analysis of opportunities. This step begins with an analysis of the production units. For any pre-selected opportunity for improvement the team discusses possibilities of implementation, which is most often done through the resource analysis, methods, equipment, and technology, as well as people who will be involved in the implementation of the proposed options. After the analysis, the team can provide a final evaluation in terms of resolution of the problem and suggest a certain number of alternatives.

Development of alternatives. Consideration of alternatives is the process of selection of the proposed solutions in order to choose the solution with the best possible outcomes.

Selection of alternatives. After consideration of the alternatives, it is necessary to select them based on the analysis defined in a System approach to the implementation of sustainable energy management.

Implementation of the Pollution Prevention Program

The first step in the implementation of sustainable energy management is the development of an action plan for all the alternatives recommended in the previous stage. The action plan is an essential part of a sustainable energy management strategy for each year and must be carefully considered prior to application. It is best to reconsider the relevant steps of quality models that are evaluated during the first phase of the program, which is crucial to the success of pollution prevention.

The action plan should be revised at least quarterly. After one year, a complete internal and external evaluation should be done. Only after a complete evaluation is the pollution prevention program ready for implementation.

Following Up on the Strategy of Sustainable Energy Management

Throughout the implementation phase of sustainable energy management it is necessary to understand it thoroughly in order to obtain data needed for the analysis and summary of results. The criterion result (the seventh criterion of the quality model) examines the actual situation in the organization and provides the results of the strategy for sustainable energy management. In addition, it is recommended to consider the levels and trends and the ways in which they refer to the impact on the environment, health, and security of employees, the requirements of other interested parties, and the main financial indicators.

It is important to stick to the rule that only what has been measured has been achieved. A pollution prevention program cannot be based only on the expression of wishes. The achieved results should serve as progress indicators, and not only as the data that will be presented to the stakeholders. After reporting on the results, the action plan can be developed for the next year.

Measuring Progress in Achieving Environmental Energy Excellence

After several years, the strategy of sustainable energy management can be considered in terms of the contribution provided, i.e., the achievement of organizational goals. The results can be summarized by the proposal envisioned by the *Green Zia* program. In this way, the insight will be included in the pollution prevention program for the entire organization. The organization will also be able to compare the achievements of its competitors and corrections can be made as needed.

5.6 LIFE CYCLE ANALYSIS

Life cycle analysis (LCA) is currently the most common method of designing environmentally friendly products and technologies and evaluating their impact on the environment. This method can also be successfully used for the implementation of certain aspects of the strategy of sustainable energy management.

This method is governed by the international series of ISO 14000 standards, which is also the reason why it is widely used in relation to other methods of designing for the environment. LCA helps companies realize the extent of the impact of their products, processes, and other activities on the environment. The goal of LCA is not only to provide an answer to the question of how serious the harmful effect is, but also to enable strategic planning of future activities.

LCA history dates back to the early 1960s, when scientists became concerned with the exhaustion of fossil fuels. LCA methodology was initially developed in order to get insight into the effect of production, distribution, and consumption of the fuel on the environment. From 1970–1975, due to the oil crisis, several analyses of profile resources and the impact of their exploitation on the environment (called Resource and Environmental Profile Analysis) were conducted.

During this study the protocol was developed as well as the standard methodology for conducting such studies. After considering the impact of energy production and consumption in the late 1970s, LCA focused on the analysis of the life cycle of waste.

Basic Steps of Life Cycle Analysis

LCA is by definition a systematic approach to evaluating the environmental consequences of a particular product, process, or activity from the acquisition of raw materials to final disposal. The basic steps of LCA are given in Figure 5.17.

Raw materials acquisition. Each product or service is made by using certain types and amounts of energy. Therefore, each production process takes from

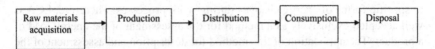

FIGURE 5.17   Basic steps of the life cycle analysis method

nature certain materials, and the exploitation of energy is one of the most intensive forms of exploitation of raw materials in general. Regardless of the type of activity that a company deals with, it uses certain amounts of energy. A great deal of energy needs are met with conventional energy resources, whose exploitation and consumption are usually connected to a lot of pressure on environmental quality, destruction of huge areas of land, and other forms of pollution. Every company has to take into account the damage that is caused to the natural environment because of its energy needs.

Production. In the broadest sense, every production process involves a certain amount of energy consumption, and the amount of energy used greatly depends on the type of activity, as well as the quality of technology that is used, the intensity of its use, and the skills of the producers. Estimation of energy consumption in a production system gives managers insight into the (non-) compliance of sustainable development and is one of the key areas for the implementation of improvements.

Distribution. Most goods in contemporary business are produced at some distance from the place of consumption and use. Therefore, the distribution is associated with huge amounts of energy for transport. Speed and distance from the destination, the way of transport, and the use of alternative energy sources for these purposes are not taken into account. Therefore, the estimation of the specific energy consumption for the needs of distribution is of great importance, because opportunities for saving energy are limited.

Consumption. Each product is spent for a particular purpose in a certain way and thereby requires more or less energy consumption. The estimation of the energy efficiency of certain products is particularly important in the category of products, one the larger consumers of energy. The intention to produce goods that use less energy is the imperative of contemporary business and market requirements.

Disposal. After being used, the product that is not consumed as a whole, or after the consumer decides to stop using it, needs to be disposed, having lost the purpose for which it was obtained. The possibility of safe disposal is certainly of special interest and is subject to numerous efforts to create products that can be disposed of in a manner that does not harm the environment and does not require intensive consumption of energy in that process.

LCA is a broad term that can be applied to a number of activities, but on the other hand, its comprehensiveness is limited. Namely, it usually does not provide data of satisfactory quality and the quantity of data is often questionable. Additionally, accurate measurements are not provided, and attention is not paid to specific forms of energy changes during work processes. Therefore, LCA is well accepted exclusively as a method for the assessment of the energy load of a product. The application of this method for description and assessment of the production cycle is extremely impractical and inefficient.

The Main Components of Life Cycle Analysis

LCA is used to assess the impact of each part of the process of obtaining a certain product. The basic components of LCA are:

- Inventory analysis;
- Estimation of impact; and
- Improvement possibility analysis.

Inventory Analysis

Analysis of inventory is considered the most objective component of LCA, because it is based on data obtained by measuring the amount of energy and raw materials, air emissions, liquid effluents, solid waste, and all other processes in the life cycle of products, technology, or activities. By doing inventory analysis it is possible to:

Establish a set of all system requirements for energy resources;
Identify the components of the process on which reduction of energy resources consumption could be reduced;
Assist in developing new products or processes in which needs have been reduced in terms of resources and the amount of emission into the environment;
Compare alternative materials, products, or activities; and
Compare internal inventory information at other producers or products in the same or other company.

Inventory analysis within LCA can be schematically presented as shown in Figure 5.18.

Given the above scheme, which is very simplified, it is necessary to know that the system of inventory analysis includes the collection of a large number of diverse data, where one should always keep in mind that the data will not be fully and accurately collected because it is influenced by a very large number of factors.

Acquisition (extraction) of raw materials is the first step in any production process, and therefore it is the first to be analyzed. Input in each production system represents a large number of components, so each input component that accounts for less than 5% can be neglected in the analysis. Otherwise, the analysis of inventory will be even more complex, more expensive, and harder to use. Properly conducted analysis of inventory will provide the ratio of the environmental impact of raw material acquisition. When analyzing the acquisition of

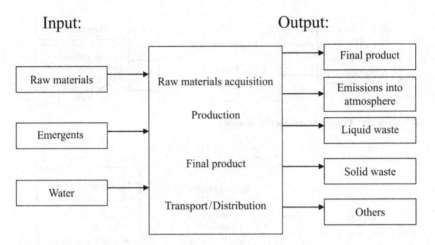

FIGURE 5.18   Inventory analysis

raw materials inventory it is necessary to specifically monitor the input materials and energy, as shown in Figure 5.19.

Production, as the next stage, consists of a series of special processes (subsystems) so that each process can be separately analyzed. Often in the production process there is one major and several minor systems. In the analysis of inventory at this stage the data is obtained on the movement of the substance and energy through the production process, as well as data on emissions into the environment. In addition to the analysis of the manufacturing process, this part of LCA involves analyzing the environmental impact of packaging as well as storage and distribution (Figure 5.20).

Transport and distribution are analyzed separately for a number of environmental impacts. Environmental impacts are caused by fuel consumption during loading, repacking, transport, and distribution to the point of delivery. In this analysis, the transport distance is measured as well as the contents of cargo and the protection of the environment during transport. The transport and distribution system to be analyzed can be represented as shown in Figure 5.21.

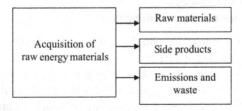

FIGURE 5.19   Acquisition of raw materials

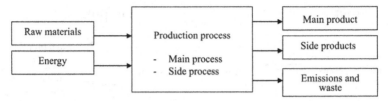

FIGURE 5.20   Process of production analysis

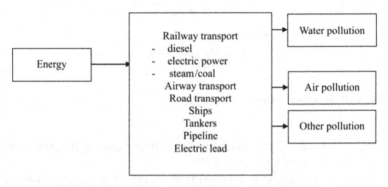

FIGURE 5.21   Analysis of transport and distribution

Consumption and disposal of products are observed and analyzed together and refer to a part of the system that arises from the moment when the product reaches the stage of consumption until it has been disposed after use. The inventory analysis phase aims to review all energy and other aspects of this stage of LCA, as the products are consumed in different ways and thus behave differently. In addition, there are different reactions during the proper or improper use of the product. The main activities of consumers covered by this analysis are the use of the product and its storage, preparation, handling, maintenance (repair), and reuse. The main issues discussed at this stage include:

The time the product was used prior to its rejection;
The effects of the use of the product;
The input parameters necessary in the maintenance process;
The frequency of repairs; and
The potential possibilities of reusing the product.

The system of consumption and disposal of products after use can be summarized as shown in Figure 5.22.

During the implementation of these activities there is energy consumption and therefore an impact on the environment. Only products that can be completely recycled in a closed system of recycling (in which no amount of waste is created) can be said to have zero effect on the environment.

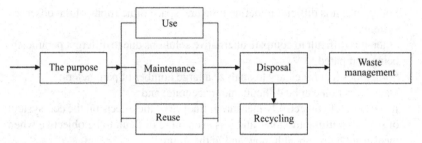

FIGURE 5.22   Analysis of consumption and after use disposal

Estimation of Impact

After collecting data during analysis of the inventory, the estimation of its impact on the environment is conducted. The environment is complex, and the impacts may be numerous and varied, so this stage of LCA is particularly complex. However, this stage of analysis tells how certain stages in the product life cycle have harmful effects on the environment and gives concrete suggestions as to what can be done to improve the situation.

It should be noted that this phase of LCA is associated with a number of objective difficulties in implementation, because the estimation of the environmental impact itself is very complex and often there are no valid and reliable data on specific impacts. Nevertheless, the evaluation of the impact of the effects on the environment should be carried out at least for those parameters whose effect has already been studied and well known.

Analysis of Improvement Possibilities

Based on data collected in the inventory analysis and evaluation of their impact on the environment, at this stage suggestions are given to improve the life cycle of a possible product or process, or to improve the environmental identity of companies. There are many different ways to improve the energy image of a company, which later results in a number of positive financial and non-financial effects. Suggestions for improvements are determined depending on priorities, possibilities of implementation, and environmental and economic cost benefits.

Main Problems in Implementation

As discussed, LCA is the oldest and most accepted method of designing for the environment, and is regulated by the international system of standards of environmental protection. Nevertheless, the application of this method has certain limitations, and there are many difficulties in its implementation. The most common problems in the application of LCA include:

Increased costs of application in small enterprises;
Time needed to master this method can surpass time of the life cycle of the product;

Sometimes it is difficult to define time and space dimensions of the observed system;
Often it is difficult to compare alternative solutions due to different parameters being compared;
Sometimes it is necessary to analyze a large number of parameters;
Data collection can be difficult and inaccurate; and
It is difficult to objectively measure influence of the effects on the eco-system of some harmful emissions, and it is even more difficult to be objective when measuring human health, now and in the future.

All these difficulties and limitations should be kept in mind when deciding on the implementation of LCA, but its advantages should never be ignored and it should be applied whenever it is possible and economically justified.

5.7 GAP ANALYSIS

For the purpose of sustainable energy management gap analysis is suggested as a very efficient technique. Gap analysis is one of the techniques of strategic management and forecasting used to determine the opportunities for growth and development in specific areas. This method is based on a number of estimates and assumptions, and its simplified schematic diagram is shown in Figure 5.23.

In traditional terms, gap analysis is used to project long-term goals and determine the gap between projected goals and objectives that the company will probably meet if no new management activities are undertaken in relation to the current ones. Successful implementation of gap analysis involves a complex and comprehensive process of analysis and assumptions. For the purpose of sustainable energy management these requirements are further defined in accordance with the specific issues being addressed. Gap analysis is particularly suitable for sustainable energy management in a particular company but is less suitable for energy management in a particular area. In general, gap analysis includes the following steps.

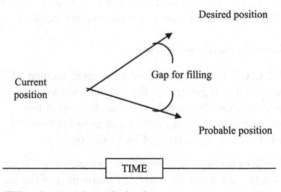

FIGURE 5.23   Filling the strategic gap in development

Determining the current position of the company, based on the estimates of internal indicators. The analysis is performed based on the perception of the current energy status of the company. This stage of analysis can be very comprehensive and complex, especially for companies that use large amounts of energy. It is necessary to determine the following:

Amount of the total energy consumed in the company for a certain period of time. Quantities should be monitored and expressed at least annually, because in this way certain seasonal fluctuations in consumption can be identified;

Amount of total energy consumed in certain periods during the year. Depending on the type of activities, significant fluctuations in the quantity of energy consumed are possible;

Amount of total energy consumed in certain sectors of the company, where it is necessary to separate the amount consumed exclusively for the production of energy from the energy used for administrative and side activities;

Types (origin) of energy used in the company, as well as the amount of energy from certain sources. In fact, in many cases companies use different types of energy (electricity, heat, coal energy, oil, gas), and it is necessary to accurately monitor the consumption according to certain types of energy resources;

Amount of energy consumed in the company that was obtained from the company's own resources or alternative sources. There are many technological solutions that enable companies to produce a certain amount of energy themselves, using some of the energy recycling technologies;

Costs that the company has according to the types of energy resources, i.e., savings achieved by exploiting their own sources of energy;

Energy costs included in the price of the final product or service;

Share of energy costs in total costs the company has recorded;

Energy efficiency of technology to be used, depending on the type of technology, training of employees, the proper management of the work process, control, and many other factors;

Current emissions in the environment that result from exploitation of energy sources, as well as the efficiency of measures taken in terms of environmental protection;

Company commitment to change in the field of energy management, which requires compliance with the concept of sustainable development;

Type, shape, and strength of environmental factors that affect energy management policy, reviewing existing regulations and treaties ratified by the country in which the company or its business unit operate;

Type and the strength of the impact of consumers who have specific requirements in terms of environmentally responsible and energy-saving operations; and

Existing requirements of banks, insurance companies, and all other financial institutions with which the company is in contact, and which to some extent evaluate the environmental and energy aspects of the business.

Based on all the above, it is possible to get insight into the three basic parameters that need to be defined in the process of sustainable energy management: energy resources consumption, energy consumption costs, and energy efficiency of business.

Determining where the company is likely to be at a specified time should the current direction and pace of growth and development continue is the next group of data and is based on the extrapolation of goals based on historical data and trends in the past. At this stage, it is necessary to:

Consider the overall quality of business and the position that the company has had in the past few years. The period for which the analysis is made is determined by the length of operations of the company itself as well as opportunities to get accurate data;

Consider the energy consumption in previous years, both in terms of type and quantity;

Determine energy costs that were recorded in the previous period;

Define the production capacity that has been made on the basis of specific energy consumption;

Assess the pace of growth expected to continue in the future, which will be based solely on the characteristics of the business. Namely, it is assumed that the company will developed at the same pace, which means it will use the same amount and type of energy resources, or, if the trend is such, will increase or decrease the amount of energy consumed depending solely on the capacity of production to be achieved. These estimates are based only on the business policy of the company, which is usually market conditioned;

Estimate the amount and type of energy the company will spend in the next period, without any special changes or adjustments, neglecting the needs for sustainable energy management;

Evaluate the energy costs that will be recorded if the company continues developing at the desired pace;

Estimate the future share of the cost of energy resources used in the unit price of the final product;

Estimate the possibilities for providing the desired types and quantities of energy in the future; and

Estimate the future emissions in the environment that will be created by energy resource consumption based exclusively on the economic principles of the business.

The basic data to be obtained by this analysis include the estimation of the amount of future energy consumption, estimation of future costs of energy consumption, and assessment of future energy efficiency. The estimation of the condition in which the company will probably find itself is a process related to a certain degree of uncertainty, since it depends primarily on the stability of general business conditions in the country. Also, the precision of this estimation can be considered acceptable for a period of 5-10 years; however, in countries with unstable business conditions this period may be shorter.

Determining the position in which the company wants to be with the use of sustainable energy management, based on desired goals. The company that wants to become and remain a part of the modern world market has to adapt its business to the requirements of sustainable development, which means that, among other things, it must incorporate in its development plans sustainable energy management. Recognizing this, companies can assess their own future after the implementation of sustainable energy management measures based on the following:

Estimation of future total energy consumption based on constant savings, rationalization, and optimization;

Estimation of future costs of energy incurred as a result of reduced consumption of energy or energy-efficient operations;

Estimation of investments required to achieve savings in energy consumption, which involves estimating the size of the investment and determining the economic and non-economic effects of investments in this area;

Estimation of future levels of energy efficiency that can occur as a result of the implementation of sustainable energy management measures;

Definition of the share of energy consumed from its own or other renewable sources in the total amount of energy consumed in the company;

Estimation of financial and non-financial effects that may occur by reducing harmful emissions into the environment;

Estimation of the condition and reaction of the environment to environmentally responsible and irresponsible behavior of the company;

Estimation of compliance of energy-efficient operations of the company with the requirements of legislation, both domestically and globally; and

Estimation of the situation in which the company wants to be if it applies sustainable energy management, which is a particularly sensitive category in the process of forecasting in general.

The future is in essence unpredictable and there is always a degree of risk. On the one hand, companies are faced with the trend of environmentally responsible business in general, and on the other there is a clear awareness on the existing energy resource limitations. All this points to the need to develop specific tools that will enable more effective predicting in this field.

Today, there are very effective criteria for assessing economic effects and justification of investments, but currently there are no mechanisms by which the effects of a long-term implementation of sustainable energy management could be predicted with great precision. Companies will introduce it into their business because it is a requirement of sustainable development and an imperative of civilized business, but predicting its effects over the long term is currently at the level of assumptions. It is necessary to distinguish between an unstable and stable environment and adjust the techniques underlying the control of the strategy. If the environment is stable, then it is relatively easy to predict, and the role of control is to observe the progress of companies. In an unstable environment, when prediction is difficult, the task of control is to assess the assumptions, goals, and components of the strategy.

Determining the size of the gap between the extrapolated and projected goals includes determining the difference that will occur in future business if certain changes are or are not introduced. Any change in itself is a process that has its causes and effects, including changes to sustainable energy management:

The gap will be greater if the company deals with the activities connected with high energy consumption;

The gap will be greater if the company uses only non-renewable energy sources;

The gap will be greater if the company recorded low energy efficiency indicators; and

The gap will be greater if the company is planning big changes in terms of transitioning to a sustainable way of managing energy.

The whole process of gap analysis is linked to a number of assumptions, and defining the gap is the last step of the method and is often impossible to define precisely. Properly conducted analysis of the gap that exists between the state that would be realized without introducing changes and the one that can be achieved by using sustainable energy management must provide the following information:

The total change in the amount of energy spent in the company, since the process of sustainable energy management results in a decrease in the amount of energy consumed;

The changes in the quantity of energy consumed, depending on the type of energy, because it may happen that the implementation of the process of sustainable management of energy reduces consumption of only one type of energy resource, which usually depends on the specifics of the technological process;

The size of the change in energy efficiency, since the process of sustainable energy management leads to reduced energy consumption for producing a product unit; and

The financial impact of implemented changes, which are related to showing changes in revenues and expenditures that inevitably occur in the process of implementation of the sustainable management of energy, by which increased costs in the form of investments in energy-efficient operations can be expected. Savings will later be registered in terms of positive effects of implemented investments.

From the point of view of proper assessment of economic efficiency of the system of sustainable energy management it is particularly important to consider financial aspects and the time in which the above effects can be expected. In this way the company will be able to predict in the long term and implement the necessary allocation of resources and anticipate the moment in which to expect the first positive financial effects. In this way financial instability can be avoided or significantly reduced.

Establishing programs to fill the gap involves a whole series of measures and activities directly related to the need to surpass the instability that arises as a consequence of all business transactions, which can be very strong if the sustainable energy management is introduced in companies whose business is connected with large energy consumption. At this point it is necessary to predict the following:

> Time frame required to move from traditional to sustainable energy management, which must be coordinated and integrated with other management plans (usually a minimum period of five years is required);
> Financial resources required for implementation of the measures and activities that lead to full implementation of the process of sustainable management of energy. When using internal sources the company must anticipate the intensity of the financial effects that may occur by investing in this aspect and neglecting other aspects of business. If the financial resources are provided from external sources, it is necessary to use adequate methods to predict expected cash flow and current liquidity problems, as well as measures to solve these problems; and
> Changes in the sphere of human resources, which must be determined both by the number and quality. The implementation of sustainable energy management is associated with changes in traditional business, which sees energy solely as an expense. If the company is able to pay for it, there are no obvious reasons to make any changes. Sustainable energy management looks at energy as a resource that should be spent optimally, regardless of the company's financial strength. It is necessary to provide training for employees at various levels, with possible involvement of people who are specially trained for the introduction, implementation, and monitoring of sustainable energy management.

At this point the company is faced with the need to cover changes that are imminent or the gap that is identified by planning and management in the sphere of its three basic resources: time, money, and people. For these resources, adequate analysis, development plans, and ways of monitoring and control are necessary because it is essential that these three resources effectively support all activities required for the complex operations of transition to a sustainable energy management system, as shown in Figure 5.24.

Determining the mechanisms of control, or a set of control points to meet the gap. The introduction of sustainable energy management is a big, strategic change in the way of thinking and operating, and must therefore be regarded as a kind of strategic management process. The process of strategic management

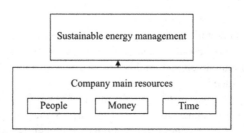

FIGURE 5.24 Basic resources of a company in the function of sustainable management of energy

involves iteratively connected steps that are aimed at ensuring efficient and effective business. On the other hand, a strategic decision-making process usually requires long-term engagement of the company, during which the company changes the way it operates. Accordingly, assessment, control, and auditing are needed to determine whether the strategy as a product of strategic management is in accordance with, above, or below expectations and planning. For these reasons the audit itself should involves several areas including:

Evaluation of internal harmonization of strategic goals with other goals of the company, as a result of the need to harmonize changes in the field of energy management with other plans of the company, which should all be integrated together into a single management system. The goals of business are numerous, but they all lead to opportunities for income generation, by which the company fulfills its mission and the purpose of its existence. On the other hand, modern companies determine social responsibility as one of their goals that promotes business in accordance with the concept of sustainable development,

Evaluation of the quality of the process of analysis, which is the basis for the development of the strategy, including a realistic assessment of the goals set as desired or likely in the process of gap analysis, and according to which further assessments and management plans are made. At the time of setting goals their supposed value has been set as well as their size or some other characteristics, and, after some time, it is possible to validate a company's ability to achieve certain predictions. Modern business is increasingly associated with predictions rather than analysis of previous events, so attention has to be paid to the selection of the required methods for predicting;

Evaluation of the very content of the strategy (direction, pace, and method of growth), which was selected as the content for the gap analysis. Depending on the goals and the size of the estimated gap, the company chooses its own way for the implementation of sustainable energy management. The above process can be implemented in whole or in part, faster or slower, with fewer or more costs, but requires monitoring of changes, as well as other enterprise business strategies that exist in the given period of time;

Evaluation of the company's opportunities to achieve the formulated strategy requires a remarkable ability to assess if the company is capable of filling the gap and reaching the set goals, after the internal and external analysis, or after setting goals and defining the gap. Sustainable energy management is often a desired goal, but its implementation is associated with a number of changes and investments, so most companies are not able to immediately implement a complete process of sustainable energy management. The gradual transition to a new way of energy management through the implementation of smaller, easily achievable goals is the safest way for most companies, although it takes more time; and

Evaluation of results brought about by the implementation of the strategy is a necessary step in assessing the effectiveness of the method of gap analysis. The gap analysis method is based on results, and its main advantage is that it provides the opportunity to precisely measure and assess whether the use of energy management has led to the desired goals. If so, the company gets valuable experience, and if goals are not achieved, or are not fully realized, the company is able to find out why it happened and to identify where the error occurred. In this way it can correct its own management plan. The most common errors that occur in the process of transition to sustainable energy management are due to misjudgment of abilities and non-objective setting of goals.

Gap analysis is based heavily on projections and assumptions. It is extremely demanding in terms of a company's ability to assess its own abilities. Most companies have a positive attitude toward the implementation of sustainable energy management, but the first problems arise even during the internal analysis of capabilities to implement something like it in practice. If the problems are ignored and companies overestimate their own strength, their goals will be too high and a gap will be present, which can cause the company to give up the sustainable management of energy because it may seem unachievable, or the company will engage in a long period of changes based on non-objective assumptions.

For the reasons stated above, companies need to understand the complexity of the process of introducing sustainable management of energy and observe it as a big change that requires a very serious approach. The application of available methods and management techniques will reduce the possibility of error and increase the likelihood of successful implementation. No company achieves all its goals, but a modern company should not be resistant to changes in the field of energy management. The methods for implementing sustainable energy management are all specific, and there are several implementation methods to choose from. The factor that in most cases affects the choice of a particular method is the activity that the company deals with, but above all, its efficiency and experience in managing change, i.e., the transition from traditional to sustainable energy management.

5.8 CASE STUDY – USA

ENERGY MANAGEMENT PATHFINDING: Understanding Manufacturers' Ability and Desire to Implement Energy Efficiency

Abstract

Manufacturers are scrambling for relief from today's energy expenses and price volatility. Most industry decision makers believe the solution is to seek the lowest available energy prices. Too often, managers fail to grasp the opportunities offered by *energy management*, which focuses on both consumption and prices. Industry can be resistant to energy management for a variety of reasons. Simply put, energy management has no traditional place in the typical manufacturer's chart of organization, job descriptions, and performance accountabilities. While technology is fundamental to energy efficiency, it is people who make it work in an organizational context. DuPont, Frito-Lay, Unilever, and Kimberly-Clark are a few of the forward-thinking companies that have found ways to build energy management into their daily operations to positive effect. The Alliance to Save Energy is documenting these companies' experiences in a series of case studies that reflect the organizational and behavioral aspects of corporate-wide energy management. Case studies show that energy management motives and approaches are somewhat varied – there is no "one size fits all" solution. The Alliance offers a typology of industrial energy management strategies to illustrate the range of opportunities available to industry. Ultimately, it is a manufacturer's organizational character that determines its ability to manage energy consumption. A checklist included in this study allows the reader to diagnose a manufacturer's aptitude for undertaking various energy management strategies.

1 ENERGY EFFICIENCY, ENERGY MANAGEMENT, AND BUSINESS IMPACTS

From the manufacturer's perspective, fuel and power are merely catalysts that refine raw materials into finished products. *Heat and power optimization* are the real value propositions behind energy efficiency. As an organizational *process*, "energy management" contributes to the *outcome* of improved business performance. "Energy efficiency" refers to practices and standards set forth in an energy management plan. Energy efficiency initiatives are selected for their potential to reduce expenses, build revenue capacity, and contain operating risk.

For manufacturers:

Activities that instill energy efficiency	→ Control over heat and power resources	→ Increased reliability of operations	→ Ability to fill more orders faster and with less expense	→ Greater productivity, more revenue

Unchecked energy expenditures are like a tax burden imposed cumulatively on each stage of production. Plants of all types, sizes, and locations use energy, so the potential for energy-driven productivity gains is everywhere. Energy management is an ideal opportunity to improve competitiveness through productivity improvement. The benefits only begin with reduced energy bills. Other impacts include greater capacity utilization, reduced scrap rates, more effective emissions and safety compliance, and enhanced risk management.

Efficiency should not be confused with *conservation*. As opposed to conservation (sacrifice), energy efficiency is an indispensable component of any effort to improve productivity. Ultimately, energy efficiency contributes to wealth.

American industry continues to waste energy. No one knows that better than Frito-Lay, Unilever, DuPont, 3M, Kimberly-Clark, and other manufacturers that have implemented the most aggressive energy management programs. This is more than a "hippies, beads, and flowers" issue. At stake is the viability of manufacturing facilities that employ people and sustain local communities. For this reason, the Alliance to Save Energy, with support from the U.S. Department of Energy's (DOE) Industrial Technologies Program, has compiled ten corporate energy management case studies to date. Most of these companies used information resources developed by the U.S. DOE to facilitate their accomplishments to varying degrees. The intent of this case study series is to encourage industry observers to learn from their peers.

Many efficiency proponents believe that if you show the projected dollar savings or payback for energy improvements, top managers will accept these proposals. That's not always true. Organizational size and complexity pose formidable hurdles to capturing efficiency opportunities. Manufacturing enterprises have organizational structures, accountabilities, and incentives that are designed to make products and get them out the door. While most companies will express a desire to "reduce costs," waste is not fully recognized in day-to-day practice. Control of energy waste requires cross-functional authority and communications that don't exist in most facilities. Given this reality, energy waste will continue no matter how financially attractive a project looks on paper.

A fully developed industrial energy management program is a work plan for continuous improvement. This plan will engage human, technical, and financial resources, and its progress will be monitored for attainment of certain goals. Criteria for action will reflect input from engineering, maintenance, financial, and utility staff. Staff will be held accountable for outcomes. The only energy improvements undertaken are those that provide business value to the organization.

2 A SAMPLE OF ENERGY MANAGEMENT LEADERS

Energy management is practiced to varying degrees by manufacturers throughout industries. No one industry dominates the practice. While it is easier to identify energy management leaders among Fortune 500 companies, there are also small, privately held companies that excel at stewardship of energy and other resources. An overview of ten companies' accomplishments is as follows:

3M. This diversified manufacturer seeks to reduce energy consumed (Btus) per pound of product by 20% over the 2000-2005 time frame. This goal will require 3M's tier-1 plants (52 facilities worldwide) to achieve 3M's own "World Class" energy management label. 3M has already surpassed that target and uses its energy performance in its product marketing. Superior energy cost control at 3M reduces the embedded energy cost that 3M's customers will normally absorb. Notable feature: 3M's executive management believes that resource stewardship makes good business sense. Energy management goals and results are routinely communicated to Wall Street analysts. 3M, and the manufacturers that purchase inputs from 3M, are responding to markets that increasingly demand products with low environmental impacts.

C&A Floorcoverings. Based in Georgia, this privately held, five-plant company demonstrates successful energy management by a mid-sized manufacturer. C&A has implemented a management system for matching energy efficiency initiatives with business goals. After two years, C&A achieved 10% savings on an annual natural gas expenditure of $824,500. **Notable feature**: C&A adopted Management Standard for Energy (MSE) 2005, an ANSI-certified standard for energy management developed by Georgia Tech, as a template for an in-house energy management program. By the end of 2004, C&A was close to becoming the first organization to become fully certified per the MSE 2005 standard.

Continental Tire North America. A lack of corporate involvement effectively puts energy management in the hands of facility managers. Continental has begun shutting down certain North American facilities due to energy waste and other cost inefficiencies. One Illinois-based facility became proactive at energy management, and was rewarded by getting a larger share of overall production quotas. The Illinois plant used a combination of energy consultants and in-house management structures to achieve a 31% reduction in energy consumption per tire. Notable feature: Continental successfully partnered with an energy services company (ESCo) to design and implement energy management procedures that are self-sustaining after the ESCo's tenure concluded.

DuPont. With over 100 plants in 70 different countries, energy management practices at DuPont are supported by two top-level strategies. The first is designating energy conservation as a high priority corporate issue. The other is applying "Six Sigma" methodology to the energy management process. Notable feature: Through 2002, DuPont applied Six Sigma to behavioral tasks, including plant utility management. Over 75 energy improvement projects, many requiring no capital, were implemented across their global operations. The average project netted over $250,000 in annual savings.

Frito-Lay. This leading snack food manufacturer's energy management features aggressive energy reduction goals with a focus on results. This demands a high degree of monitoring, measurement, and communication. Frito-Lay organized the required engineering talent as its Resource Conservation Group. While surpassing intermediate targets on the way to even larger savings, Frito-Lay's efficiency initiatives have returned over 30% on investment. **Notable feature:** Large and challenging energy reduction goals were used to rally and motivate staff to generate results.

Kimberly-Clark Corporation. This personal care products manufacturer has a broad mandate for environmental stewardship. KCC's global portfolio of over 165 plants practice energy conservation, air emissions abatement, wastewater treatment upgrades, process water user education, packaging reduction, land-fill elimination, toxic chemical elimination, and Environmental Management System implementation. Five-year plans help coordinate benchmarking efforts across a global facility network. KCC's energy conservation efforts are currently in the middle of a second five-year plan, which seeks to expand on the success of the first plan (1995-2000). The first plan led to a Corporate-wide, 11.7% reduction in energy use per ton of product. Notable feature: A large, global population of mills allowed KCC to generate its own proprietary energy benchmarking discipline. Sharing best practices across plants prevents "reinventing the wheel."

Merck & Co. Inc. This pharmaceutical products and services corporation seeks to improve the productivity of existing assets while reducing energy expenses. A corporate energy program is mobilized by goals that hold site managers accountable for annual performance targets. Energy costs at manufacturing sites are on a growth-adjusted pace to be cut 22% between 2001-2005. This equates to at least 250,000 tons of avoided carbon emissions and 11.5% energy expenditure savings. Notable feature: Energy efficiency was employed to boost the production capacity of existing assets, thus avoiding the need to finance new capital assets.

Mercury Marine. This manufacturer of marine propulsion systems consolidated energy decisions under the authority of a Central Facilities Manager (CFM), and implemented a power monitoring system that permits electricity costs to be tracked and billed to individual cost centers. Valuable energy flow data gives the CFM leverage in gaining corporate approval of energy technology upgrades. The centerpiece of these efforts in 2004 was the installation of a new, centralized compressed air system that carved roughly half a million dollars from an annual electricity bill of $7 million. Notable feature: Simple and effective energy management (1) placed the authority to make energy improvements in a single manager, (2) assigned cost control responsibility to production units, and (2) used information technologies to monitor energy flows and to directly bill production units for their actual energy use.

Shaw Industries. Concerted efforts to manage energy at Shaw Industries got underway in mid-2004. By primarily using the U.S. Department of Energy's (DOE) plant audit methods and Best Practices reference material, a newly hired Demand-Side Engineer documented potential energy savings at a rate of $1 million per month for the first six months of his tenure. Notable feature: U.S. DOE resources were effectively adopted by in-house personnel to drive their energy auditing and remediation activities.

Unilever HPC. Unilever's Health and Person Care Division's energy management program coordinates 12 facilities by combining energy-use targets with an energy service outsourcing strategy. A simple budget-to-actual spreadsheet compares energy performance at 14 facilities. Notable feature: Because its use resulted in a saving of $4 million on energy and another $4 million in avoided costs, the spreadsheet has captured the attention of individual facility managers and Unilever's Board of Directors as well.

The energy management features exhibited in the ten Alliance case studies are summarized in Tables 5.1 and 5.2 below. Table 5.1 compares each company's tactics, approaches, management tools, functions, and modes of organization and communication. Table 5.2 summarizes authority, leadership, and accountability profiles for each company.

Energy management at all ten companies includes:

Leadership of energy improvements provided by a key manager or "champion"; and
Technical planning, evaluation, and assistance rendered by an in-company energy team.

Features that are frequent but not universal include:

Performance goals and metrics specific to energy use (nine cases);
Routine audits or baseline assessments of plant-wide energy use (seven cases); and
A database to archive energy performance benchmarks and/or project profiles (six cases).

There are both *project-based* approaches and *behavioral* approaches to energy management. The project approach concentrates on hardware upgrades. The behavioral/procedural focus enhances efficiency awareness and decision making among production personnel. Note that:

Four of the ten companies combine a project-based approach with a behavioral/procedural approach;
Three focus primarily on behavior; and
Three companies take a projects-only approach.

Half of the companies studied:

Implement a multiyear planning horizon for coordinating their efforts;
Use budget-to-actual comparisons that incorporate energy consumption data;
Publish their results for investor relations purposes; and
Participate in government-sponsored energy events, programs, and collaboratives.

Some leadership and accountability lessons are evident in Table 5.2, including:

TABLE 5.1 Common features of in-house industrial energy management programs (Based on a sample of ten case studies)

	3M	Continental tire	C&A floorcovering	DuPont	Frito-Lay	Kimberly-Clark Corp.	Merck & Co	Mercury marine	Shaw industries	Unilever
Tactics & approaches										
Performance goals and metrics	×	×	×	×	×	×	×	×		×
Project-based approach	×			×	×	×		×	×	×
Behavioral/procedural approach		×	×	×	×	×	×	×		
Multiyear planning horizons	×			×	×	×	×			
Make prominent use of ESCOs	×	×								×
Employ a standardized protocol for energy or quality control			×	×					×	
Energy stewardship supports marketing strategy	×		×						×	
Tools & functions										
Energy performance reflected in budget-to-actual comparisons	×				×	×	×			×
Database to archive energy performance metrics and/or projects	×	×	×	×		×	×			

(Continued)

TABLE 5.1 Continued

	3M	Continental tire	C&A floor-covering	DuPont	Frito-Lay	Kimberly-Clark Corp.	Merck & Co	Mercury marine	Shaw industries	Unilever
Routine auditing or self-assessment of energy consumption	×	×	×		×	×	×		×	
Use DOE analytical software and related reference material				×			×	×	×	
Corporate energy implementation guide	×		×							
Easier financial criteria for energy improvement investments					×					×
Direct billing of energy costs to teams within a plant								×		
Organization & communications										
Corporate energy coordinator or "champion"	×	×	×		×	×	×	×	×	×
In-company energy team offers technical evaluation & assistance	×	×	×	×	×	×	×	×	×	×

(Continued)

| Energy performance results released in investor publications |
| Energy performance communicated to all employees |
| Plant-level teams and/or supervisors support energy improvements |
| Participate in government-business energy efficiency collaboratives |
| Improvement suggestions filtered up through staff |
| Internet or intranet workshops, online peer networking |
| Categorically recognize high-performance plants |
| Energy awareness events for employees (past or planned) |

Source: Alliance to Save Energy

TABLE 5.2 A comparison of corporate energy management styles Authority, Leadership, Organization, and Accountability (A sample of ten case studies)

3M

Authority: Broad corporate goal to reduce overall energy consumed per volume of product.
Leadership: Corporate leaders regularly review all plants' energy performance.
Who Decides to Act? Plant managers act, with influence from plant-based energy teams.
Organization: A corporate energy management team provides technical assistance and evaluation. Plant-based energy teams pursue implementation.
Accountability: Energy stewardship is one of many variables used to annually evaluate plant performance.

C&A floorcoverings

Authority: Top management periodically reviews energy performance.
Leadership: An energy coordinator leads all functions required by the MSE 2005 standard for energy management. Top management stands behind this standard.
Who Decides to Act? Key individuals decide to act per the accountabilities set forth in the MSE 2005 standard.
Organization: An in-house, cross-disciplinary team was assembled to initiate MSE 2005.
Accountability: Once implemented, the MSE 2005 standard sets roles and accountabilities for key personnel.

DuPont

Authority: Broad, five-year, corporate-wide goals require reduced energy consumption, increased use of renewable energy, and reduced carbon emissions.
Leadership: Corporate direction requires use of Six Sigma quality control methodologies for virtually all procedures at DuPont.
Who Decides to Act? A Six Sigma culture at DuPont is the incentive for all staff to seek improvement projects.
Organization: A corporate energy management team assists plants by providing technical assistance, documentation, and communication to build and replicate knowledge of energy solutions.
Accountability: Personnel promotions at DuPont are contingent upon gaining proficiency in Six Sigma. This drives DuPont's professionals–including energy utility engineers–to improve operations through application of Six Sigma.

Continental tire

Authority & Leadership: A Facilities Engineer takes nominal leadership of an in-plant energy team. Key supervisory engineers enforce energy discipline largely through personal influence and leadership. Corporate officers have no role in goal setting or progress reviews.
Who Decides to Act? The Facilities Engineer acts on the consensus of the in-plant energy management team.
Organization: A cross-disciplinary energy management team discovers, evaluates, and prioritizes energy improvement opportunities.
Accountability: Plant personnel generally observe in-plant leadership. While corporate officers play no day-to-day role in energy management, their long-term decisions regarding plant closure usually include energy cost performance.

TABLE 5.2 Continued

Frito-Lay

Authority: Aggressive corporate goals specify desired reductions in energy and water. Goals are pro-rated across plants. A Senior VP for Operations reviews comparisons of plants' progress.

Leadership: A Group Leader for Energy & Utilities coordinates corporate-wide discovery and evaluation of improvement opportunities.

Who Decides to Act? Plant managers and personnel make implementation decisions.

Organization: Several tiers of energy leadership are involved: a corporate tier provides technical assistance, a regional tier coordinates audit functions, and site champions assume implementation details.

Accountability: Corporate comparison of plants' budget-to-actual energy performance is the mechanism for ensuring compliance.

Kimberly-Clark

Authority: Corporate-wide five-year plans impose goals for energy savings.

Leadership: The VP for Energy and Environment ultimately leads technical support, benchmarking, databasing of results, and corporate-wide communication/promotion of success stories.

Who Decides to Act? Individual plant managers make actual implementation decisions.

Organization: A corporate energy management team provides technical support, energy auditing, benchmarking, and documentation services. Plant staff perform implementation.

Accountability: Energy performance is integral to plant and plant manager performance evaluations.

Merck & Co.

Authority: Five-year plans establish corporate-wide goals for energy cost reduction. The Sr. VP for Manufacturing monitors reported energy performance.

Leadership: The Senior Manufacturing Head and Energy Manager coordinates corporate-wide energy management functions. Facility representatives participate in developing a 4-point strategy for strategic planning, reporting, best-practice identification, and awareness development.

Who Decides to Act? Each facility's General Manager makes ultimate implementation decisions.

Organization: At the corporate level, "Global Energy Management" is led by an Energy Reduction Initiative Team, which is in turn comprised of a core team (for monthly review and guidance) and an expanded team (of in-house subject experts called upon as needed). Team subcommittees each represent many functions, including engineering, benchmarking, procurement, etc. Facility representatives identify improvement opportunities for their individual sites.

Accountability: Energy performance is a line item in each general manager's performance evaluation.

TABLE 5.2 Continued

Mercury marine

Authority & Leadership: A Central Facilities Manger assumes responsibility for all energy improvement decisions, including discovery, evaluation, and technical assistance. There are no energy-saving goals or corporate reviewers.

Who Decides to Act? Individual unit managers make energy improvement decisions per the advice of the Central Facilities Manager.

Organization: Personnel with a variety of professional disciplines form an in-house energy management team to identify improvement opportunities and assist with implementation. Unit managers petition the Central Facilities Manager and energy team for assistance as needed.

Accountability: Unit managers have cost control responsibilities. An in-plant power metering system permits direct energy cost assignment to unit managers. Energy management is therefore integral to cost performance.

Shaw industries

Authority: Senior management issues a general directive to "get some energy savings". Leadership: A Demand-Side Engineer leads a corporate Energy Management Department.

Who Decides to Act? An individual site's plant engineer or maintenance supervisor takes responsibility for action.

Organization: The six-person corporate Energy Management Department supports plants with energy accounting, acquisitions, monitoring, and technical assistance (auditing and evaluation).

Accountability: Individual plant managers are influenced by the Energy Management Department. The Demand-Side Engineer communicates success stories to boost awareness and encourage greater responsiveness to recommendations.

Unilever

Authority: All energy management results are reviewed by a Senior Vice President. Leadership: The Energy & Environmental Manager leads a corporate energy team that advises staff and energy service vendors.

Who Decides to Act? Plant managers make the ultimate decision to implement improvements.

Organization: Plant managers approve a budget that incorporates planned energy consumption. Budget input comes from various stakeholders in each plant. Energy service vendors are contracted to do much of the implementation.

Accountability: Quarterly budget-to-actual energy performance comparisons hold plant managers accountable for results.

Source: Alliance to Save Energy

There is no singular approach to energy management;

Every organization has a unique, established balance of authority and influence (1) between corporate headquarters and its subordinate facilities, (2) across facilities, and (3) among personnel within facilities. Creativity and initiative are extremely helpful in tailoring an energy management program to fit each company's unique circumstances;

Companies rarely compel their plants to make energy-efficient choices. Instead, accountability is usually indirect through (1) general cost control responsibility, or (2) regular plant/personnel evaluations that position energy as one of many areas for performance credit; and

In-company energy support networks provide crucial assistance, but implementation is usually exercised *at the plant manager's discretion.*

Some programs are relatively complicated, involving overlays of management teams and detailed reporting metrics (Merck, Frito-Lay). Others are amazingly simple, yet equally effective:

Mercury Marine puts energy decision-making authority in the hands of a central facilities manager. Process managers must follow his lead with respect to energy decisions. Also, an investment in power monitoring equipment allows Mercury to accurately bill power costs to substations within the plant. Cost accountability is all the motivation needed for subunit managers to enforce smart energy behavior on the part of their staffs;

Unilever routinely circulates a plant-by-plant comparison of energy performance, comparing each plant's budget-to-actual performance. Pride, competitive spirit, and perhaps a bit of shame are all that's needed for laggard plants to seek assistance coordinated by the corporate energy manager; and

Continental Tire's "corporate" energy policy is to leave energy management up to individual facility managers. There is no corporate officer who actively monitors energy performance. Their Illinois facility has effective energy management tactics, and enjoys an increasing share of corporate production quotas. Continental facilities with poor efficiency records are being shut down.

Motivations for pursuing energy management are surprisingly varied. Perhaps the most obvious reason is to "control energy expenditures," although this is far from being the only reason. Some companies put a premium on resource stewardship, for both public relations and risk management purposes. Other companies wish to sustain and replicate operational improvements that will be otherwise lost in complex, multifacility environments. Table 5.3 summarizes the motivations for undertaking energy management, as expressed by the ten companies in the case study series.

The summary of motivations in Table 5.3 clearly reflects the multipurpose nature of energy management, as follows:

Energy expense control and management of energy price volatility;

Non-energy expense control, such as avoided capital expenditure;

Increased revenue potential through replication of capacity improvements;

Improved product marketing through visible resource stewardship; and

Risk mitigation related to environmental liabilities and operational reliability.

TABLE 5.3 Motivations for initiating in-house industrial energy management programs (Based on a sample of ten case studies)

	3M	Continental tire N.A.	C&A floorcovering	DuPont	Frito-Lay	Kimberly-Clark Corp.	Merck & Co.	Mercury marine	Shaw industries	Unilever
Dealing with the volatility and complexity of energy markets				×	×			×	×	×
Consolidating, coordinating, and replicating improvement knowledge			×	×		×		×		
Expense reduction opportunity	×	×	×	×				×	×	
Environmental compliance			×	×		×				
Energy management leads to new product/revenue opportunities	×								×	
Disenchantment with ESCO services					×					
Improve plant capacity, performance		×					×			

Source: Alliance to Save Energy

Federal, state, and trade association programs attempt to boost general industry awareness of energy management principles through workshops, industry conferences, and trade press. Industry's response is at best lukewarm. Many companies are frankly intimidated by the prospect of implementing energy projects, much less day-to-day energy management processes. After all, competitive pressures have stripped manufacturers to the point where surviving staff are over-tasked in simply "keeping the car on the road," much less finding time to monitor and adjust performance. Also, despite every effort to reach industry's empowered decision-makers, awareness outreach too frequently attracts the wrong audience. This is because "energy efficiency" is almost always perceived as a technical or maintenance pursuit, so it is delegated to maintenance staff that are uninterested in, or unprepared to tackle, the organizational measures needed to make meaningful energy improvements.

3 LESSONS LEARNED: TEN CASE STUDIES

A comparison of the ten case studies presented here suggests that industrial energy management is not prescriptive in nature. It is tempting to argue that some companies' approaches are stronger than others. Upon further thought, it is useless to suggest that Company A is somehow "better" at energy management because it achieved greater relative energy reductions than Company B. After all, one company may have already been somewhat more efficient to begin with. The structure of authorities within companies is a major factor. So too are market conditions and asset management strategies. Is energy management helped or hindered by corporate policies regarding investment, human resource development, and outsourcing? The answers are unique for every company.

It is clear that a corporation approaches energy management with a strategy that reflects the company's organizational characteristics. Among the leading determinants are the degree of corporate authority and involvement, depth of in-company technical support, and capacity to express energy performance's contribution to business goals.

4 THEORY: CORPORATE RECEPTIVENESS TO ENERGY MANAGEMENT

The purpose of this section is to propose a typology of corporate "aptitudes" for energy management. This discussion is based on the Alliance to Save Energy's observations and research. Until these theories can be properly tested, readers are asked merely to consider this persuasive argument.

Human, technical, and financial criteria all contribute to a robust energy management program. Collectively, these attributes constitute a "culture" and receptiveness not only to energy management but to operational efficiency in general.

The following is a listing of organizational attributes that enable energy management. Manufacturers will enjoy a wider range of energy management options (moving up on a continuum from "do nothing" to sustained, daily energy management) by adopting as many of these attributes as possible. How can the presence of these organizational attributes be determined?

5 ORGANIZATIONAL ATTRIBUTES THAT DETERMINE AN APTITUDE FOR ENERGY MANAGEMENT

Fundamental business viability. Companies that are the subject to merger or acquisition, labor disputes, bankruptcy, or severe retrenchment may have fundamental distractions that will interfere with the attention that energy management deserves. A preponderance of such conditions indicates management turmoil that makes energy management impractical.

Replication capacity. Logical attributes for replication include (1) a multi-plant organization, and (2) general consistency in process activities and products across plants. Staff's ability to cooperate across sites and functional boundaries is crucial. Organizations must simultaneously engage many different professional disciplines and accountabilities to maximize their energy management potential.

Energy leadership (or "champion"). Successful energy improvements are usually led by an "energy champion," a manager that (1) understands both engineering and financial principles, (2) communicates effectively both on the plant floor and in the boardroom, and (3) is empowered to give direction and monitor results.

Energy market capability. This dimension is straightforward: Does the corporation wish to purchase energy through ongoing market activity? If so, the corporation should be prepared to maintain sophisticated search and verification procedures to support its contracting activities. Purchasing decisions should reflect the collaboration of procurement, production, and plant utilities personnel.

Leadership intensity. Quality of operations should be demanded, facilitated, and recognized by top officers of the corporation. Adoption of professional and industry standards are helpful in attaining this attribute. Energy-smart operations will hold employees accountable for adherence to energy management goals and other quality standards.

Pride intensity. Energy efficiency is as much dependent on behavior as it is on technology. A positive, can-do attitude on the part of staff is helpful in attaining potential energy savings. Rewards and recognition can be harnessed to good effect.

Fiscal protocol. The finance question is not always *how much.* Are purchase decisions made on first cost or life cycle costs? Who in the organization pays, and who claims the savings? Do savings count only fuel bill impacts, or do they include the value of material waste minimization and greater capacity utilization? What criteria determine adequate payback?

Engineering protocol. Successful energy management depends on an ability to understand energy consumption. This requires benchmarking, documenting, comparing, remediating, and duplicating success stories. Internal skills, procedures, and information services are engaged. The likelihood of building value through energy efficiency varies directly with the depth of these technical capabilities.

In the absence of an energy management process, energy expense control is reduced to one-dimensional efforts. Many manufacturers (either wittingly or not) settle for something less than full energy efficiency potential due to a lack of time, interest, or understanding. The approach taken by individual manufacturers is very much a function of their organizational attributes and business culture.

6 ENERGY MANAGEMENT STRATEGIES

The aim of this section is to present the range of energy management options available to industry. Every manufacturer employs SOME energy management strategy, even if the choice is to do nothing about energy consumption. Consequently, every manufacturing organization adopts one or more of these strategies:

1. DO NOTHING. Ignore energy improvement. Just pay the bill on time. Operations are business-as-usual or "that's the way we've always done it." The result is essentially "crisis management," in that energy solutions are induced by fire-drill emergencies and undertaken without proper consideration of the true costs and long-term impacts.

   WHO DOES THIS? Companies that do not understand that energy management is a strategy for boosting productivity and creating value. Or, companies that are subject to merger, buy-out, bankruptcy, union disputes, relocation, or potential closure. Or, companies that are extremely profitable and don't consider energy costs to be a problem.

   PROs: You don't have to change behavior or put any time or money into energy management.

   CONs: You don't save anything. Income is increasingly lost to uncontrolled waste. Because you don't inventory your energy usage, you are exposed to volatility in energy markets. You are less prepared to adapt to evolving emissions compliance agendas, and you are less capable of spotting opportunities presented by new technologies. Because you don't monitor anomalies in energy flow data, you are more susceptible to lapses in mechanical integrity and plant reliability.

2. PRICE SHOPPING. Switch fuels, shop for lowest fuel prices. No effort to upgrade or improve equipment. No effort to add energy-smart behavior to daily O&M procedures.

WHO DOES THIS? Companies that "don't have time" or "don't have the money" to pursue improvement projects. Or, these companies truly believe that fuel price is the only variable in controlling energy expense.

PROs: You don't have to bother plant staff with behavioral changes, or create any more work in the form of data collection and analysis.

CONs: Lack of energy consumption knowledge exposes the subject company to a variety of energy market risks. You don't know where your waste occurs, nor do you identify opportunities to boost savings and productivity. You are also exposed to energy market volatility and emissions and safety compliance risks.

3. OCCASIONAL O&M PROJECTS. Make a one-time effort to tune-up current equipment, fix leaks, clean heat exchangers, etc. Unable/unwilling to make capital investments. Revert to business-as-usual O&M behavior after one-time projects are completed.

WHO DOES THIS? Companies that are insufficiently organized to initiate procedural changes or make non-process asset investments. They cannot assign roles and accountabilities for pursuing ongoing energy management.

PROs: You spend very little money when just pursuing quick, easy projects.

CONs: Savings are modest and temporary because you don't develop procedures for sustaining and replicating your improvements. Familiar energy problems begin to reappear. Energy bills begin to creep back up.

4. CAPITAL PROJECTS. Acquire big-ticket assets that bring strategic cost savings. But beyond that, day-to-day O&M procedures and behavior are business-as-usual.

WHO DOES THIS? Companies that lack the ability to perform energy monitoring, benchmarking, remediation, and replication as a part of day-to-day work. However, they have the fiscal flexibility to acquire strategic assets that boost productivity and energy savings.

PROs: Obtain fair to good savings without having to change behavior or organize a lot of people.

CONs: Forfeit savings attributable to sustained procedural and behavioral efforts. Also, savings from the new assets may be at risk if adequate maintenance is not applied.

5. SUSTAINED ENERGY MANAGEMENT. Merge energy management with day-to-day O&M discipline. Diagnose improvement opportunities, and pursue these in stages. Procedures and performance metrics drive improvement cycles over time.

WHO DOES THIS? Companies with corporate commitment to quality control and continual improvement, well-established engineering and internal communications protocol, and staff engagement through roles and accountabilities.

PROs: Maximize savings and capacity utilization. Increased knowledge of in-plant energy use is a hedge against operating risks. Greater use of operating metrics will also improve productivity and scrap rates while reducing idle resource costs.

CONs: You need a lot of in-house talent, cooperation, and a capable energy "champion" to do this.

It is beyond the scope of this study to comment on which strategies are predominantly encountered in industry. Anecdotal evidence suggests that all industrial energy management strategies can be categorized per one of these five selections. It is also possible for firms to practice multiple strategies simultaneously, for example "price shopping" for low-priced fuel commodities in concert with a "capital projects" focus.

It should be noted that most of the ten of the experiences documented in the Alliance's case study series can be categorized as "sustained energy management." As such, these companies integrate energy management with day-to-day operating procedures and accountabilities.

7 ENERGY MANAGEMENT PATHFINDING: MATCHING STRATEGIES WITH CORPORATE ATTRIBUTES

This section will build on the theory of corporate receptiveness to energy management, as presented above. The energy management strategies available to a manufacturer are a function of its organizational attributes, as summarized in Table 5.4. Note that this is currently presented as theory.

8 EXAMPLES FOR INTERPRETING THIS TABLE

A manufacturer should have attained the attributes of "fundamental viability," "leadership intensity," "fiscal protocol," and "engineering protocol" in order to effectively pursue *capital projects* as a single-site energy reduction strategy.

Alternatively, a manufacturer that has attained "fundamental viability," "replication capacity," "leadership intensity," "pride intensity," "engineering protocol," and has an "energy champion," should be capable of pursuing both the *occasional O&M projects* and *sustained energy management* strategies across multiple sites. In this instance, the company may wish to start with the lesser strategy (O&M projects) and evolve into the practice of sustained energy management.

This typology presumes that energy management for multisite organizations is more demanding than for single-site companies. Accordingly, adoption of a certain strategy by a multisite organization requires all the organizational attributes that a single-site organization will be expected to muster, plus the capacity to replicate.

Managers that are contemplating improved energy management are encouraged to consider the case study results and theory presented in this study. To act on this information, the steps are:

TABLE 5.4 Theory: matching corporate attributes to energy management strategies

Organizational attributes

	Fundamental viability	Replication capacity	Energy champion	Energy market capability	Leadership intensity	Pride intensity	Fiscal protocol	Engineering protocol
Strategies for single-site energy reduction:								
Do Nothing								
Price Shop				REQUIRED				
Capital Projects	REQUIRED				REQUIRED		REQUIRED	REQUIRED
Occasional O&M Projects	REQUIRED				REQUIRED	REQUIRED		
Sustained Energy Management	REQUIRED		REQUIRED		REQUIRED	REQUIRED		REQUIRED
Strategies for replicating energy reduction at multiple sites:								
Do Nothing		REQUIRED		REQUIRED				
Price Shop		REQUIRED	REQUIRED					
Capital Projects	REQUIRED	REQUIRED	REQUIRED		REQUIRED		REQUIRED	REQUIRED
Occasional O&M Projects	REQUIRED	REQUIRED	REQUIRED		REQUIRED	REQUIRED		
Sustained Energy Management	REQUIRED	REQUIRED	REQUIRED		REQUIRED	REQUIRED		REQUIRED

Source: The Alliance to Save Energy

Determining an Organization's Aptitude for Energy Management. Note that organizational attributes have been substantially attained by the subject company.
Compare the attained attributes to the information in Table 5.4. The presence (or absence) of certain attributes determines that energy management strategies are available to the subject company.
Use these findings to understand what the subject organization can or cannot achieve in terms of energy management.

Keep in mind that this study indicates what a manufacturer can expect from energy management, given its *current* organizational attributes and business culture. There may be a desire to evolve to a higher level of energy management than what the current organization allows. What if a manager wants to advance energy management in his or her organization? There are windows of opportunity. An obvious example is when energy market turmoil brings top management's attention to fuel costs. Also, take advantage of annual planning sessions or strategic reorganizations to propose the kind of organizational processes needed to practice sustained energy management. Remember that energy cost control is as much dependent on people as it is on technology. Learn from the case studies shared here.

9 CONCLUSION

Volatile energy markets are here to stay. So are competitive and regulatory pressures. Energy price movements will put some manufacturers out of business, while others will decide to move offshore. Surviving manufacturers will not only provide superior products and service, they will maximize value through operating efficiencies. Energy efficiency is an indispensable component of wealth creation.

Energy procurement strategies such as shopping for low energy prices and supply contracts are only partial solutions to soaring energy expenses. Management of consumption is an underappreciated opportunity. While technology is the foundation for managing consumption, it is the human dimension that makes technology work. Organizational procedures, priorities, and accountabilities are crucial to energy management.

A few forward-thinking companies have allowed their energy management experience to be documented for industry's wider benefit. Frito-Lay, DuPont, and Kimberly-Clark are among these companies. The "best of the best" companies' energy management programs feature corporate accountabilities, a mechanism for providing technical support, and a "champion" to manage energy improvement efforts.

A manufacturer's ability to manage energy consumption is ultimately a function of organizational attributes and corporate culture. This study advances "energy management pathfinding" concepts. While sustained, day-to-day energy management is recommended for providing the greatest and most durable value, it is also the most demanding in terms of operational character. Many companies will find that they are suited for strategies that are less challenging but may also provide less value. The same management diagnostic presented in this study serves as a pathfinder for matching organizational characteristics with appropriate energy management strategies.

Chapter 6

Energy Management – Control

6.1 ELEMENTS OF STRATEGIC CONTROL

Control may be conditionally considered as the last step in the process of sustainable energy management. Its essence is to compare the achieved with the planned, identify possible deviations, and undertake corrective activities. Control is hence only in theory perceived as the last strategic step, while, essentially speaking, it represents a starting point for making necessary corrections to the energy management process, planning it again, and performing it with no interruptions and with constant improvements. Control therefore focuses on goals stemming from an enterprise's mission. Definition of the basis that can be used for contrasting is one of the main issues. A well-conceived control offers information on negative deviations and their removal at early stages, prior to the conclusion of envisioned activities. Control is a system that has to be explicitly and clearly set, based on its contrasting processes and unambiguous principles.

Control System

Control may be defined as "a process ensuring that implemented activities match the envisioned ones [59]." Objectives and their implementation are defined in the planning stage. Control should measure fulfillment of a plan, i.e., the level of achieving set goals, establishing deviations, and proposing corrective activities.

Control is a systematic effort, and envisioned goals become controlling standards: defining the feedback and comparing real results with previously defined standards, identifying size, and importance of deviations. Control should also deal with taking appropriate actions that use all available resources to reach objectives in an effective and efficient manner. Such a definition implies that the control process consists of four stages, as shown in Figure 6.1.

The four control process stages include defining standards and result-measuring methods, results measuring, contrasting achieved results with standards, and undertaking corrective measures. A strategic energy management control system has a number of specificities.

Sustainable Energy Management. http://dx.doi.org/10.1016/B978-0-12-415978-5.00006-0

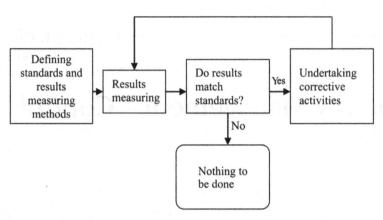

FIGURE 6.1   Control process

Defining Standards and Result-Measuring Methods

As has already been underlined, planned objectives are most often taken for standards and basic elements of control. Objectives are defined and at the same time result-measuring methods and responsibility of entities (managers most often) are established. If standards are not set, managers may respond to small deviations or, more seriously, fail to react to significant and alarming deviations. The motivation function and comparison "threshold" function are two important functions of standards in the control process. Objectives will give better results and control will be efficient, provided objectives are motivational and not too low. Most employees feel better and are more satisfied if they manage to respond to challenges and be successful in execution of high standards. However, if they are too high and cannot be objectively carried out, it would be counterproductive and bring about motivation decline. In this case, it is better to spend more time and set up new standards that objectively fit business terms and conditions.

Standards should be comprehensible for those that execute them, which is one of the main prerequisites. It is necessary to articulate adequate result measurements so that an objective can become a control standard and result-measuring methods reliable. For example, when discussing services, the waiting time may be set up as a control standard. In manufacturing, measures should synthesize quality control indicators and production and financial indicators as well.

Results Measurement

Unlike the previous stage of control, results measurement is a repetitive activity, the frequency of which depends on the activity being measured. For example, the level of greenhouse gases a chemical or metallurgy plant emits in the environment should be measured on a daily basis, while the growth rate of an enterprise may be measured once a year. Selection of activities to be measured should be established in line with the management principle of *exception based*. A manager who is responsible for a certain process should only be informed about the activities if significant deviations from standards are recorded. In this case, the manager should react so that activities are adjusted and an adequate solution that would result in compliance with a standard for a certain activity is found.

Most systems function according to the feedback principle. This is particularly relevant to result-measurement systems, since by measuring results, obtained information is communicated to decision makers.

Comparison of Achieved Results with Standards

This stage of control is, in practice, the easiest step of the controlling process, since if results are in line and at the level of set standards, no intervention or corrective activity is needed. Setting standards in an adequate way and measuring results is much more complex than determining whether there are any deviations from such standards.

Undertaking Corrective Activities

If the previous stage proved there are deviations from standards, the control process stage is needed. The corrective activity itself might be simple and involve changing only one activity, or it may include a number of activities or alternations of previously defined standards. The control process is a dynamic process, and if managers apply it afterwards, i.e., *ex ante*, and follow it only through its results, it can be said that they only follow results but fail to "exercise control." Control should be *exercised*, so that results get closer to standards in a constructive manner before a concrete activity is undertaken.

Control Parameters

Frequency, control subject and object, and quantity and quality of information provided through implementation of the control are the main control parameters in the process of sustainable energy management.

The *frequency of control* depends on a number of factors, and may be defined at the very beginning, at the planning stage of energy management process. However, changes in frequency of control quite often take place during the process itself. In most cases, frequency of control is under the influence of the following factors:

- The activity of an enterprise. Controls will be much more frequent in companies that are large energy users. In such cases, both theory and practice urge for everyday controls of implementation of the energy management process, which primarily refers to following energy consumption levels and recording relevant factors. In the course of time, having gained insight into certain regularities in energy consumption, these controls might be reduced, however, they should still be performed at least once a month;

- The range of sustainable energy management the organization wants to carry out has an impact on frequency of control. Sustainable energy management may be conducted in a different range, from partial and symbolic up to comprehensive and full. Naturally, implementation of partial enhancements does not require frequent controls since these measures are often determined only once in the observed period (for example once a year) with effects visible following the introduction. Implementation of comprehensive sustainable energy management is an extremely complex and long-lasting process requiring the introduction of frequent, precise, and, in a special way, planned controls, and parameters under control may in this case be numerous and varied;

- Experience in implementation of the management process as a factor that largely and quite often defines frequency of control, despite being based on immeasurable influence. Namely, companies with certain experience in the implementation of a management system in different areas (not only energy management) may believe their experience allows for control to be performed less often since they may be reassured that experience overcomes mistakes resulting from insufficient and imprecise controls. On the other hand, companies with less experience in implementation of the management process very often fail to attach any importance to the control itself, not even in the planning phase; rather they have mainly been devoted to the implementation. In other cases, insufficiently experienced companies may plan for extremely rigorous and frequent control that may burden the system with a great quantity of information that is often useless, and important information may go unnoticed or be insufficiently reconsidered. Introduction of a sustainable energy management process is in most cases a huge change and companies quite often fail to understand that the sustainable management approach differs greatly from traditional approaches. It is thus recommended due to the complexity of change and insufficient knowledge of sustainability mechanisms that companies have objective frequency of controls regardless of experience. Application of other companies' experiences (benchmarking) could give good guidelines in this area;

- The degree of changes coming from the outside, as a result of the fact that the sustainable energy management process has been carried out for years. It very likely that during this period certain changes will take place in internal and external environments that may influence the implementation process and control results as well. The most common changes impacting control standards are those pertaining to adoption of certain legal and international obligations and agreements, significant changes in prices of energy substances, and internal changes in the structure and liquidity of the company itself; and

- Perception of the listed changes by control stakeholders (subjects). Most frequent controls of the implementation of sustainable energy management process have been carried out in companies feeling those changes might have a huge impact on both company and the environment. Furthermore, some companies often fail to conduct any review or plan any changes in the control system, which leads to irregular and inaccurate and hence useless results. Less-experienced companies often demonstrate insufficient sensitivity toward changes taking place in the environment that may affect an energy management system. It is therefore important for a company to have an elaborated system of providing responses to changes and the capacity to assess the ongoing potential level of influence of changes on expected results and control of implementation at the same time. Any changes in price of energy substances, for companies that are large energy consumers, have an immediate effect on output parameters and the results of the application of sustainable energy management, so an adequate system for financial aspects assessments has to be provided in this sphere in particular.

Apart from the above, the frequency of control in the sustainable energy management process does not depend on every factor acting individually, but most often on the effect created by their mutual interaction. It is therefore very important to be familiar with the rules pertaining to the factors' mutual relations. Significant changes in the energy market may cause significant consequences if a company is run by insecure and inexperienced management. On the other hand, there are companies that spend significant quantities of energy substances but believe the costs may be refunded through product prices, in which case the sustainable energy management process becomes irrelevant and marginalized by both management and employees.

The *object of control* is of special interest in the control process of implementation of the sustainable energy management process. This control is based on measurement of certain parameters, which are then contrasted to the values stipulated in the goals planning phase and give a success measure of the implemented strategy.

Identification of an object of control, i.e., control items, first has to be harmonized with previously set objectives. Each objective set in the planning phase has to be measureable, so that it can be contrasted in the control phase. Setting objectives in immeasurable units in sustainable energy management is useless.

Apart from harmonizing control items with previously set objectives, control items have to be selected so they provide data relevant for further measurement in the sustainable energy management process. A control item may be any place at which one may obtain certain data within the company. However, the following controls are considered necessary for sustainable energy management:

- Measuring the total energy consumption following a certain period of time defined by the control plan (day, week, months), which provides data on consumption trends (increase, decrease, stagnation) with definition of clear reasons for the spotted trend, aiming at arriving at some regularities;
- Specifying the total energy consumption by sectors in a company, with the comparison of the measured consumption with the planned one, registering all parameters that may have influence on the measured level of consumption (capacity, type of products, technology, season, obstacles);
- Establishing the biggest energy consumption area in a certain time frame, aiming at defining areas suitable for the most significant savings;
- Measuring consumed energy by a product unit following a certain control period and comparing that quantity with the quantity defined as a goal, with the purpose of establishing whether conducted measures lead toward the goal at the desired pace;
- Specifying the quantity of energy obtained from the company's own or other energy renewable resources. Following a certain control period it is possible to establish the quantity of energy replaced by the energy from renewable resources and compare it with the defined goals. It is possible in this way to establish in a timely manner whether measures have been enough to meet this goal (new technological solutions, use of new energy resources, creating alternative energy resources, etc.). Often, companies set unrealistic goals and envision quick and abundant replacement of traditionally obtained energy with energy from renewable resources; and
- Defining financial effects of sustainable energy management measures implementation is of great interest in the overall process since it could largely define survival of implementation of the process pertaining to transfer to sustainable business. The comprehensive transfer to sustainable energy management principally implies increased costs in the very beginning, so that companies using a lot of energy have to be ready to accept initial negative financial effects. Investing into energy wise efficient technologies and building one's own capacities for production of energy from alternative sources are financially demanding investments, and the return period is often longer than five years. Company should expect the positive effects only after a certain period of time. Defined by application of adequate economic measures for specifying justifiability of investments. Therefore financial reports are one of the control items, monitoring general and incomes and expenses situation in the company, since investments for energy sustainable business are inseparable from other financial business indicators. In some cases current business results are sufficient to compensate for initial negative financial effects resulting from larger investments in energy solutions. However, quite often positive financial effects cannot be resolved without adequate help from additional investments by foreign investors, special funds, or banks.

The above list proves that energy quantity, efficiency, and financial indicators are the most common control subjects. A combination of these indicators may provide insight into the given circumstances and conduct of the overall system, on the basis of which one may conclude whether the company is moving in the sustainable energy management direction or not. Timely control of the given indicators leaves enough time for review of goals and redefinition of the strategy, provided measured activities prove to be unsatisfactory. Figure 6.2 shows the relationships among consumption, energy efficiency, emissions in the environment, and financial effects as main control subjects.

Control of sustainable energy management is a particularly complex process that does not actually have an end result. Control is the process based on measuring the realization of objectives, as well as their effects, but control is performed continuously and does not end after the measurement of certain effects. The essence of control lies in the fact that all measured values are reduced to a single measure, which is suitable for conducting the required analysis, mathematical operations, and comparison with targets. The most appropriate unit for expressing all subjects of control is money.

Energy consumption is monitored and measured on several grounds, depending on whether the consumption is measured in household consumption,

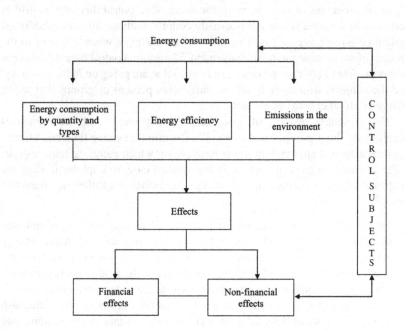

FIGURE 6.2 Interrelation among control items in the process of sustainable energy management

the company, or any particular country, but is most often expressed through consumption by quantity and types of energy sources. All values can be expressed in a certain monetary value. In addition, power consumption can be expressed by certain parameters of energy efficiency that can be expressed in different units, but the most appropriate is the unit of income. Control of emissions into the environment can be expressed in specific measurement units depending on the type of contaminants, but these values can be converted into a certain value in money that takes into account the amount of penalty to be paid if there is a certain over-pollution, or as savings (income) that occurs if the pollution is within the allowed limits. Also, at this point the sum attained on behalf of the premium for the reduction of pollution can be expressed in money.

The controller is an issue of special interest that is often not given enough attention. More specifically, companies often approach the process of energy management sporadically, and if there is a total commitment to solving this problem, the issue of entities that initiate and control the whole process and are therefore responsible for conducting of the control cannot be solved in a way that could be considered satisfactory. At this point it should be noted that the main aim of control is the provision of sufficient information to indicate the correctness of or discrepancies in the whole process. Information obtained by control is used in making important decisions in the process of further planning. Therefore, the issue of control is of special interest.

In all processes of management, the issue of responsibility and control is particularly sensitive because it is usually coupled with the already-established system of governance that may or may not be adequate when it comes to the process of sustainable energy management. Therefore, control over this process shares the fate of all other processes of control that are going on in the company, and the subjects who carry it out are most often persons or groups that are in charge of all other areas of control.

The presented process of sustainable management, with its numerous peculiarities, long process, and especially the initial negative financial effects, requires a special approach in the control phase, which therefore requires specially oriented subjects of control. In the optimal case, to implement adequate control a person or persons should be engaged, having the following characteristics:

- Professional competence, which includes education, skills, and abilities necessary to monitor and control processes such as sustainable energy management, is the first prerequisite in this process. People must have the necessary theoretical and practical knowledge in the field of energy and sustainable development, as well as basic knowledge in management science. It is necessary that these persons are able to identify problems, to distinguish between important, less relevant and irrelevant events in the system, and to react accordingly. It is expected that the person in charge of control is capable of independent and team work.

- Legal competence related to the position that the person or persons responsible for the control of the process of sustainable energy management already take in a particular company. Usually, these are people who are in high management positions, which normally have insight into control of all the processes of management. On the one hand, it is desirable that the sustainable energy management is integrated in other management processes in the company, but it is not good if, because of that, energy management becomes marginalized in some way. The process of sustainable energy management is complex and requires full engagement of people who cannot have other responsibilities.

From the above, it can be concluded that in order to implement the stage of control it is necessary to engage people who were engaged in the process of planning the strategy of sustainable energy management. Moreover, these people should have been engaged in creating the basic objectives. Apart from that, in companies that consume more energy, the process of transition to the new energy management should be managed by a person who has only that task to do.

Preferably the process should be controlled by a person working in the company, because such a person is aware of the technological processes and the ways of doing business in general, although there are no restrictions regarding the hiring of expert consultants. Although outsourcing can bring many advantages, because the people who are engaged as consultants are very experienced professionals familiar with good or poor business practices in this sphere, and are especially objective because they have no stake in the internal interests of the company, which each management process requires.

The amount of information that can and must be gathered by controlling the implementation of the process of sustainable energy management can greatly vary. Too much information in most cases cannot provide a clear picture of the implementation of the strategy of sustainable energy management, because in most cases it overloads the control system and increases the possibility of error. An insufficient amount of information, on the other hand, does not provide enough information on implementing the strategy, and thus it is not possible to make the right decisions on further actions. The optimal amount of information is a parameter that is passed in accordance with experience and in accordance with the specific features of the company in which control is performed and the ability of the system of control that is in use.

Quality of information is a more important parameter of control and indicates the reliability of information or its ability to present the real situation. Measurable information is usually the information of the greatest degree of quality and it is most frequently obtained in the energy sector, so the quality of information in this case does not present a particular problem provided that the measuring instruments are adequate and precise and readings are regularly made.

Frequency of control depends on the activities of the company and control models. The choice of subjects and frequency of control along with the definition of the basis for comparison are the main problems of control implementation.

As the modern environment is characterized by rapid and intense change, it becomes more difficult to set a basis for comparison. Response of the company to changes often increases the degree of diversification and decentralization. On the other hand, decentralization brings new demands. These requirements refer to the coordination and determination of responsibility for decisions with a greater number of participants. In order to more easily respond to the consequences of this change, it is necessary to continually test and timely modify the previously issued decisions.

6.2 CONTROLLING FINANCIAL EFFECTS OF IMPLEMENTATION

The process of sustainable energy management is complex, time consuming, and demanding on several grounds. One of the key problems that slows down the process of transitioning from traditional to sustainable energy management is exactly the problem of planning, monitoring, and evaluation of financial effects of the application of the stipulated strategy. For companies involved in energy-intensive and demanding activities, these problems are severe and therefore require special attention and the development of acceptable methods for assessment. Great difficulties in practice often cause uncertainties in the field of cost-effectiveness and problems in monitoring financial effects during the implementation of sustainable energy management.

In most cases, the transition to sustainable energy management involves significant financial investment in equipment, resources, training, implementation of a series of standards, and engagement of experts and consultants. All this represents a significant cost to the company. In addition, in the transition to sustainable energy management in the initial phase delays occur most commonly, as well as incompatibility of equipment and people, the problems of energy supply, and the like, which inevitably lead to loss of value of production, thus decreasing revenue. Therefore the investment in sustainable ways of using energy requires careful monitoring of financial parameters.

To assess feasibility of investment in sustainable energy management, certain methodology should be used that provides consistency in assessment of various investments, thus allowing testing and comparison of indicators over several years. The basic algorithm of estimation of profitability of investment in sustainable energy management involves the following activities:

- Assessment of the developing capacities of the company – which requires real estimation of the current business performance and analysis of the development strategy, new markets, changes in the process of work, estimation of the risk in the environment, and all other economic, financial, social, technical and technological, and other environmental conditions;
- Estimation of current status and the necessary changes – necessary to analyze conditions in the field of energy management in the company, and above

all assess in detail the parameters of energy supply (quantity, type, price, conditions given by the supplier) to determine the way in which energy is consumed (type, quantity, energy efficiency, energy waste) and based on that set a list of priorities that require investment in order to transfer to a sustainable energy management;

- Analysis of procurement options – if the company finds out it is necessary to implement certain changes or invest in new sources of energy, it should estimate opportunities for procurement of necessary equipment for carrying out such an enterprise, because the markets in which the equipment can be found are sometimes too distant and prices and quality are different;
- Estimation of the value of the necessary investment – having chosen the type and quality of the necessary equipment, the total value of investment should be estimated, which provides insight into the total amount of money necessary for carrying out certain investment activity;
- Determining sources of funding – investments in sustainable forms of energy management often require engagement of bank capital, so it is necessary to analyze bank offers and special funds that provide finance for this purpose;
- Estimation of future income – transition to sustainable energy management is often connected with certain problems in the initial stage that inevitably lead to reducing the value of production and thus reducing income that should help companies survive, pay its liabilities, and settle loan installments;
- Estimation of future expenditures – transition to sustainable energy management often requires investments during several years that increase usual expenditures of the company and, apart from that, the expenditures entail liabilities toward sources of funding;
- Static estimation of the cost effectiveness of investment – comparison of revenue and expenditures in the representation year of investment (the year in which the full capacity of production has been achieved and liabilities toward sources of funding are still active) can provide insight in cost effectiveness of the stipulated investment;
- Dynamic estimation of the cost effectiveness of investment – the first impression of cost effectiveness of investment is created by financial indicators during the whole period of exploitation of the stipulated investment (which estimates in which year the investment would pay off), method of net current value (by applying the method of discounting), and method of internal rate of return (which estimates if the investment would pay off under certain bank conditions);
- Sensitivity analysis – respects the fact that during the whole period of the project certain changes may occur in the environment that would directly influence financial indicators of return. Therefore it is necessary to estimate to which extent the investment is able to stand them without jeopardizing liquidity. The following has been estimated: changes in the price of energy raw materials, changes in the price of other input, reduction of revenues,

increase of expenditures, and other parameters that are considered to be of special importance; and

- Final financial estimation of investment feasibility – carried out based on previously determined parameters, by which it is necessary to ensure that the investment fulfills all stipulated methods of estimation in order to be considered acceptable.

All investments in the systems and facilities for sustainable energy management are to a certain extent subsidized and supported by the state authorities but are still a huge cost for the company and therefore special attention should be paid to estimating financial aspects.

6.3 CONTROLLING NON-FINANCIAL EFFECTS OF IMPLEMENTATION

Improvements in the field of energy are the consequence of a long-term unplanned exploitation of natural resources and pollution of the environment caused by their exploitation and consumption. The energy problem has been widely recognized as one of the priorities of sustainable development, and it can be solved by implementation of different activities at all levels. All projects in the field of quality management of sustainable energy development attract the interest of huge groups of stakeholders, such as:

- Companies, basic units of business and consumers of energy, which get a series of advantages by implementation of adequate methods of management that refer to creating positive image, profit, cost reduction, and provision of certain subsidies;
- Beneficiaries, consumers of goods and services provided by the companies, that want to support energy saving companies, and therefore put pressure on institutions and are willing to ensure a healthy environment for themselves and their descendants;
- Citizens, who as consumers of energy stimulate energy saving and investments in renewable energy sources, thus gathering certain advantages, favorable loans, and tax reductions;
- Legislation, as a necessary framework for functioning of the system, should regulate the sustainable energy management and should be changed depending on the changes in requirements imposed by world integration;
- Investors, who want to direct their capital toward companies that are doing business responsibly towards the environment, and that invest in the projects for saving energy or production of energy from renewable sources, because they know that such companies have a more secure future and more secure capital;
- International communities, who are interested in efficient energy management, since energy stability is one of the key issues in terms of the survival of the world;

- Generations that have shown interest in regulating very complex problems of energy management, regardless of the fact that full effects in this field can be expected in the long-term; and
- Future generations, who are the biggest beneficiaries of the changes in the field of production and consumption of energy, and who have the right to have at their disposal the same quantity of resources that the previous generations had at their disposal, which is a key principle of sustainable development, called the principle of intergenerational justice.

6.4 CASE STUDY – SERBIA

Review of the Economic Viability of Investing and Exploiting to a Biogas Electricity Plant in Vizelj, Serbia

Abstract

Construction and operation of plants that produce energy from renewable energy sources is the subject for discussion in all countries that have accepted the sustainable development concept and the Kyoto Protocol as their own development direction. Enlargement of renewable energy production is clearly an imperative, but only economically viable construction and operation can result in long-term sustainability, which is initially the goal when deciding upon such investments. In line with this goal, this study presents the estimation of the economic viability of constructing and operating a biogas electricity plant on a farm in Vizelj, Serbia. The time frame for this estimation is from 2011 until 2020. This study also presents all parameters necessary for performing this estimation, respectively, analysis of revenues and expenditures, projection of economic and financial flow, ratio analysis, dynamic and static analysis, and analysis of sensitivity of the project, i.e., the impact of the changes in prices and raw material on the overall performance of the project. The observed investment was predicted to be financed by the EBRD's (European Bank for Reconstructing and Development, London, UK) credit line for renewable energy in the Western Balkans, and the total investment was estimated at 958,000 Euros. After the economic assessment was performed, the conclusion was that the observed investment in biogas electricity plant is very acceptable and can serve as a role model for similar investments in the region.

Keywords: electricity plant, biogas, economic viability, case study, Serbia

1 INTRODUCTION

The objective of this study is to research the economic indicators of constructing and operating a biogas electricity plant. It is assumed that the energy is produced from the slaughterhouse confiscate, from its own meat processing facility, IMES from Padinska Skela. The capacity of the envisioned plant was 250 kW/h. Except for the heat, it was estimated that the capacity utilization would be 75%. This project also estimated that the heat generated from the biogas electricity plant would be used for heating the nearby farm. During the

TABLE 6.1 Estimated annual sales of electricity

Products	2011	2012	2013	2014-2020
Electricity (MWh)	271	1,359	1,365	1,369
Heat (Sm3)	26,961	61,657	61,943	62,424
CERs	2,149	8,818	8,823	8,827
Removing slaughterhouse confiscate (t)	90	450	472	486

TABLE 6.2 Prices of the projected outputs from the given plant

Product	Unit	Price
Electricity (MWh)	MWh	158.20 €
Heat (Sm3)	Sm3	0.30 €
CERs	Kom.	11.00 €
Removing slaughterhouse confiscate (t)	t	47 €

regular operation process of the biogas electricity plant, 51.88% of the heat was consumed for heating the plant itself. The rest, 48.12%, i.e., 1,037.6 MWh, was at the outside consumer's disposal. The financial data is given in Euros.

In line with the Law on Energy of the Republic of Serbia, producers of energy from renewable sources have the status of privileged producers. Therefore, buying this energy is obligatory for the state-owned electricity sector, and this cycle is under the state control. Feed-in tariff for the given case on the Vizelj farm is c € 15.82 per kWh during the 2011-2020 period. The initial phase in the evaluation of the economic viability of such a venture was to define basic production parameters, which are estimated sales and estimated prices. Table 6.1 gives an overview of estimated sales (annual quantities). Table 6.2 gives an overview of the projected prices.

2 REQUIRED INVESTMENTS AND FINANCING

Investing in the biogas electricity plant was estimated at the total amount of 958,950 Euro. This amount consists of the investments into fixed and current assets (Tables 6.3 and 6.4).

Out of the total investment, the loan was envisioned in the amount of 809,214.69 Euros. The investor would finance part of construction works from its own sources, which would amount to 149,735.31 Euros (15.61% of the total investment) (see Table 6.5).

TABLE 6.3 Required fixed assets

Fixed assets	Dec. 31st 2010	Mar. 31st 2011	May 31st 2011
Real estate	0,00	149,735	149,735
Technological equipment	139,752	111,802	27,950
Mechanical equipment	32,201	25,760	6,440
Electrical equipment	150,701	75,350	25,116
IT equipment	0,00	32,201	32,201
Total investment (Eur)	**322,655.44**	**394,850.40**	**241,444**

TABLE 6.4 Required current assets

Current assets	Days	Coef.	Amount 2011	Amount 2012	Amount 2013	Amount 2014
I Business-related current assets			10.387	52.281	49.232	49.436
Buyers			8.193	39.140	39.402	39.566
Electricity (MWh annually)	60	6	7.167	35.836	36.003	36.102
Heat (Sm3)	30	12	674	1.541	1.548	1.560
Confiscate removal (t)	30	12	352	1.762	1.850	1.903
Cash	10	36	2.193	13.140	9.829	9.869
II Sources of current assets			1.468,25	5.727,99	6.106,24	6.121,99
Suppliers			588	2.207	2.234	2.249
Confiscate	30	12	105	525	551	567
Maintenance of the investment	30	12	358	1.432	1.432	1.432
Other expenditures – CER (CER – Carbon Emission Reduction) fees	30	12	125	250	250	250
Salaries	30	12	880	3.520	3.872	3.872
Amount of continuous current assets (I-II)			8.919	46.553	43.125	43.314
Difference year by year				37.633	–3.427	188

TABLE 6.5 Overview of the investments

Type of investment	Amount (Euros)	Own sources	Loan
Real estate	299,470	149,735	149,735
Technological equipment	279,505	-	279,505
Mechanical equipment	64,402	-	64,402
Electrical equipment	251,168	-	251,168
IT equipment	64,402	-	64,402
Total investment	958,950	149,735	809,214
TOTAL INVESTMENT	100%	**15.61%**	**84.39%**

TABLE 6.6 Financing conditions

Period	**5 years**
Annual interest rate	6.5%
Processing fee	1%
Moratorium	9 months
Grace period	9 months (Sept. 30th 2011- Jun. 30th 2012)
Payback type	Equal installments June 30th 2012 – December 31st 2015
Payback dynamics	Monthly
Success fee upon completion 15%	€ 121,382.20 (2012)

TABLE 6.7 Long-term and short-term liabilities at the end of the year (period 2011-2015)

Date	Dec. 31st 2011	Dec. 31st 2012	Dec. 31st 2013	Dec. 31st 2014
Short - term liabilities	119,498.83 €	238,997.66 €	238,997.66 €	238,997.66 €
Long - term liabilities	716,992.99 €	477,995.33 €	238,997.66 €	-

The dynamics of settling liabilities towards financing sources was based on the financing conditions of the appropriate EBRD's credit line. Table 6.6 provides the financing conditions.

For financing this biogas electricity plant, the loan will last 5 years, with a grace period of one year. The payback plan is presented in Table 6.7.

3 REVENUES AND EXPENDITURES

Defining the economic viability of the exploiting electricity plant based on bio-gas conversion began with the estimation of expected revenues and expenditures.

Total Revenues

Total revenues were estimated for the project life cycle, and the results are presented in Table 6.8.

When looking at Table 6.8, it can be seen that the smallest revenue of 20% is expected in the first year. During the second year, somewhat larger revenue is expected, and it is supposed to result in receiving the success fee for finishing the project. The following revenues, over the next 8-year period, will be stable and will amount to 355,292 Euros annually.

Total Expenditures

Expected expenditures in running the biogas electricity plant are raw material, workforce, maintenance, certification, approval of the lowering emission, interest, other expenditures, and taxes (see Table 6.9).

Other costs of doing business include the fees for selling CERs, which amount to 3,000 Euros annually.

The corporate profit tax has been estimated at 10%.

4 SYNTHETIC FINANCIAL REPORT

Estimating the economic viability of doing business was conducted by preparation and analysis of the P/L statement, balance sheet, and cash flow and free cash flow. These are given in Tables 6.10–6.13.

The P/L statement clearly shows that the EBRD funds will come to the project when they are most needed – during the second year of the project. At that time, the company's funds will already be heavily exploited by investing, while the revenues will still be small and unstable. After analyzing the P/L statement, cash flow was observed with all its elements, and is given in Table 6.11.

Cash flow analysis shows expected changes for a project of this kind. Clearly, during the first two years the flow is positive, and this is primarily the result of the EBRD investment. After that, from the third until the fifth year, the project enters the critical phase and records negative net flows. After the fifth year, the positive values show up and remain until the end of the observation period.

When looking at the balance sheet, it is clear that most company's property is in fixed assets. On the other hand, long-term liabilities are almost four times larger than short-term ones. This is of course quite favorable from the investor's point of view. After the year 2014, the project is free of liabilities.

TABLE 6.8 Expected revenues of the biogas electricity plant (2011-2020)

2011

Assortment	MU	Sales	Price/ MU	Total revenue
Electricity	MWh	271	158	43,004
Heat	Sm3	26,961	0.30	8,088
CERs	kom	2,149	11	23,648
Removing slaughterhouse confiscate	t	90	47	4,230
Total revenues				**78,971**

2012

Electricity	MWh	1,359	158	215,020
Heat	Sm3	61,657	0.30	18,497
CERs-Approval of lower emissions	kom	8,818	11	97,007
Removing slaughterhouse confiscate	t	450	47	21,150
Success fee		1		121,382
Total revenues				**473,057**

2013

Electricity	MWh	1,365	158	216,018
Heat	Sm3	61,943	0.30	18,583
CERs	kom	8,823	11	97,063
Removing slaughterhouse confiscate	t	472	47	22,207
Total revenues				**353,872**

2014-2020

Electricity	MWh	1,369	158	216,616
Heat	Sm3	62,424	0.30	18,727
CERs-Approval of lower emissions	kom	8,827	11	97,107
Removing slaughterhouse confiscate	t	486	47	22,842
Total revenues				**355,292**

TABLE 6.9 Estimated annual expenditures

Raw material	Price	2011	2012	2013	2013-2020
Confiscate supply	14.00 €/t	1,260	6,300	6,615	6,804
Workforce-related expenditures					

Job	Education	No			
Total	High school	5	42,240		

Type of expenditure	Amount of the investment	Life expectancy	Annual cost of maintenance		
Real estate	299,470	25	1.0%	2,994	
Processing equipment	279,505	10	2.0%	5,590	
Mechanical equipment	64,402	20	1.5%	966	
Electrical equipment	251,168	20	2.0%	3,767	
IT equipment	64,402	10	5.0%	3,220	
Total	958,950			16,538	

Depreciation

Type of the investment	Amount	Interest	Bank fee	Total initial amount	
Real estate	-	5,047	1,497	306,015	
Processing equipment	299,470	9,421	2,795	291,722	
Mechanical equipment	279,505	2,170	644	67,217	
Electrical equipment	64,402	8,466	2,511	262,147	
IT equipment	251,168	2,170	644.02	67,217	
Total	64,402	27,277	8,092	994,319	

Type of investment	Total initial amount	Period	Annual interest		
Real estate	306,015	25	4%	12,240	
Processing equipment	291,722	10	10%	29,172	
Mechanical equipment	67,217	20	5%	3,360	
Electrical equipment	262,147	20	5%	13,107	
IT equipment	67,217	10	10%	6,721	
Total	994,319			64,602	

TABLE 6.9 Continued

Raw material	Price	2011	2012	2013	2013-2020
Costs of certifying CERs	15,000				
Cost of interest					

Year	2011	2012	2013	2014	2015
Cost of interest	13,204	51,244	38,354	23,264	8,173

5 FINANCIAL ANALYSIS

Based on the projected synthetical financial reports, it was possible to quantify and analyze the indicators of business performances of the biogas electricity plant. Table 6.14 gives an overview of the most common ratio analysis.

Keeping in mind that activating fixed assets was envisioned for October 1, 2011, investments were discounted to the net present value for September 30, 2011. The chosen discount rate was 8% (see Table 6.15).

As a base for calculating dynamic indicators of efficiency, the net inflows were also discounted, as presented in Table 6.16.

Net Present Value (NPV)

The difference between the investment and net inflows from the project is positive, and it amounts to **893,006.31 Euros**. Based on the criterion NPV>0, this investment is acceptable for financing.

Profitability Index

$$PI = \frac{NPV}{SVU} = \frac{893,006}{85,437} = 104.55$$

Internal Rate of Revenue (IRR)

The present value of the expected net economic flow was equalized with the NPV of the investment at the discount rate of 22.13%. Keeping in mind that the IRR was greater than the initial discount rate, this project is acceptable for financing.

The project resulted in an average rate of annual yield over the engaged capital of 22.13%.

TABLE 6.10 Balance sheet

Revenues	2011	2012	2013	2014	2015	2016-2020
Electricity	43,004	215,020	216,018	216,616	216,616	216,616
Heat	8,088	18,497	18,583	18,727	18,727	18,727
CERs	23,648	97,007	97,063	97,107	97,107	97,107
Confiscate removal	4,230	21,150	22,207	22,842	22,842	22,842
EBRD donation	-	121,382	-	-	-	-
Total revenues	**78,971**	**473,057**	**353,872**	**355,292**	**355,292**	**355,292**
Expenditures	2011	2012	2013	2014	2015	2016-2020
Raw material	1,260	6,300	6,615	6,804	6,804	6,804
Labor	10,560	42,240	46,464	46,464	46,464	46,464
Depreciation	16,150	64,602	64,602	64,602	64,602	64,602
Maintenance	4,134	16,538	16,538	16,538	16,538	16,538
CER verification	15,000	15,000	15,000	15,000	15,000	15,000
Other expenditures	1,500	3,000	3,000	3,000	3,000	3,000
Interest	13,204	51,244	38,354	23,264	8,173	-
Total expenditures	**61,809**	**198,925**	**190,575**	**175,673**	**160,583**	**152,409**
Gross profit	17,162	274,131	163,297	179,619	194,709	202,883
Profit tax (10%)	1,716	27,413	16,329	17,961	19,470	20,288
Net profit	**15,445**	**246,718**	**146,967**	**161,657**	**175,238**	**182,595**
% of net profit	**19.56%**	**52.15%**	**41.53%**	**45.50%**	**49.32%**	**51.39%**

TABLE 6.11 Cash flow

Years	2011	2012	2013	2014	2015	2016	2017	2018	2019	2020
Inflow from sales	78,971	473,057	353,872	355,292	355,292	355,292	355,292	355,292	355,292	355,292
Residual value										396,743
Total inflow	**78,971**	**473,057**	**353,872**	**355,292**	**355,292**	**355,292**	**355,292**	**355,292**	**355,292**	**752,036**
Raw material	1,260	6,300	6,615	6,804	6,804	6,804	6,804	6,804	6,804	6,804
Labor	10,560	42,240	46,464	46,464	46,464	46,464	46,464	46,464	46,464	46,464
Maintenance	4,134	16,538	16,538	16,538	16,538	16,538	16,538	16,538	16,538	16,538
CER verification	15,000	15,000	15,000	15,000	15,000	15,000	15,000	15,000	15,000	15,000
Other outflows	1,500	3,000	3,000	3,000	3,000	3,000	3,000	3,000	3,000	3,000
Interest	13,204	51,244	38,354	23,264	8,173	0.00	0,00	0.00	0.00	0.00
Loan installment		119,498	238,997	238,997	238,997	0.00	0,00	0.00	0.00	0.00
Profit tax	1,716	27,413	16,329	17,961	19,470	20,288	20,288	20,288	20,288	20,288
Total outflow	**47,374**	**281,234**	**381,299**	**368,030**	**354,449**	**108,094**	**108,094**	**108,094**	**108,094**	**108,094**
Net outflow	**31,596**	**191,822**	**-27,427**	**-12,737**	**843.78** €	**247,198**	**247,198**	**247,198**	**247,198**	**643,941**
Cumulative net inflow	31,596	223,419	195,991	183,254	184,097	431,295	678,493	925,691	1,172,889	1,816,831

TABLE 6.12 Balance sheet

Years	2011	2012	2013	2014	2015
ASSETS (A+B)	1,020,152	1,189,266	1,094,186	1,017,050	953,291
A. Fixed assets	978,168	913,565	848,962	784,360	719,757
Facilities	302,955	290,714	278,473	266,233	253,992
Technology equipment	284,429	255,257	226,085	196,912	167,740
Equipment works	66,376	63,016	59,655	56,294	52,933
Electrical equipment	258,870	245,762	232,655	219,548	206,440
IT equipment	65,536	58,815	52,093	45,371	38,649
B. Current assets	41,984	275,700	245,223	232,690	233,534
Receivables	8,193	39,140	39,402	39,566	39,566
Cash	2,193	13,140	9,829	9,869	9,869
Cash from inflow	31,596	223,419	195,991	183,254	184,097
LIABILITIES (C+D+E)	1,020,152	1,189,266	1,094,186	1,017,050	953,291
C. Equity	182,206	466,599	610,139	771,985	947,224
D. ST liabilities	120,953	244,670	245,049	245,064	6,067
Suppliers	574	2,153	2,179	2,195	2,195
Labor	880	3,520	3,872	3,872	3,872
Part of the loan	119,498	238,997	238,997	238,997	-
E. LT liabilities	716,992	477,995	238,997	0.00	0.00
Part of the loan	716,992	477,995	238,997	-	-

Discount Payback Period (DPP)

The present value of the net inflows from economic flow reaches the amount of NPV of the project in 2015.

$$DPP = 4.25year + \frac{13,324.38}{(13,324.28 + 151,708.41)} = 4.33year$$

TABLE 6.13 Economic flow

Economic flow	0	2011	2012	2013	2014	2015	2016	2017	2018	2019	2020
Sales		78,971	473,057	353,872	355,292	355,292	355,292	355,292	355,292	355,292	355,292
Residual value											396,743
Total Inflow		**78,971**	**473,057**	**353,872**	**355,292**	**355,292**	**355,292**	**355,292**	**355,292**	**355,292**	**752,036**
Raw materials		1,260	6,300	6,615	6,804	6,804	6,804	6,804	6,804	6,804	6,804
Labor		10,560	42,240	46,464	46,464	46,464	46,464	46,464	46,464	46,464	46,464
Maintenance		4,134	16,538	16,538	16,538	16,538	16,538	16,538	16,538	16,538	16,538
Verification of CERs		15,000	15,000	15,000	15,000	15,000	15,000	15,000	15,000	15,000	15,000
Other		1,500	3,000	3,000	3,000	3,000	3,000	3,000	3,000	3,000	3,000
Interest		13,204	51,244	38,354	23,264	8,173	0.00	0.00	0.00	0.00	0.00
Profit tax		1,716	27,413	16,329	17,961	19,470	20,288	20,288	20,288	20,288	20,288
Total outflow		**47,374**	**161,736**	**142,301**	**129,032**	**115,451**	**108,094**	**108,094**	**108,094**	**108,094**	**108,094**
Net flow		**31,596**	**311,321**	**211,570**	**226,260**	**239,841**	**247,198**	**247,198**	**247,198**	**247,198**	**643,941**
Investment	**994,319**	-	-	-	-	-	-	-	-	-	-
Cumulative inflow	−994,319	−962,722	−651,401	−439,831	−213,570	26,270	273,468	520,666	767,864	1,015,062	1,659,003

TABLE 6.13 Continued

Economic flow	0	2011	2012	2013	2014	2015	2016	2017	2018	2019	2020
Return period	**4.14**										
Discount factor	8.00%										
Accumulation factor		1.02	1.10	1.19	1.28	1.39	1.50	1.62	1.75	1.89	2.04
Discounted inflow		30,994	282,767	177,930	176,189	172,931	165,032	152,808	141,489	131,008	315,992
Investment – discounted amount	**854.137**										
Discounted cumulative inflow	-854,137	-823,143	-540,376	-362,445	-186,255	-13,324	151,708	304,516	446,005	577,013	893,006
NPV	**893,006**										
Discounted return period	**4.33**										
	-854,137	31,59	311,321	211,570	226,260	239,841	247,198	247,198	247,198	247,198	643,941
IRR	**22.13%**										
Profitability index	**104.55%**										

TABLE 6.14 Ratio analysis

Indicators	2011	2012	2013	2014	2015
Liquidity					
General liquidity ratio	0.35	1.13	1.00	0.95	38.49
Strict liquidity ratio	0.35	1.13	1.00	0.95	38.49
Net current assets (EUR)	−78,969	31,029	174	−12,374	227,466
Activity					
Turnover	9.64	12.09	8.98	8.98	8.98
Average delayed receivables	37.87	30.20	40.64	40.65	40.65
Turnover of suppliers	107.58	92.39	87.44	80.03	73.15
Average delayed payments	3.39	3.95	4.17	4.56	4.99
Financial structure					
Share of borrowed capital	82.14%	60.77%	44.24%	24.10%	0.64%
Long-term sources/ total sources	88.14%	79.43%	77.60%	75.90%	99.36%
Outflow/interest	2.30	6.35	5.26	8.72	24.82
Profitability					
Business profit rate	38.45%	68.78%	56.98%	57.10%	57.10%
Net profit rate	19.56%	52.15%	41.53%	45.50%	49.32%
Efficiency	127.77%	237.81%	185.69%	202.25%	221.25%
Yield/total assets	1.56%	22.33%	12.87%	15.31%	17.79%
Yield/own assets	9.31%	76.05%	27.30%	23.39%	20.39%

The investment would be paid off after 4 years, as shown in Table 6.17.

$$PP = 3,25year + \frac{213.570,98}{(213.570,98 + 26.270,47)} = 4,14year$$

The investment payback period is 4 years and 51 days.

TABLE 6.15 Dynamic efficiency indicators

Investment Date	Amount	Accumulation factor/Discount factor	NPV Sept. 30th 2011
December 31, 2010	8,092	1.059	8,572.975
March 31, 2011	149,735	1.039	155,609.498
July 31, 2012	19,916	0.938	18,679.243
August 31, 2012	19,916	0.932	18,559.829
September 30, 2012	19,916	0.926	18,441.178
October 31, 2012	19,916	0.920	18,323.285
November 30, 2012	19,916	0.914	18,206.146
December 31, 2012	19,916	0.908	18,089.756
January 31, 2013	19,916	0.902	17,974.110
February 28, 2013	19,916	0.897	17,859.204
March 31, 2013	19,916	0.891	17,745.031
April 30, 2013	19,916	0.885	17,631.589
May 31, 2013	19,916	0.880	17,518.872
June 30, 2013	19,916	0.874	17,406.876
July 31, 2013	19,916	0.868	17,295.596
August 31, 2013	19,916	0.863	17,185.027
September 30, 2013	19,916	0.857	17,075.165
October 31, 2013	19,916	0.852	16,966.005
November 30, 2013	19,916	0.846	16,857.543
December 31, 2013	19,916	0.841	16,749.774
/	,	.	
/	,	.	.
September 30, 2015	19,916	0.735	14,639.201
October 31, 2015	19,916	0.730	14,545.614
November 30, 2015	19,916	0.726	14,452.626
December 31, 2015	19,916	0.721	14,360.232
Total	**994,319**		**854,137.962**

TABLE 6.16 Table of discounted net inflows

Investment date	Net inflow (EUR)	Accumulation factor/Discount factor	Discounted net inflow 30/09/2011	Cumulative discounted net inflow
September 30, 2100	–854,138	1.00	–854,138	–854,138
December 31, 2011	31,596	1.02	30,994	–823,143
December 31, 2012	311,321	1.10	282,767	–540,376
December 31, 2013	211,570	1.19	177,930	–362,445
December 31, 2014	226,260	1.28	176,189	–186,255
December 31, 2015	239,841	1.39	172,931	–13,324
December 31, 2016	247,198	1.50	165,032	151,708
December 31, 2017	247,198	1.62	152,808	304,516
December 31, 2018	247,198	1.75	141,489	446,005
December 31, 2019	247,198	1.89	131,008	577,013
December 31, 2020	643,941	2.04	315,992	**893,006**

TABLE 6.17 Investment payback period

Investment date	Net inflow	Cumulative net inflow
September 30, 2011	–994,319.28	–994,319.28
December 31, 2011	31,596.56	–962,722.71
December 31, 2012	311,321.47	–651,401.24
December 31, 2013	211,570.17	–439,831.07
December 31, 2014	226,260.09	–213,570.98
December 31, 2015	**239,841.45**	**26,270.47**
December 31, 2016	247,198.01	273,468.48
December 31, 2017	247,198.01	520,666.49
December 31, 2018	247,198.01	767,864.51
December 31, 2019	247,198.01	1,015,062.52
December 31, 2020	643,941.39	1,659,003.91

Breakeven Point

To determine the breakeven point, 2014 was used as the representative year because of the maximum capacity utilization. To calculate the breakeven point, costs were separated into two groups – fixed and variable costs (Table 6.18), and also, the rate of profit contribution was determined.

Based on the given parameters, the value and capacity breakeven points were determined:

Value Breakeven – VB

$$VB = \frac{FC}{PCR} = \frac{165,869.66}{97.28\%} = 170,576.57$$

where:

FC – fixed cost
PCR – profit contribution rate

TABLE 6.18 Breakeven parameters

Costs	2014
Raw material	6,804
Other costs – fees for selling CERs	3,000
Variable costs	**9.804**
Labor	46.464
Depreciation	64.602
Maintenance	16.538
CERs	15.000
Interest	23.264
Fixed costs	**165.869**
Total costs	**175.673**
Total revenues	355.292
Variable costs	9.804
Profit contribution	345.488
Fixed costs	165.869
Gross profit	179.619
Rate of variable costs	2,76%
Profit contribution rate	**97,24%**

Capacity Breakeven – KB

$$KB = \frac{VB}{TI} = \frac{170,576.57}{353,872.13} = 48.01\%$$

Sensitivity Analysis

Exploiting this type of plant was observed over a ten-year period. In the given time frame, certain changes can happen and these can more or less affect the functioning of the project. Therefore, it was important to reach certain values of changes that the project could sustain. These changes in parameters are given in Table 6.19.

TABLE 6.19 Sensitivity analysis

Parameter	% change	IRR	NPV	Payback Period
Basic value	0%	22.13%	893,006.31	4 years, 4 months
Selling prices	+5%	23.65%	994,263.77	4 years
Selling prices	+10%	25.14%	1,095,521.23	3 years, 9 months
Selling prices	−10%	19.00%	690,491.39	5 years, 1 month
Selling prices	−5%	20.58%	791,748.85	4 years, 8 months
Raw material prices	−5%	22.16%	894,910.40	4 years, 4 months
Raw material prices	−10%	22.19%	896,814.49	4 years, 4 months
Raw material prices	+10%	22.08%	889,198.13	4 years, 4 months
Raw material prices	+5%	22.10%	891,102.22	4 years, 4 months
Selling prices and raw material prices	−5% & −3%	20.60%	792,891.31	4 years, 8 months
Selling prices and raw material prices	−10% & −6%	19.04%	692,776.30	5 years, 1 month

TABLE 6.20 Eliminating criteria

Indicators	Criterion	Project
IRR	> 8%	22.13%
NPV	> 0	893.006
Payback Period	< 10 years	4.33 years

6 OVERALL CRITERIA

In the process of estimating this project's viability, all inputs and outputs were analyzed. The overall opinion about the project was based on three groups of criteria: eliminating, functional, and descriptive.

Eliminating Criteria

This group of criteria defines acceptable values of the dynamic efficiency indicators, whereby the IRR criterion was crucial for decision making. The achieved values are given in Table 6.20.

Based on this set of criteria, the project was considered profitable and acceptable.

Functional Criteria

Project assessment based on the functional criteria entails liquidity and sensitivity analysis. The liquidity criterion requires paying all invoices on time. If the project is not liquid in each year of its life cycle, then the project's functionality is secured by the cumulative net inflow being positive. In the financial analysis it was observed that the project will have negative net flow during 2013-2014, but during the previous years enough cash would be generated to cover all expenses. According to all functional criteria, basic conclusion about project acceptability can be described as follows:

- Taking into account that cumulative net flow during the whole life cycle was positive, the project was declared acceptable;
- Sensitivity and risks of the project were determined within the breakeven analysis and sensitivity analysis, and they were both acceptable; and
- Based on the given facts, it can be concluded that investment in the biogas electricity plant was acceptable.

Descriptive Criteria

The investor has organizational, know-how, and human resources to implement planned investments, and planned activities, schedule, and predicted deadlines for preparation and execution of the project are realistic. Expected business results are satisfactory. Besides positive financial effects, execution of such project has very positive effects upon the environment.

7 CONCLUSION

This project will be executed in line with the priorities of the Republic of Serbia that relate to the Sustainable Development Strategy and the Kyoto Protocol. Also, the importance of the project was proven by the fact that EBRD's credit line for the Western Balkans will be used for financing. By executing this project, i.e., by turning slaughterhouse confiscate into electricity, it will be possible to self-heat the slaughterhouse, while the rest of the electricity will be supplied to the electricity company of Serbia. This company, as a public company, is obliged to buy RES electricity at privileged prices by the decree on the status of privileged electricity producers. Besides the given environmental indicators, detailed financial analysis was done. All the usual financial indicators were estimated as very positive during the 10-year life cycle. Also, after the sensitivity analysis was done, it was determined that there are no threats to the stability of this project whatsoever.

In line with the given opinion, this project can be considered viable and completely acceptable, but its development will be carefully monitored. Keeping in mind numerous historical, industrial, geographical, and energy-related similarities among the neighboring countries, this project could easily become a useful example to other investors in the region as well.

Chapter 7

Strategic Priorities of Sustainable Energy Development

FIRST STRATEGIC OBJECTIVE – EXPLOITATION OF RENEWABLE ENERGY SOURCES

Global development of society in the future will largely depend on the situation in the field of energy. Problems faced by most countries in the world are associated with the provision of energy and environmental protection. Explosion of human population on the earth has caused a constant increase in the demand for energy, particularly electricity. The growing trend in energy demand is 2.8% per annum. On the other hand, the current structure of primary energy sources cannot meet such an increasing demand for electricity on a global scale. The reason for this also lies in the environmental problems caused by the direct combustion of fossil and nuclear fuels the present generation of electricity is based on. In addition, the current dynamics of consumption of fossil fuels will lead to the disappearance of reserves in the near future.

The direct consequence of these opposite conditions of production and consumption is a steady increase in electricity prices, which already at the current level create an environmentally and economically justifiable need for the inclusion of alternative energy sources in a global energy development strategy. These energy flows have forced highly developed countries to invest huge capital and employ a large number of experts in the development of alternative renewable energy sources (wind power plants, solar power plants, biomass and biogas power plants, geothermal energy, etc.). As a result of such an investment, the technology industry has adopted and developed technically reliable conversion of some primary renewable sources. In addition, international protocols and commitments to reduce CO_2 emissions (the Kyoto Protocol) and local environmental problems have forced many governments to stimulate, through various subsidies, the construction of environmentally clean power plants using renewable sources of electricity.

These energy flows have forced highly developed countries to invest in the construction of electrical power plants that use wind power and solar power, currently the most promising alternative renewable sources of electricity.

Sustainable Energy Management. http://dx.doi.org/10.1016/B978-0-12-415978-5.00007-2

7.1 SOLAR ENERGY

The Potential of Solar Energy

The energy of the sun's rays coming to the earth's surface, i.e., the potentially usable radiation of the sun, is about 1.9×10^8 TWh per year. This energy is approximately 170 times greater than the total energy of coal reserves in the world and when compared to the energy needs of mankind, it amounts to 1.3×10^8 TWh per year. This means that the solar energy that reaches the earth's surface during only 6 hours is sufficient to meet all the needs of the world on an annual basis. In order to get better insight into these data, an average household in some of the most developed countries spends about 10,000 kWh of electricity a year and it would take about 100,000 years to spend 1 TWh.

About 37% of world energy demand is met by the production of electricity that during 2008 amounted to 17,000 TWh. If this energy is generated by photovoltaic (PV) systems (systems that convert solar energy into electricity) of modest annual power output of 100 kWh per square meter, a surface of 150 x 150 km² would be necessary for the accumulation of solar energy. Much of the absorption surface could be placed on roofs and walls of buildings and would not require additional surface area on the ground.

The energy of solar radiation is sufficient to produce an average of 1,700 kWh of electricity per year per square meter of soil, and the greater the radiation at a site, the higher the generated energy. Tropical regions are more favorable in this respect from other regions with moderate climate. Secondary radiation in Europe is about 1,000 kWh per square meter, while, by comparison, it is 1,800 kWh in the Middle East.

Application of solar energy can be achieved in two ways:
Converting solar energy into heat; and
Converting sunlight into electricity.

Solar systems for heat production are used in households, industry, agricultural buildings, and other facilities, that, for example, use large quantities of sanitary water. However, in the last ten years, PV conversion of solar energy has become the primary industry of solar devices due to the large number of technological advantages in relation to the conversion into heat and due to the rapid development of relevant technologies and their projected abilities.

Production of PV devices has doubled every two years, with an average annual increase of 48% since 2002, so this economic sector shows the fastest growth in the world compared to all other branches of energy technology. From an economic point of view, the price of electricity produced from solar energy is continuously decreasing as a result of technological improvements and growth in mass production, while it is expected that fossil fuels will become more expensive in the near future.

It is now justifiable to encourage the use of solar energy for the production of heat and electricity in the area of households, industry, and some branches of agriculture due to lower investment costs. However, in the long run, the future conversion of solar radiation is in PV technology and its integration with other branches of technology, which is in accordance with the attitudes, plans, and current status in the European Union and other leading economic countries.

The Technology of Solar PV Devices

The direct conversion of solar energy into electricity, the so-called PV effect, was perceived almost two centuries ago, but it was only after the development of quantum theory at the beginning of the twentieth century that this phenomenon was explained and understood. The first solar PV cell was developed at Bell Laboratories in 1954. With the development of space exploration, PV cells made of semiconductor silicon quickly became the main source of electricity for satellites, primarily because of its reliability, while the price was of minor importance. The importance of terrestrial use became topical in the world energy crisis of the early 1970s. Today, PV conversion implies a high-technology production of electricity from solar energy.

PV systems consist of modules made of solar cells. PV systems are modular so their strength can be designed for virtually any application. Additional parts to increase the output power are easily adapted to existing PV systems, which is not the case with conventional energy sources such as power plants and nuclear power plants whose economic viability and feasibility require multi megawatt installation. PV modules contain a number of serial- or parallel-connected PV cells used to obtain the desired voltage or current. Solar cells are laminated between two protective layers. On one side, there is a specially tempered glass with low iron content and on the other Tedlar protective plastic material or another layer of glass (Figure 7.1). In a typical solar module, solar cells are integrated and laminated with the laminating plastic (EVA). Such a laminated PV module is protected from adverse environmental impacts to extend service life. The typical

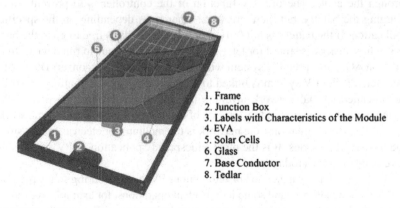

1. Frame
2. Junction Box
3. Labels with Characteristics of the Module
4. EVA
5. Solar Cells
6. Glass
7. Base Conductor
8. Tedlar

FIGURE 7.1   The structure and layout of PV modules

manufacturer's warranty on a PV module is 25 years. PV panels include one or more modules that can be used individually or in groups to form a modular system, together with the supporting structures and other necessary components. Systems can be fixed in a certain position toward the sun or mechanically adjusted continuously toward the sun (adjustable supporting systems).

PV latest generation of solar modules consist of so-called thin-layer PV cells and panels in which the thickness of the PV material or layer is about 2 μm (micrometers or microns), almost 100 times less than the traditional PV cells made of crystalline silicon. This reduction in the quantity of used semiconductor material reduces the cost per unit area, the price of the generated power (expressed in W), and the cost per installed kWh and increases profitability. However, the total price (balance of costs) of a PV system is still somewhat higher for thin-layer PV modules because of the additional costs of supporting structures required to install these modules, and to a lesser extent because of the costs of inverters and connecting to the distribution network. Also, operation and maintenance is more expensive because of the greater surface area they cover in relation to the classical PV modules, so these additional costs must be compensated by the benefits arising from savings of materials. In the past several years the price, related to larger module surface, has been constantly falling due to substructure optimization, higher module efficiency, and better quality of electrical connections. During 2005-2007 the total cost of thin-layer PV modules decreased by 60%, and this trend is expected to continue in the future.

The standard components of PV systems are:
PV modules;
Controllers and battery charge regulators;
Accumulators or batteries;
Cables and mounting systems; and
AC/DC power converters – inverters (autonomous and network).

Direct current produced in solar cells or modules leads to the controller through the cable. The primary function of the controller is to prevent overcharging the battery, but there are some other roles depending on the specific application. If the battery is not fully charged, electricity is free to go to the battery, where energy is stored for later use. If the system needs to run devices that work on AC, part of the PV system will also be inverters that convert DC to AC electricity. If the PV system is linked to an electric power distribution network, the so-called special network inverters that allow synchronization of the PV system with the network and restoration of electric power back into the network are used. In these situations, the network is the medium for electric energy storage instead of batteries. It is the most widespread application of PV systems in developed countries today.

The excess energy generated in autonomous PV systems during sunny periods is collected in batteries, and some independent operations, for example, for direct pumping of water or running other engines, do not require the use of batteries.

The water is pumped when the sun is shining and is directly stored in a reservoir located at a higher level for later pumping through the effects of gravity. Other PV systems convert DC to AC power and inject excess electricity into the electrical distribution network, while taking energy from the grid during the night, when there is no sunlight. This is an example of the operation of PV systems connected with a distribution system network. The three typical configurations of PV systems are the autonomous system, the system connected to the distribution network, and the hybrid system. The autonomous and hybrid systems can be used independently, so they are not connected to the electrical distribution network and are often used in physically remote areas. PV systems connected with electricity distribution networks are one of the ways to create decentralization of the electrical network. Electrical energy is generated by these systems closer to the locations where there is demand, not only by power plants, nuclear power plants, or large hydropower plants. Over time, these systems will reduce the need for increased capacity of transmission and distribution lines.

High Temperature Solar Collectors

In solar thermal power plants, solar radiation through lenses and mirrors is concentrated to achieve higher temperatures. This technique is called a concentrated solar energy. Depending on the temperature of the medium in the solar power plants, appropriate engines are used. Steam turbines are used up to 600°C temperature of the working medium, while gas turbines are used for achieving higher temperatures. Very high temperatures condition the use of different materials and techniques. For temperatures over 1100°C, liquid fluoride salts should be used as the working medium and multistage turbine system, which can be achieved, and the degree of efficiency to 60%. High operating temperatures enable a large savings of water, which is particularly important in deserts where the construction of such plants is expected.

As the storage of thermal energy is cheaper than electricity, solar thermal power plants are usually performed with thermal tanks. This enables the production of electricity at night. Solar power plants can be a reliable source of electricity in locations with good solar radiation. Reliability can provide additional power to fossil fuels (see Figure 7.2).

Advantages of Solar Energy Exploitation

The advantages of using solar energy are numerous. First, the source of energy is practically inexhaustible. The sun always shines on certain land, and at any time half the planet isn't illuminated by sunlight. Sunlight is free and as such, accessible to all. In addition, regardless of the brightness disparity, there is the possibility of a relatively accurate prediction of the amount of radiation, and thus of the amount of energy that can be obtained in this way. The great advantage of using solar energy is that the energy obtained in this manner can be stored in batteries and used when needed.

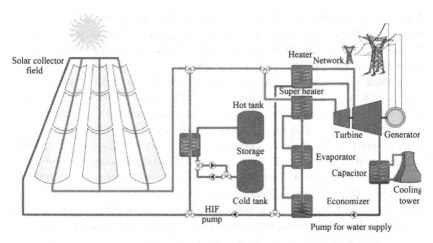

FIGURE 7.2 Power plants with a parabolic reflectors channel

Disadvantages of Solar Energy Exploitation

The main disadvantage of the production and use of solar energy is its price. Namely, the production, installation, and maintenance of the equipment needed to convert solar energy into electricity is relatively expensive. The production of energy from the sun is a few times more expensive than the production from other alternative sources. In addition, the energy obtained in this way is not suitable for transmission over long distances, nor is it enough to run high power devices.

Application of PV Systems

PV systems are very diverse; they can be smaller than a coin, larger than a football field, and can provide power for any device, from clocks to entire settlements. With the ease of handling, these factors make them particularly attractive for a wide range of applications.

The recent increase in production of low-cost PV cells has opened many new markets with a large number of different applications. Applications such as lighting, telecommunications, cooling, pumping water, and providing electricity for a whole settlement, especially in remote areas, have proven to be competitive and profitable in comparison to existing technology. Additionally a relatively new application of these systems with very high potential – PV façade systems (BIPV – building integrated PVs) – has appeared.

Facade PV Systems

The more current aspect of electricity generation is the conservation of the environment in which energy is generated and consumed. Solar electricity can contribute to available energy and at the same time help in preventing global climate change.

Approximately 75% of energy used in the developed world is spent in cities, of which about 40% is consumed in buildings. PV systems can be embedded in almost any building or structure, from bus shelters to large office buildings and even gardens, parks, and so on. Although the exact forecast of PV performance in buildings requires a careful analysis of various factors such as quantity of solar radiation coming to the surface of the building, stability and quality of electrical installation, electrical and distribution networks, etc., it is easy to see that this technology has great potential. Even in climatic conditions characterized by moderate sun radiation, the roof of a building of a household is large enough to set a PV system that can provide enough electricity during the year. PV modules and generators are traditionally placed on special supporting structures, but they can also be placed on buildings or they can become integral parts of buildings. The use of PV systems can significantly reduce the consumption of electrical energy from power plants. The buildings can even be converted into small manufacturers and distributors of electrical energy that can be of general benefit.

The problem of architectural integration of PV technology requires an interdisciplinary approach. This requires not only collaboration and the presence of highly specialized experts on the project team, but also consideration of sensitive issues related to their social, economic, and energy aspects. For example, the facade of the building not only has to protect against precipitation and regulate heat losses, but it must also regulate the flow of sunlight, provide sound insulation, provide ease of maintenance, and must also meet architectural and aesthetic criteria.

As facade PV modules can replace traditional building materials, the difference in cost between the solar elements per unit area and materials that can be replaced is of particular importance. As the cost per unit area of a facade PV system, connected to the distribution network, is almost the same as the price of the highest quality materials, such as marble or ornamental stone, the additional benefits of PV are practically free.

Potential of PV Devices

Solar PV technology was previously used mainly in space programs or in remote locations, and thus was marginal and exotic at first. However, in the last ten years it has become the main technology for the production and distribution of electric power in urban areas with the potential to become equally cost competitive with the prices of energy obtained and distributed by conventional technologies. Since 1990, the PV conversion industry has shown consistent annual economic growth of over 20%, beginning in 1997. In 2000, the total installed capacity worldwide exceeded 1000 MW, and from then until today, the trend growth has exceeded 40% annually. In developing countries more than a million households use electricity produced by PV systems.

An increasing number of companies and organizations are actively involved in the promotion, development, and manufacture of PV systems. Companies that produce and distribute electric power in cooperation with manufacturers of solar equipment, urban organizations, and supporting national and international funds are planning and implementing increasingly large projects. The market value of the PV industry currently amounts to over $5 billion a year, and it is expected to increase to over $10 billion per year by 2010. Representatives of the PV solar industry are some of the leading global companies, such as Sharp, Mitsubishi, Sanyo, BP, Shell, Kyocera, etc. At the end of 2009 the total installed PV capacity worldwide amounted to 15 GW, which exceeds predictions made at the beginning of the twenty-first century by almost 50%.

Ten years ago, it was expected that the two most promising applications of PV systems would be in the sector of large power plants of several megawatts, connected to the distribution network, and in ten million solar home systems in developing countries. However, the situation is somewhat different today and the market is dominated by urban (residential) PV systems connected to the electricity network. In the period from 2000 to 2005 there was a relative increase of 50% per year in the number of networked solar PV power plants. Moreover, studies of the European Photovoltaic Industry Association (EPIA) and the Greenpeace organization predict that half the capacity of 207 GW in 2020 will be systems that are connected to the electrical distribution network, out of which 80% will be installed in residential buildings. The PV industry is increasingly present in national energy strategies of more and more countries.

Development of PV Technology

In addition to 27 national research and development programs, the European Union (EU) has also been funding research (The Directorate-General for Research and Innovation-DG RTD) and development projects (The Directorate-General for Transport and Energy-DG TREN) within the FP (Framework program) since 1980. Funds for these projects are an important incentive for the European agenda on PV technologies. A large number of research groups, ranging from small research groups at universities to the teams of major research centers, are included in the program, which includes research related to PV, from semiconductor materials to industrial optimization processes. During the FP 6 program (Framework program for period 2002–2006), a platform for PV technology was established, which aimed to mobilize all researchers who agree to the long-term European research programs related to PV technology. The platform has developed the European strategic agenda for PV research for the next ten years in order to retain the leadership role of Europe in the PV industry. A special role in the new FP 7 program (Framework program for period 2007–2013), which began in 2007 and lasts until 2013, is given to basic research related to PV technologies.

It is anticipated that, with technological development, the price of PV electricity that is connected to the distribution network should be between 0.10 and 0.25 €/kWh, compared to the current price of between 0.25 and 0.65 €/kWh, depending on solar radiation and local conditions on the market. It is expected that the results of research and development should enable reduced consumption of materials, higher efficiency of solar devices, and improvement of production processes based on compliance with environmental standards and cycles.

7.2 BIOMASS ENERGY

Energy plays a crucial role in modern life. It is needed for heating, lighting, and cooking in households and for virtually every industrial, commercial, and transport activity. At the global level, consumption of energy grew steadily by around 2% a year in the decade 1990-2000 and probably more in 2000-2020. Fossil fuels (coal, gas, and oil) currently account for about 79% of world energy consumption, nuclear energy for 7%, and renewable energy sources for 14%. Global consumption of fossil fuels grew in line with overall energy consumption during the 1990s and is expected to grow faster than overall consumption in the period up to 2020. Fossil fuels have two main disadvantages. First, when they are burned they emit pollutants, including the greenhouse gases that are causing climate change. Second, countries without adequate reserves of fossil fuels are facing increasing risks to the security of their energy supplies. Using renewable energy sources in place of fossil fuels reduces emission of greenhouse gases and other pollutants, improves security of supply by boosting diversification of energy production, and encourages the creation of new jobs and businesses. In the 25 countries of the EU, renewable energy sources provided about 6% of total energy requirements in 2002. Currently, nearly two-thirds of all energy from renewable sources used in Europe comes from biomass, and this source is set to play a significant role in meeting the 2010 target [15].

What is Biomass?

Biomass covers a wide range of products, byproducts, and waste streams from forestry and agriculture (including animal husbandry) as well as municipal and industrial waste streams. A definition adopted by EU legislation is: "...the biodegradable fraction of products, waste and residues from agriculture (including vegetal and animal substances), forestry and related industries" Biomass thus includes trees, arable crops, algae, other plants, agricultural and forest residues, effluents, sewage sludge, anures, industrial byproducts, and the organic fraction of municipal solid waste.

Photosynthesis and the Carbon Cycle

All life on earth is based on green plants that convert carbon dioxide (CO_2) and water from the atmosphere into organic matter and oxygen, using the energy of the sunlight [17]. This process is called photosynthesis. Expressed simply, the overall chemical reaction is as follows (in this example glucose is the product):

$$6CO_2 + 6H_2O = C_6H_{12}O_6 + 6O_2$$

In plants, the energy from the sun is conserved in the form of chemical bonds. This chemically stored energy can be used by the plants themselves or by animals and human beings.

When biomass is burned or digested, the organic carbon is recycled in a complex global process known as the carbon cycle. In this process, the CO_2 that was absorbed as the plants grew is simply returned to the atmosphere when the biomass is burned. Therefore, if the cycle of growth and harvest is maintained, there is, broadly speaking, no net release of CO_2. This is why biomass is regarded as an energy source that does not emit CO_2 into the atmosphere when burned (see Figure 7.3).

Fossil fuels, of course, are also organic matter. However, in their case the matter has been transformed and stored over a long period of time under heat and pressure in the absence of oxygen. When we burn fossil fuels we release

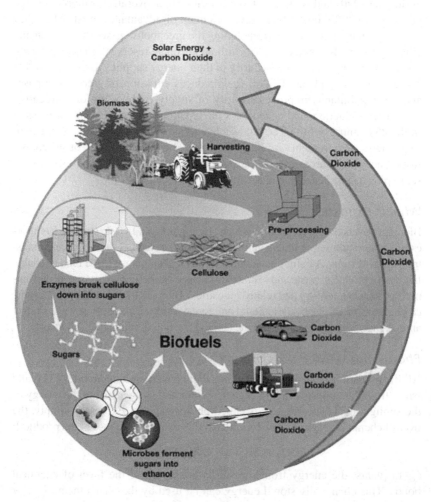

FIGURE 7.3   Life Cycle Analysis of CO_2

in a short period a quantity of CO_2 that has been locked up in plants and their follow-up products over millions of years.

Biomass has many advantages as an energy resource. It can be used to produce a wide variety of product types – heat, electricity, solid fuels, liquid transport fuels, gaseous fuels, and other products. Biomass raw materials come in a range of forms that are abundant in most parts of the world, including Europe. Their use does not increase the CO_2 content of the atmosphere. As a result of work done over the last decade, some excellent modern processes now exist for converting raw materials to usable products such as the biodiesel and bioethanol blends that can be used in today's vehicles without engine adaptation. In addition, some high-technology equipment has been developed especially for biomass product use – for example, fully automated boilers suitable for burning wood pellets. A research framework is in place that will allow further technological progress to be made. Some legislative measures to support the use of bioenergy, i.e., energy derived from biomass, have already been introduced. Other policy steps needed to allow widespread use of bioenergy products are being developed.

Of course, not all the existing biomass resource can be used for energy purposes. Food, timber, paper, and board and certain high-value chemicals are also derived from biomass. Therefore, bioenergy production must be integrated with other priority applications. Biomass must be used in a wise and sustainable way.

The overall aim is to show where and how bioenergy products can be used right now, the high technology nature of today's processes and equipment, the role of research in developing these, the benefits of using biomass for energy purposes, and what needs to be done from the policy and research viewpoint to ensure full development of this sector in the future. Following this introduction the section is arranged in three main subsections covering biomass resources, conversion processes, and products. These focus on the European situation but, where appropriate, also refer to activities in other parts of the world.

Biomass Resources

Biomass comes in various forms, each of which has specific properties, uses, and advantages. The main sources are wood from conventional and short-rotation forestry, other energy crops, residues from forestry and agricultural production, and byproducts and wastes from industrial and municipal processes. The energy that can be obtained from a particular resource depends on its chemical composition and moisture content (see Table 7.1).

Availability and Potential

Total EU land area is around 385 million acres. Forests and woodlands cover 137 million acres and crops 178.5 million acres. Once the requirements of the food, wood products, and paper sectors have been met, the biomass resources from these trees and crops could provide around 8 EJ (1 exajoule = 1.0E+18 joules) energy a year – about 11% total annual EU energy consumption. In practice, we are exploiting less than a quarter of available resources.

TABLE 7.1 Examples of biomass resources

Category Examples	Examples
Dedicated plantations	Short-rotation forestry (eucalyptus, willow) Perennial crops (miscanthus) Arable crops (rapeseed, sugarcane, sugarbeet)
Residues	Wood from forestry thinning Wood felling residues Straw from cereals Other residues from food and industrial crops (sugarcane, tea, coffee, rubber trees, oil, and coconut palms)
Byproducts and wastes	Sawmill waste Manure Sewage sludge Organic fraction of municipal waste Used vegetable oils and fats

If biomass is to play its expected role in achieving the EU target for renewable energy in future much more will need to be done both to exploit the existing bioenergy resources and to establish new ones. Integrated production of wood and biomass could be introduced where the trees are thinned to maximize the value of the wood produced and the thinnings used for biomass. Increased areas of land could be used for cultivating energy crops, i.e., short rotation forestry or non-wood crops. Finding the land for growing energy crops is an important issue. Various woody and non-woody plant species are available that are suitable for different qualities of land. In recent years, surplus food production in the EU has led to cropland being left fallow. In 2001, for instance, as much as 15% (i.e., 5.7 million acres) of the EU-15 (The EU15 comprised the following 15 countries: Austria, Belgium, Denmark, Finland, France, Germany, Greece, Ireland, Italy, Luxembourg, Netherlands, Portugal, Spain, Sweden, United Kingdom) cropland was under voluntary set aside. Other options may exist, such as growing non-food crops. Recent reforms of the EU's agricultural policy encourage production of energy crops. An extra subvention of 45 Euros per acre can now be given for energy crops for a guaranteed area of 1.5 million acre. In addition, fallow land can be used for energy crops (see Table 7.2).

Table 7.3 gives an idea of the crop and energy yields that can be expected from different types of woody and non-woody energy crops. Ranges of yield data are given because actual yields depend on land type and climate. From the information in the table it can be seen that to achieve the 1.2 EJ a year solid biomass. This is little more than the 5.7 million acres set aside in the EU in 2001.

TABLE 7.2 Heating values of various types of biomass

Biomass type	Higher heating value in GJ/tonne
Dry lignocellulosic	18
Wet cellulosic	9
Oils and fats	36
Ethanol	26

TABLE 7.3 Crop and energy yields from some energy crops

Crop	Yield (dry tonnes/ hectare/year)	Energy yield (GJ/ acre year)
Woody crops		
Wood	1-4	30-80
Tropical plantations (with no fertilizer and irrigation)	2-10	30-180
Tropical plantations (with fertilizer and irrigation)	20-30	340-550
Short rotation forestry (willow, poplar)	10-15	180-260
Non-woody crops		
Miscanthus/switchgrass	10-15	180-260
Sugarcane	15-20	400-500
Sugarbeet	10-21	30-200
Rapeseed	4-10	50-170

In that year, only 0.9 million acres of the set aside were dedicated to non-food crops. It should not be difficult to increase that figure in the future.

Wood

Wood is a renewable raw material that can be used not only for making timber-based products, pulp, and paper but also as a source of energy. In traditional professionally managed forests, the normal life cycle of the tree includes plantation, a stage of rapid height growth followed by a stage of steady growth in diameter, height, and volume. The point of harvesting depends on the species, but it is generally reached after 30-80 years. Twenty-five to forty-five percent of the wood harvested each year is in the form of residues, i.e., the wood from forestry thinning and the residues from felling. Therefore, most of the woody biomass resources available from forests and woodlands for energy production consists of residues. In the EU, woody residues are estimated to have the potential to provide 3.8 EJ energy annually.

For certain species, the use of short-rotation techniques can reduce the life cycle of the tree to 3-15 years. Thus plantations dedicated to short-rotation forestry can provide an important additional source of woody biomass for energy purposes. The species most commonly used in this way include poplar, salix, willow, and eucalyptus.

The key to creating an economic energy-from-woody-biomass scheme is to establish effective logistical systems for harvesting, recovering, compacting, transporting, upgrading, and storing the wood. Harvesting and transport, in particular, can have significant impacts on energy balance and costs. The trend is to move towards greater mechanization of harvesting for reasons of economy and safety. Because firewood and forest residues are low-value commodities, transport costs constitute the most important part of total production costs. Care must be taken, therefore, to choose an appropriate method of transport and locate the conversion plant as near as possible to the woody biomass source [23][24][26].

Non-woody Energy Crops

Non-woody plants most often used for energy purposes are wheat, barley, rye, sugarcane, sugar beet, leguminous plants (e.g., alfalfa or lucerne), grass (e.g., miscanthus and switch grass), and oil crops (e.g., rapeseed). Many other species have been studied with respect to agronomy, yield optimisation, harvesting, storage, and processing. These include high-yield fibrous plants such as giant reed (arundo donax) and a form of globe artichoke (cynara).

These plants can provide biomass suitable for direct combustion or thermochemical or biological conversion. Wheat, barley, rye, sugarcane, and sugar beet, for instance, are generally converted to ethanol. Leguminous plants and grasses can be processed together with manure or waste to obtain biogas for heat, electricity, or fuel. Oil crops can be used to produce biodiesel.

There are also plants that can be processed to give liquid biofuels and cellulosic materials simultaneously. An example is sweet sorghum, which yields both bioethanol and dry cellulosic material for other bioenergy use. Some of the plants listed above are perennials. Others are annuals. All can fit into conventional agricultural practice. Cultivating crops for energy use does not preclude a farmer from growing food crops as well – or vice versa. Food and energy crops can usefully be grown hand-in-hand to maximize the overall efficiency of the farm. The advantages of energy crops over food crops are that they do not require the best land and need significantly less care, water, and fertilizers. This is because it is the quantity, rather than the quality, of the product that is important. As discussed earlier, it is estimated that it should be possible for some 2 EJ energy a year to be produced in the EU from new woody and non-woody energy crops by the year 2010.

Agricultural Residues and Byproducts

As indicated above, the residues from forestry thinning and felling constitute a major biomass resource. To this must be added the different types of residue and byproducts that come from processing wood, e.g., sawdust, bark chippings, wood shavings, plywood residues, and black liquor. (The latter, a byproduct of paper manufacturing, can be burned or gasified to produce fuels for transport.)

Residues from the harvesting and processing of food and other crops – cereals, sugarcane, tea, coffee, rice, cotton, rubber trees, and coconut palms – are also important. It should be noted, however, that only about 20% of total straw can be used for energy purposes, to ensure that the demands of the agricultural sector and other markets can be fulfilled. Animal manure is another useful resource from the farming sector.

Together, these resources have the potential to provide the EU with 6.75 EJ energy a year – 56% from wood residues and byproducts, 8% from straw, 25% from other crop residues, and the rest from animal manure. It should be practicable to mobilize 1.5 EJ of the 6.75 EJ potential annually by 2010.

Marine Biomass

Plankton, algae, and other marine-based organisms constitute a biomass resource that has not yet been exploited. This area is, however, the subject of continued research. Bearing in mind the volume of the sea, this resource could provide a major carbon-neutral source of energy for the future.

Wastes

Municipal solid waste (MSW) is primarily waste produced by domestic households, although it also includes some commercial and industrial wastes that are similar in nature. Each EU citizen produces on average of more than 500 kg of MSW a year. The total MSW resource in the EU is therefore of the order of 225 million tons a year.

The organic fraction of MSW has a significant heat value. Typically, MSW has a heat value of 8-12 MJ/kg, about a third of the heat value of coal. A ton of MSW will give about 2 GJ of electricity.

Decisions on whether to use MSW as an energy resource are linked to local and national waste management policies and the views of the public on, say, recycling and incineration.

The choice of MSW treatment for a particular locality must take into account, among other things, the composition and properties of the input waste, the available technologies, and the market for the various outputs. The whole process must be integrated to avoid conflicting claims on the different streams (see Figure 7.4).

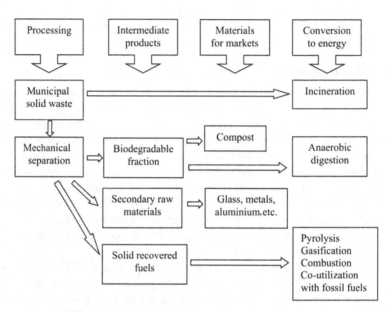

FIGURE 7.4   Pathways for MSW treatment and conversion

In general, for wastes with a high energy content, the major part will be used for power and heat production. Incineration offers one route for doing this. In 2002 there were approximately 340 incineration plants in the EU handling between them 50 million tonnes of waste a year. Recent installations tend to be efficient in generating power or producing combined heat and power. Another option is to convert the waste to solid, liquid, or gaseous fuels that can be more easily transported and used to produce heat or power or to drive vehicles. The biodegradable part of MSW can be used to produce compost or digested along with other suitable wastes to give biogas. Biogas is another useful source of energy. It can be recovered from landfill sites or produced by digesting not only parts of MSW but also sludge from sewage treatment, livestock manure, and suitable agricultural and agro-industrial effluents. It has been estimated that the total energy content of all the above resources capable of producing biogas in the EU exceeds 3.35 EJ. It should be possible to mobilize 0.63 EJ by 2010 and 0.75 EJ a year by 2020. Actual production in the EU-25 region (Austria, Belgium, Bulgaria, Cyprus, Czech Republic, Denmark, Estonia, Finland, France, Germany, Greece, Hungary, Ireland, Italy, Lithuania, Luxembourg, Netherlands, Poland, Portugal, Romania, Slovakia, Slovenia, Spain, Sweden, United Kingdom) in 2004 was 0.17 EJ – 24% higher than in 2003. Biogas is largely methane – one of the greenhouse gases. More widespread exploitation of biogas would also be in line with European environmental policy since it would reduce emission of methane, a powerful greenhouse gas, to the atmosphere.

Biomass Conversion

Except for cases where straightforward combustion is appropriate, it is not usually possible to use biomass raw materials as they stand. They have to be converted in some way to solid, liquid, or gaseous fuels that can be used to provide heat, generate electricity, or drive vehicles. This conversion is generally achieved by some type of mechanical, thermal, or biological process.

Mechanical processes are not strictly conversion processes since they do not change the nature of the biomass. They are commonly used in the treatment of woody biomass and waste. The sorting and compaction of waste, the processing of woody residues into bundles, pellets, and chips, the cutting of straw and hay into pieces, and the squeezing of oil out of plants in a press are all examples of mechanical processes. Such processes are often used to pre-treat a biomass resource for further conversion. They are therefore discussed, where relevant, later in this section in conjunction with other methods of conversion, or when describing the final products.

Combustion, gasification, and pyrolysis are examples of thermal processes. They produce either direct heat or a gas or oil, such as bio-oil. The gas can be used to drive a motor or a fuel cell. The bio-oil can be further transformed into gaseous and liquid fuels.

Fermentation and digestion are examples of biological processes. These use microbial or enzymatic activity to convert sugar into ethanol, or biomass to solid fuels or biogas. The following sections highlight some major thermal and biological conversion processes (see Figure 7.5) [49–53].

Thermal Processes

Processes where biomass conversion is achieved by heat are the most commonly used technologies.

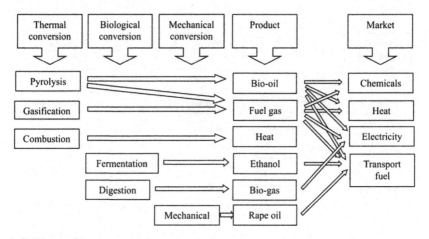

FIGURE 7.5 Biomass conversion pathways

Combustion is the most ancient and frequently applied way of using biomass as an energy source because of its low cost, ease of handling, and high reliability. The biomass can either be fired directly (as when firewood is burned for heating or waste is incinerated) or cofired with fossil fuels. Modern coal-fired plants are increasingly being designed for cofiring in order to reduce carbon dioxide emissions.

Factors to be considered when designing a biomass combustion system include the characteristics of the fuel to be used, environmental legislation, the cost and performance of available equipment, and the output required. During combustion, a biomass particle first loses its moisture at temperatures up to 100°C using heat given off by other particles. Then, as the dried particle heats up, volatile gases containing hydrocarbons, carbon monoxide (CO), methane (CH_4), and other gaseous components are released. In the combustion process, these gases contribute about 70% of the heating value of biomass. Finally, the char oxidizes, leaving ash. The technical aspects of heating and large-scale combustion are discussed in more detail in the section on heat and power production from biomass, as follows (see Figure 7.6).

The basic thermochemical processes are gasification and pyrolysis. Both processes involve heating the feedstock in the presence of less oxygen than is required for complete combustion and produce a mixture of gas, liquid, and char. Yields of the various outputs depend on the nature of the biomass used, the rate of heating, the highest temperature reached, the way in which off-gases react with hot solids, the amount of water (as steam), and the presence or absence of other substances that may act as catalysts. At one extreme, processes can be optimized to produce charcoal. At the other, they can be designed to produce a mixture of hydrogen and carbon monoxide (synthesis gas) suitable for use in the catalytic formation of a variety of liquid fuels.

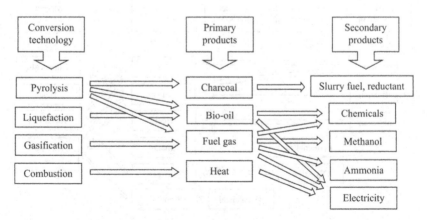

FIGURE 7.6 Thermochemical conversion products and uses

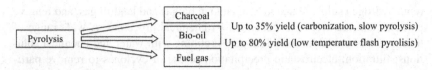

FIGURE 7.7   Pyrolysis products

In pyrolysis, an external source of heat is used with no oxygen. Heat causes the biomass molecules to break down to form water (steam) and highly-reactive low-molecular weight fragments, many of which combine to form char. Hot char will react with steam to produce carbon monoxide and hydrogen, giving a gas with a high heating value. Some of the low-molecular weight compounds may be swept from the reactor, recombining to form tars as they cool. Fine particles of ash and partly carbonized biomass may also be carried in the gas stream. For these reasons, the gas requires advanced cleaning before it can be used in a combustion engine or turbine (see Figure 7.7).

Fast pyrolysis is a high-temperature process in which small particles of biomass are burned rapidly in the absence of oxygen causing them to decompose to give vapors, aerosols, and some char. After cooling and condensation, a dark viscous liquid (bio oil) is obtained. It has a heating value about half that of conventional fuel oil and can substitute for the latter in combustion systems or engines for heat or power generation. Further processing of the oil by hydrogenation or using a catalyst will give a higher-quality product close in specification to petroleum-derived fuels oils. This product can be used in diesel-fired vehicle engines.

Liquefaction is a low-temperature, high-pressure thermochemical conversion process carried out in the liquid phase which has the potential to produce high-quality products. It requires the use of either a catalyst or hydrogen under high pressure. Gasification is a high-temperature thermochemical process carried out under conditions that lead to a combustible gas, rather than heat or a liquid. Modern gasifiers can use a large variety of feedstocks. The process involves partial combustion of the biomass feedstock with a restricted supply of air or oxygen at temperatures in the range 1200 to 1400°C. Ideally, pure oxygen would be used because this would lead to a product of high heating value consisting mainly of carbon monoxide, hydrogen, methane, and carbon dioxide. However, most biomass gasifiers use air for costs reasons so the output is diluted with nitrogen and therefore has a lower heating value. Whichever process is used the product can, after appropriate treatment, be burned directly or used in gas turbines or engines to produce electricity or mechanical work. The process can be varied to give a hydrogen-rich gas or synthesis gas that can be used to make other fuel products.

Gas Cleaning

This is a critical step in both combustion and gasification systems. The aim is to reduce emissions in flue gases, reduce the level of damaging contaminants (e.g., hydrogen sulphide and mercaptans) in biogas and landfill gas, and remove particles and tars from gas generated by chemical processes. A wide range of techniques is available, including gas scrubbing with water and chemical solutions; filtration; electrostatic precipitation or use of cyclones to remove particles; use of molecular sieves or chilling to remove water and other impurities; and use of iron, calcium, or zinc oxide or chemical reduction to remove sulphur compounds. In particular, tars may be cracked by passing the gas stream back through the gasifier bed, or through a second stage, with external heating. Scrubbing with water or various proprietary liquids may also be used to reduce levels of carbon dioxide. Most of these processes are commercially viable when used on a large scale, as they are in the petrochemical industry. They may not, however, be economically feasible when applied to small biomass-based facilities (see Figures 7.8 and 7.9).

Non-Thermal Processes

A number of processes are available that use some type of biochemical process, rather than heat, to achieve biomass conversion.

In anaerobic digestion, a process which takes place in the absence of oxygen, a mixed population of bacteria catalyzes the breakdown of the polymers found in biomass to give biogas (MHV gas). This primarily consists of methane and carbon dioxide but may also contain ammonia, hydrogen sulphide, and mercaptans, which are corrosive, poisonous, and odorous. The process takes place in several stages. First, polymers such as cellulose, starch, proteins, and lipids are hydrolyzed into sugars, amino acids, fatty acids, etc. These are then converted to a mixture of hydrogen gas, low molecular weight acids (primarily

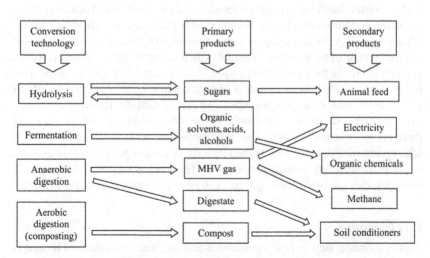

FIGURE 7.8 Biochemical conversion product and uses

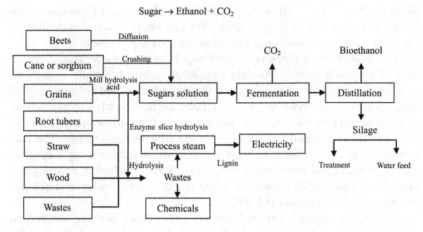

FIGURE 7.9   Fermentation

acetic acid), and carbon dioxide, in the process of acetogenesis. Finally, these products react together to generate methane, in methanogenesis. Millions of small anaerobic digesters have been built in rural areas worldwide with methane generation as their primary aim.

Fermentation is one of the oldest biological processes used by mankind. It normally uses yeast (an organism that secretes catalytic enzymes) to initiate chemical reactions that lead to the desired outputs – ethanol and carbon dioxide. Ethanol is used not only for alcoholic drinks but also as a solvent, additive, and fuel. Bioethanol as a transport fuel is discussed in more detail later in this chapter in the section on liquid biofuels. The main producers are Brazil (which uses sugarcane as feedstock) and the United States (which uses corn). Some fuel alcohol is produced in Europe using wheat, molasses, and petrochemical feedstocks.

Breakdown of the biomass into useful products can be speeded up by the use of enzymes other than yeast. These enzymatic conversion processes are mainly used for degradation of starch, cellulose, proteins, or lipids feedstocks.

The Biorefinery

A biorefinery is defined as a facility for achieving large-scale integrated production of fuels, power, and chemicals from biomass. It is analogous to a petroleum refinery which produces multiple fuels and products from petroleum. The possibility of integrated production of a number of products from biomass is a concept that is gaining increased attention in many parts of the world. In the EU, the developments towards a carbon-constrained economy and evolving agricultural policy make the idea particularly interesting. No large-scale biorefineries exist at present, but they are regarded as being an important element in the future of biomass.

Work has been carried out in the EU, the United States, and elsewhere on the design and feasibility of such facilities. The concept is explored in the following. A biorefinery is based on a number of conversion processes.

The biorefinery requires year-round supply of biomass feedstock, preferably of a specific quality. Possible sources of feedstock include agricultural crops and residues, wood residues, and woody and non-woody energy crops. If supplies are of heterogeneous type and quality, the feedstock has to be mechanically separated into fractions and, where necessary, pre-treated to give interim products of the required specification for the different applications. The final conversion to energy, fuels, or other products is carried out using a range of thermochemical and biochemical processes – some of which are already at a stage of commercial development, while others require further research and technological development [84–87].

A range of products is delivered with multiple end uses, including low-volume and high-value speciality chemicals that have niche uses in the food and other industries; high-volume and low-value liquid fuels for widespread use in the transport industry; heat; electricity, etc. The diversity of the products gives a high degree of flexibility to changing market demands and allows plant operators many options for gaining revenues and achieving profitability.

In addition, there are economies of scale. Advanced conversion of biomass (gasification, pyrolysis, etc.) has proved costly to date. The large-scale operations conducted in the biorefinery offer cost savings by, for example, allowing preconversion feedstock treatment facilities to be shared. A large-scale operation has greater buying power when acquiring feedstocks: it can source biomass over a larger geographical area and negotiate cheap long-term contracts. In the immediate future, biorefinery products will not, as a rule, compete in costs terms with products made from fossil fuels. Rather, their main competitive advantage comes from environmental sustainability. In particular, biorefinery products are near-neutral in terms of greenhouse gas emissions. Assessing the financial advantage this will bring in ten or more years' time is very difficult [13,14].

For biorefineries to succeed, different sectors of the economy – agriculture, forestry, agro- and wood-based industries, chemical, food, transport, and energy industries – will need to cooperate to develop processes for making new biomass-based products and bring them to market. There is a need for extensive research and technological development to test and prove the supply of biomass feedstocks, the extensive range of biorefining technologies, the end uses of the products, etc. Research institutes and universities are therefore also important stakeholders. Policymakers, regulators, and lawmakers will also play an important enabling role in establishing the biorefinery concept. As discussed later in this chapter, today's hydrocarbon-based economy could well evolve into a bio-based economy where biorefineries play a very important role (see Figure 7.10).

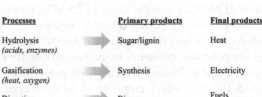

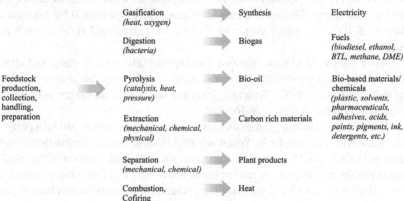

FIGURE 7.10 The biorefinery

Products from Biomass

As explained above, biomass can be converted by a variety of processes to a wide range of products – including heat, electricity, and solid, liquid, and gaseous biofuels. The following sections give more information on these products and the ways in which they can be made, supplied, and used, and the current status of their application in the EU. Solid and gaseous biofuels are discussed first because their main applications are as intermediates in the preparation of other products, i.e., heat, electricity, and liquid biofuels. At the end of this chapter a short discussion on biomass products that are not energy-related is given [16, 20].

Solid Biofuels

Solid biofuels can be derived from many biomass resources such as wood and wood residues and byproducts, agricultural crops and residues, and the combustible part of solid wastes.

The quality of solid biomass as a fuel is related to properties such as moisture content, heating value, ash content, and content of volatile matter. The higher the moisture content of a fuel, the lower its heating value. This is partly because fuels with a high moisture content have, by definition, a lower content of combustible material. It is also because some of the heat liberated during combustion is used up in evaporating the water. For a fuel to be capable of being ignited and having energy extracted from it, its moisture content must be below 55%. The moisture content of biomass sources ranges from less than 10% for straw to 70% for forest residues and wet waste. The moisture content of bark and sawmill products is in the range of 25-55%. By contrast, the moisture content of processed wood pellets is less than 10%.

Ash content is significant because it determines the behavior of the biomass at high temperatures: quantities of molten ash can, for instance, cause problems during combustion. The ash contents of biomass can range from 0.5% for wood, through 5-10% for energy crops, to 30-40% for agricultural residues such as husks.

When subjected to heat, biomass decomposes into volatile gases and char. The volatile matter content of a resource is described by the proportion that volatilizes at 400-500°C. Typically, biomass has a volatile matter content of over 80%, compared with 20% for coal.

The formation of nitrogen oxides (NOx) during combustion can be a particular issue with biomass fuels. When any fuel is subjected to combustion in the presence of air, some NOx will be formed as a result of the reaction of the nitrogen in the air with the fuel at temperatures above 950°C. In the case of biomass fuels, however, some NOx will also be formed at low temperatures because of the presence of nitrogen in the biomass itself. The quantity of NOx formed can generally be controlled by using appropriate combustion techniques.

Wood is the most commonly used solid biofuel. The raw material can be in the following forms: logs, stems, needles, and leaves from the forest; bark, sawdust, and redundant cuttings from the sawmill; chips and slabs from the wood industry; and recycled wood from demolition. These can be used directly as a fuel, where this is appropriate. Alternatively, they can be processed into forms that allow for easy transport, storage, and combustion, such as chips, pellets, briquettes, and powder. Firewood is forest fuel in the form of treated or untreated stem wood. A new technique that allows for easier handling is bundling, where branches are pressed together into log-like bundles of equal size. Wood powder consists of ground wood raw material. Wood chips are pieces of wood about1-5 cm in diameter. Pellets are short cylindrical or spherical pieces with a diameter less than 25 mm. Pellets are produced from sawdust, cutter shavings, chips, or bark by grinding the raw material to a fine powder that is pressed through a perforated matrix. The friction of the process provides enough heat to soften the lignin present. During the subsequent cooling, the lignin stiffens and binds the material together. Wood briquettes are rectangular or round pieces made by pressing together in a piston press finely ground sawdust, cutter shavings, chips, bark, etc.

The energy content of pellets and briquettes is around 17 GJ/tonne with a moisture content of 10% and a density of around 600-700 kg/m^3. To replace oil, one needs about three times its volume in pellets.

Using wood-derived biofuels instead of fossil fuels is helpful from the viewpoint of aiding sustainability, reducing greenhouse gas emissions, and improving quality in forestry. Since the major users of solid biofuels are companies concerned with heat and electricity production, better integration of the forestry industry with energy companies would undoubtedly improve the overall efficiency of the scheme. If, for instance, bulky materials with a high moisture content have to be transported far, the costs of using solid biofuels are increased considerably.

It is possible to use many agricultural crops and residues as solid fuels. Examples are straw, husks, stalks, bagasse from sugarcane, and grass. Using these residues as fuels can in addition solve the problem of how to dispose of them.

As described earlier, the initial sorting of municipal solid waste results in the recovery of combustible solids that can be used as fuels.

Gaseous Biofuels

The most commonly-used biofuels are biogas and hydrogen. Both are produced from biomass wastes by biochemical processes (see Figure 7.11).

Biogas

Chemically, biogas is comprised of a mixture of hydrocarbons (mainly methane) and other gases. It can be produced by anaerobic digestion of sewage

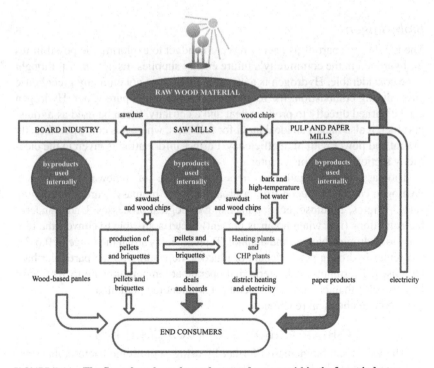

FIGURE 7.11 The flow of products, byproducts, and energy within the forest industry

sludge, grass, and other ley crops, manure, and agricultural and food wastes, including those from slaughterhouses, restaurants, grocery stores, and wastes from the pharmaceutical industry. It can also be extracted from landfills, where it is formed spontaneously and, if left, would cause environmental problems since methane is a powerful greenhouse gas.

Biogas production plants based on agricultural wastes can be found in the countryside in most EU member states. Their unique smell has given rise to the unfortunate and untrue notion that biomass is not a clean source of energy. The availability of landfill as a source of biogas will, of course, vary from member state to member state according to national policies on the use of landfill as a means of disposing of organic wastes. In some countries, landfill gas is recovered with fully industrial technologies. Frequently, biogas is used close to the place where it is produced. Its main applications are for production of heat, electricity, and combined heat and power. Further information on these are given later in this chapter. Biogas's main advantage over other biomass-derived fuels is that it can be burned directly in any gas-fired plant. It can also be injected into the natural gas network. In addition, biogas can be used as a transport fuel for vehicles adapted to run on gas. The environmental benefits of replacing petrol or diesel with biogas are considerable. However, although the cost of biogas is significantly lower than petrol per unit of energy, one has to equip vehicles with extra gas tanks to use it.

Biohydrogen

The EU devotes part of its energy research budget to exploring the possibilities for hydrogen in the community's future energy supplies. Its potential is thought to be considerable. Hydrogen is a clean fuel that does not emit any greenhouse gases during combustion: the remnant of combustion is pure water. Hydrogen can be burned directly to produce heat and electricity. It can be used as a transport fuel. It also makes an ideal input for fuel cells, where it is converted directly to heat and power with high efficiency. Further information is given in the electricity section later in this chapter.

Biomass is potentially an important source of renewable hydrogen. Hydrogen can be produced from a broad range of biomass sources containing carbohydrates, cellulose, or proteins using biological processes. Under anaerobic conditions (i.e., when no air is present) bacteria are able to convert the biomass to hydrogen, biogas, and ethanol. Typical yields are in the range of 0.6-3.3 molecules hydrogen per molecule of glucose, depending on the particular bacteria used. Thermophylic bacteria that operate at temperatures up to 70°C give higher yields than bacteria that operate at ambient temperatures.

A typical chemical reaction is:

$$C_2H_{12}O_6 + 2H_2O = 2CO_2 + 2CH_3COOH + 4H_2$$

The yields can be increased further by using phototropic bacteria that convert acetic acid to hydrogen, as follows:

$$CH_3COOH + 4H_2O = 2CO_2 + 4H_2$$

As a rule, biological processes require modest investments and are effective even on a small scale (see Figure 7.12).

Heat

In 2002, around 1.6 EJ heat were produced from biomass in the EU.

Wood-based solid biofuels are the biomass sources most commonly used in heat production. Combustion is the normal conversion method. Traditionally, of course, firewood was used for thousands of years to provide heat for domestic purposes, i.e., local heating and food preparation. Today, the availability of biomass-derived fuels in clean and convenient forms (e.g., chips, pellets, and briquettes) and of modern, automatically operating combustion equipment have created renewed interest in the use of solid biofuels for domestic heating. The use of liquid biofuels is also making headway for small-scale heating. For the commercial and industrial sectors, fixed-bed, fluidized-bed, and dust combustion equipment now available allow efficient production of heat from biofuels of different types and forms on a larger scale. Cofiring of biofuels with coal is also practicable. All these processes are discussed in more detail below.

Pre-Treatment

For domestic, commercial, and industrial combustion equipment, it is advisable to use solid biofuels that have undergone some form of pre-treatment and processing (e.g., washing, drying, size reduction, and compacting) to achieve greater uniformity and ease of handling and reduce the moisture content to an acceptable level.

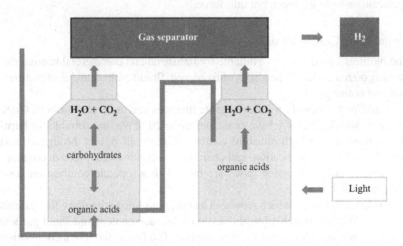

FIGURE 7.12 Non-thermal H_2 production from biomass

Small-Scale Heating

At the domestic level, appliances that burn wood and similar biofuels are popular because they not only provide heat, but also help create a pleasant atmosphere and decorative space. Commonly available appliances include fireplaces, heat storage stoves, pellet stoves, and burners, and central heating furnaces and boilers for wood logs and wood chips.

There is also a range of automatically operating appliances for wood chips and pellets on the market. The excellent handling properties of pellets make them a good fuel for domestic and other small-scale use. Good automatic pellet-fueled boilers with low emissions and high efficiencies have been on the market for about ten years now. Existing oil-fired boilers can be adapted for pellet use by replacing the existing burner by one constructed to take pellets. If this burner-boiler combination is well designed, an efficiency of over 90% can be achieved, which is comparable to the efficiency of a modern oil burner. Changing an oil burner to a pellet burner – or replacing electric heating with a pellet-burning stove – can be profitable. A pellet boiler takes a little more effort to run than an oil boiler because the chimney has to be swept and the ash removed a few times a year. Apart from this, the system is completely automatic and therefore requires less effort than a traditional wood-fired boiler. Environmentally, a good pellet burner is preferable to a fossil fuel burner. Pellet burners meet the demands of national and international eco-label schemes. Their emissions are well below the building regulations requirements of most EU member states and will not contribute to the greenhouse effect. A further advantage of moving from fossil fuels to biofuels for small-scale heating is that it will help the rural economy.

The use of liquid biofuels and blends of these with liquid fuels for heating domestic properties is also gaining ground. Standards for biodiesel for heating applications have just been put into force.

Large-Scale Combustion

The biofuels-based systems available for industrial and commercial heating can be categorized under the headings of fixed-bed, fluidized-bed, and dust combustion and cofiring.

Fixed-bed combustion includes grate furnaces and underfeed stokers. Grate furnaces, which generally have capacities up to 20 MW_{th}, are suitable for burning biomass with a high moisture content. Primary air passes through a fixed bed where drying, gasification, and charcoal combustion take place in consecutive stages. The combustion gases are burned in a separate combustion zone using secondary air.

Underfeed stokers, which represent a cheap and safe technology for systems up to 6 MW_{th}, are suitable for biofuels with low ash content and small particle size such as wood chips, pellets, and sawdust. The fuel is fed into a combustion chamber from below by screw conveyors and transported upwards to a grate.

Fluidized-bed combustion systems are suitable for large-scale applications exceeding 30 MW_{th} in size. The biomass is burned in a self-mixing suspension of gas and solid bed material (usually silica sand and dolomite) in which air for combustion enters from below. The high heat transfer and mixing encourage complete combustion. Fluidized-bed systems allow a good deal of flexibility regarding fuels, although attention has to be paid to particle size.

Dust combustion systems are suitable for biofuels, such as wood dust, in the form of small, dry particles. A mixture of fuel and air is injected into the combustion chamber. Combustion takes place while the fuel is in suspension. Fuel gasification and charcoal combustion take place simultaneously because of the small particle size. Quick load changes and efficient load control can be achieved.

Cofiring of biomass with coal in traditional coal-fired boilers is becoming increasingly popular, as it capitalizes on the large investment and infrastructure associated with existing fossil-fuel based power systems and at the same time reduces the emission of traditional pollutants (sulphur dioxide, nitrous oxide, etc.) and greenhouse gases (carbon dioxide, methane, etc.). Up to 10% biomass can be added to nearly all coal-fired plants without major modifications. Wood chips, willow chips, sawdust, and organic waste are the forms of biomass most often used.

Cofiring is normally realized by what is termed direct cofiring, i.e., firing the biomass and coal together in one combustion chamber of the power plant boiler. However, a number of other systems exist that fall into the category of indirect cofiring. One indirect system, known as a hybrid system, uses 100% biomass firing to generate steam. This steam is then mixed with the steam coming from the conventional coal-fired boiler for sending to the steam turbines for electricity generation. A second type of indirect system involves burning the biomass in a pre-furnace and feeding the resulting flue gases into the existing coal boiler. In a third type, the biomass is gasified and the resulting combustible gas fed to the coal combustion chamber.

Emissions Treatment

Appropriate techniques exist for treating all the emissions that emerge from biomass combustion plants. For large installations, flue gas cleaning is economically viable. As explained in the section on solid biofuels, NOx emissions can usually be controlled by appropriate combustion techniques. Secondary reduction measures to remove Sulphur Oxides are not usually necessary because biomass combustion does not yield as much of these pollutants as does coal combustion. Solid ash and soot particles are, however, emitted by biomass combustion. As these cause aerosol formation, additional gas cleaning is required to remove them.

Electricity

In 2002, some 43 TWh (i.e., about 0.15 EJ) electricity were produced from biomass in the EU. About a third of this is expected to be achieved via biomass-based combined heat and power plants.

The usual route for producing electricity from biomass has two stages. The biomass is converted to heat, which is then used in generation of electricity (or combined heat and power) using technology originally developed for conventional power production. A more modern approach, still at the research stage, involves the use of fuel cells. The biomass is converted to hydrogen, biogas, or methanol that are then fed to suitable fuel cells, the outputs being heat and electricity. Further information on both is given below.

Biomass-Based Processes for Conventional Power Production

In the simplest route, the biomass is converted to heat by combustion, the heat used to produce steam, and the latter used to drive steam piston engines or turbines. Basic information on the conversion of biomass to heat was given earlier in this chapter. The subsequent steps involve well-proven technologies. Suitable steam engines are available for electricity production in the range of 50 kWe to 1 MWe and steam turbines for the range of 0.5-500 MWe.

The conventional steam turbine route has a number of disadvantages. For instance, the steam boiler has to be operated at high temperatures and there can be erosion of the turbine blades due to the presence of moisture. Turbogenerators that work on the organic rankine cycle (ORC) present a useful alternative. The ORC is similar to the cycle of a conventional steam turbine except that the fluid that drives the turbine is an organic fluid that can operate efficiently at lower temperatures to produce electricity in the range 0.5-2 MWe. The organic fluid operates in a closed cycle. It is vaporized in an evaporator by an external heat source (such as biomass-derived heat). The vapor expands in the turbine and is then condensed and pumped back to the evaporator.

In a development that is applicable to small-scale electricity production, heat from biomass combustion is used as the external heat source for operation of a Stirling engine. A 30 kWe prototype plant has been built that has achieved around 20% electricity efficiency in combined heat and power production.

An example of biomass-based generation of electricity by a process similar to that used in conventional power production is the straw-burning power plant that went into operation in Sangüesa, Navarra, Spain, in 2002.

With an installed capacity of 25 MWe, it will, when operated for 8000 hours a year, consume 160,000 tonnes grain straw and produce 720 TJ electricity. It was designed to burn 100% grain straw or 50% straw and 50% wood waste.

Typical production costs for biomass-based power were found to vary in 2002 between 0.088 $ and 0.252 $ €cents, depending on feedstock and country. Feedstock collection costs increase roughly with the square of the distance from the plant. For this reason, the upper size limit of a biomass-based power plant lies somewhere between 30 MWe and 100 MWe. In general, they are below 30 MWe. This comparatively small size favors their operation as combined heat and power plants. These can meet the district heating and electricity needs of small communities [23].

Combined Heat and Power Production

Combined heat and power (CHP) plants use the waste heat from electricity production for heating purposes, normally for district or industrial heating. When biomass is employed as fuel for CHP plants, the availability of a stable and sufficient feedstock supply within a reasonable distance from the plant is essential. The use of biomass in CHP plants is one of the best options for achieving simultaneously increased bioenergy utilization and significant reductions in emissions. Renewable energy sources, mainly biomass, already account for 13% of all fuel inputs to CHP in the older member states. In the new member states the figure is just 1%, indicating that these countries have large unexplored opportunities to increase their use of biomass fuels for CHP.

Fuel Cell Processes

A research topic that is currently receiving a good deal of interest is simultaneous production of electricity and heat by fuel cells driven by biomass-derived hydrogen, biogas, or methanol. Information on methods of converting biomass to these chemicals was given earlier in this chapter. Fuel cells require very clean fuels and are sensitive to certain substances present in biomass, e.g., sulphur. Therefore, the major problem to be resolved by the research relates to gas and methanol purification.

Liquid Biofuels

The major market for liquid biofuels is in the transport sector, although products have also been developed for direct use in boilers and engines for heat and electricity production. In modern society, transport of people and goods plays a significant role in economic development, and requires increasing amounts of energy.

Currently, some 98% of transport fuels used in the EU are petroleum-derived. At the moment, the only technically viable way of using renewable energy resources to reduce EU dependence on fossil fuels in the transport sector is to increase the consumption of liquid biofuels. In 2004, the latter constituted close to 1% (corresponding to 2.4 million tonnes) of total EU petrol consumption. However, production is growing at 26% a year, and there are already a number of players in the EU liquid biofuels market. Liquid biofuels can be used neat or blended with petroleum-derived products. The possibility of using tax incentives to encourage more widespread use of biofuels has been introduced into a EU directive. The future fate of liquid biofuels will depend to a large extent on how far the member states decide to take this option – or, indeed, introduce fiscal and support measures of other kinds.

As a result of technological developments carried out over the past few years, transport biofuels come in a variety of types, notably bioethanol, biodiesel, and synthetic fuels (biomass-to-liquid or BTL fuel). Biodiesel is already well-known. BTL, on the other hand, is at the beginning of its development.

Manufacture is from agricultural resources of different kinds – currently, grain, sugar, oil crops, etc. For the future, processes are being developed to allow lignocellulosic biomass to be a major additional source. Yields vary according to feedstock: about 1.2 tonnes of bioethanol can be produced from an acre of wheat and 4.1 tons from an acre of sweet sorghum. Generally speaking, production costs today are high compared to petroleum-derived products. It costs around twice as much to make a liter of biofuels compared to a liter of petroleum-derived diesel (this ratio obviously depends on the price of crude oil), and it requires on average 1.1 liters of biofuel to replace 1 liter of diesel. The employment prospects associated with liquid biofuels manufacturing are good: around 16 jobs per 1000 tonnes of biofuel can be created, mostly in rural areas. The following sections discuss some of the different types of liquid biofuel in more detail [70, 72–74].

Bioethanol

Bioethanol, a colorless liquid, is the most widely produced biofuel in the world, with Brazil and the United States being the leading producers. In 2003, world production was 18.3 million tonnes. In the same year in the EU, 310,000 tonnes were produced – some 17.8% of total EU liquid biofuels production. (The major biofuel product in the EU is biodiesel.) Despite the EU's modest production compared with other parts of the world, steady growth has been achieved in the last decade.

Bioethanol is mostly obtained by fermentation of sugar beet, sugarcane, corn, barley, wheat, woody biomass, or black liquor. Production is generally in large-scale facilities, such as those in Abengoa in Spain. Most bioethanol production today is based on feedstocks from food crops. For the future, lignocellulosic biomass is expected to be an important feedstock and its use would reduce competition between the food and energy industries for raw materials.

Because the characteristics of lignocellulosic biomass differ from those of other forms of biomass, technologies for biofuels production have to be adapted for its use. Over the past 30 years, considerable research effort has been put into this area. The focus has been on producing fermentable sugars from the lignocellulosic material that can subsequently be converted into ethanol [58, 60, 61].

Normal vehicles can run on a 15% blend of bioethanol and gasoline. To use pure bioethanol, they need modification. Flexible fuel vehicles adapt automatically to run on fuels ranging from pure petrol to a blend of 85% bioethanol-15% petrol known as E85. The additional costs of manufacturing such vehicles on a mass scale, compared to normal vehicle manufacture, amount to 150 € a car.

Using ethanol in vehicles is beneficial to the environment because the emissions from ethanol are cleaner than those from petrol. An 85% ethanol-15% petrol blend can reduce greenhouse gas emissions by 60-80% compared with pure petrol. A 10% ethanol-90% petrol blend reduces emissions by up to 8%. The exact figure depends on the feedstock used to make the ethanol; a 10% blend using ethanol made from sugar, for instance, reduces harmful gases emissions by only 4%. Ethyl t-butyl ester (ETBE) is a biofuel, made from bioethanol, with an octane rating of 112 that can be blended with petrol in proportions up to approximately 17%. Methyl t-butyl ester (MTBE) has similar properties [18,19].

Biomethanol

Biomethanol is similar to bioethanol but is much more toxic and aggressive to the engine material. It can be produced from synthesis gas made by gasification of biomass.

Biodiesel

Chemically, biodiesel consists of methyl (or ethyl) esters of fatty acids. In response to its established market for diesel engines, the EU is the principal region of the world with a developed market for biodiesel. Growth rates have been 34% a year for over a decade. Biodiesel has the largest share of the EU's liquid biofuels market: it accounted for some 79.5% of EU total liquid biofuels production in 2004. Eight member states have production facilities. Biodiesel is produced by a chemical process – the esterification of fatty acids produced from vegetable oils. Rapeseed oil is the most commonly used feedstock because as much as 1-1.5 tonnes rapeseed oil can be produced per acre of rape. However, sunflower oil or used cooking oils are also used as feedstock. Like bioethanol, biodiesel is manufactured in large facilities.

Practically all diesel engines can run on biodiesel or blends of biodiesel with normal diesel. Using recently developed additives it is also possible to blend diesel with ethanol for use in trucks. Emissions of carbon dioxide, the major greenhouse gas, are 2.5 kg per liter less for biodiesel than they are for fossil fuel diesel. Emissions of hydrocarbons and soot are also lower for biodiesel than for fossil fuel-derived diesel. In addition, biodiesel releases fewer solid particles and, because it contains no sulphur, does not create SO_2, which contributes to acid rain. NOx emissions, on the other hand, are somewhat higher because of the presence of nitrogen in the biomass raw material.

Synthetic Fuel (BTL-fuel)

BTL fuel is a short term for biomass-to-liquid fuel. Typical examples are BTL-diesel and dimethylether (DME). BTL-diesel has exceptionally good fuel characteristics – a high cetane index and low sulphur and aromatics contents. It meets all the standards for normal diesel fuel. DME is a fuel of diesel quality with physical characteristics similar to liquified petroleum gas (LPG).

An advantage of BTL-fuel is its flexibility regarding feedstocks. It is pro-duced in a two-step process involving, first, preparation of synthesis gas (a mix-ture of carbon monoxide and hydrogen) from a biomass feedstock and, second, conversion of the synthesis gas into liquid fuel. The biomass feedstock is con-verted to bio-oil by pyrolysis as described in the section on thermochemical pro-cesses earlier in this chapter. This can be carried out in decentralized units close to the place of feedstock production. The bio-oil is then gasified under high pressure (30 bar) and temperatures (1200-1500°C) to a high-quality clean syn-thesis gas. Conversion of the latter to liquid fuel is carried out using a catalytic process known as the Fischer-Tropsch process originally developed to produce liquid fuels from synthesis gas derived from coal.

Because of the high quality of the product and flexibility regarding feed-stocks, BTL-fuel is set to make a major contribution to the EU biofuels market in the future.

Bio-Oil

As indicated in the discussion on BTL fuel above, bio-oil's main use is as a valuable intermediate for production of other products. However, it also has a direct application in boilers and furnaces for heat production and in static engines for heat and electricity generation. In the future, it may have an applica-tion as a source of hydrogen.

The yield of bio-oil from the solid biomass feedstock is about 75%. Bio-oil is much cleaner than the fossil fuel-based original because it contains 100 times less ash. As a liquid, it makes a versatile energy carrier since it can be pumped, stored, transported, and burned without difficulty. Its energy density is about 20 GJ/m^3 compared with 4 GJ/m^3 for solid biomass.

Other Products

Any consideration of biomass as an energy resource would be incomplete without a reference to its use as a feedstock for non-energy products. This was touched on in the discussion of the biorefineries earlier in the chapter. A few more details are given here. A wide range of chemicals and materials can be derived from biomass. This includes the traditional plant-based products – for example, oils, starch, fibers, and drugs – for which there are already major estab-lished industrial bases. It also includes many other possibilities. For instance, lubricants made from biomass offer significant environmental advantages over their fossil fuel-based counterparts. Printing inks, polymer additives, and poly-mers can also be made from biomass. Linoleum, for example, can be made from linseed oil. Surfactants are another group of products capable of being made from biomass. Some organic solvents can also be vegetable derived as can some pharmaceuticals, colorants, dyes, and perfumes. As indicated earlier, bio-mass has to be used wisely as an energy resource in a way that allows optimal manufacture of other priority products – not only the chemicals and materials described above, but also of food, wood products, paper and board, etc.

7.3 WIND ENERGY

History of Wind Energy

Humankind has used wind power since ancient times. The application of wind energy dates back to the time of the first civilizations, when wind energy was used to power boats on the river Nile (5000 BC).In Afghanistan, the first windmills that appeared were used for grinding grain. During the nineteenth century, thousands of windmills were set in North America and were mainly used for pumping water on farms and plantations. At the beginning of the twentieth century, small windmills began to be massively used in the United States to generate electricity, but many of them stopped working before World War II due to the intensive expansion of the electro-distributive system even to the most remote parts of the earth.

During the oil crisis in the 1970s, the production of electricity from wind recorded a sharp increase, and it was only in the 1980s that it experienced a boom upon the application of new technological achievements.

Since the 1980s, wind energy has strongly progressed in both installed capacity and in volume production. Until a few years ago, wind turbines of 500 kW were rare, but today's wind turbines with a power of 1.9-2.5 MW and a rotor diameter of 50-90 meters are standard. Figure 7.13 presents the development of wind turbine technology.

Wind as a Potential Source of Energy

Wind is an inexhaustible source of energy the global potential of which exceeds several times the world's electricity needs. The use of wind power in the production of electricity began to be used in the early 1930s of the last century when the first wind farms (plant electro-mechanical conversion of wind energy) began

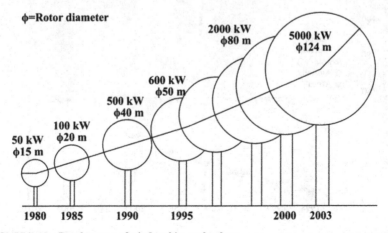

FIGURE 7.13 Development of wind turbine technology

to be built. Nowadays, the wind energy industry is considered the fastest growing, both in terms of technology and in terms of growth of wind farms in the world (see Figure 7.14).

Wind farms use the kinetic power of wind. Wind turbines transform mechanical energy of wind through electric generators, into electricity.

About 75% of all wind farms have been installed in the EU. Out of the installed 34,466 MW, the EU serves about 3% of the total electricity needs. Germany is a leading country by the amount of installed wind power of 16,629 MW, twice as high as the strength of all power plants in Serbia, but the largest share of electricity production from wind farms is in Denmark, which creates 20% of its needs.

Wind Power

Eliza rotor wind turbines rotate due to the air mass flow. The amount of energy transferred by the wind to the turbines is not directly dependent on the density of air, the rotor size, or the wind speed. The kinetic energy of the body in motion is proportional to its mass, so that kinetic energy depends on the density of air. More specifically, the thicker the air, the more energy is obtained on the wind turbines.

The density of air is equal to the amount of molecules per unit volume of air. In normal air pressure and air temperature of 15°C, the mass of air is 1.2 kg/m^3, but with the increase of the moisture content, its density decreases.

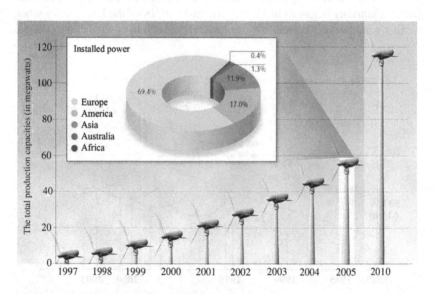

FIGURE 7.14 **Annual increase in installed wind power in the world**

Colder air is denser than warm, which is why, at the same speed of wind, a wind turbine produces more electricity in winter than in summer. Atmospheric pressure decreases with increasing altitude, because at high altitudes (mountains) the pressure is lower and the air less dense.

Wind turbines convert the kinetic energy of wind in the power of the turning propeller. The surface of the rotor determines the amount of energy that can be derived from wind turbines. Since the rotor area increases with the square of the diameter of the rotor, a turbine twice as large produces four times more energy. Increasing the rotor surface is not as simple as creating long propellers, since the increase in the surface of the rotor increases the strain of the whole system, regardless of wind speed. In order to compensate for stress, the entire mechanical system must be strengthened.

Another factor that affects the input power is that wind turbines affect the turn of wind, even before the wind is affected by the rotor blades. The rotor slows down the wind speed, i.e., the wind speed is higher in front of the rotor than behind, which means that wind turbines cannot use the whole power of wind. Conversion of the kinetic wind energy into the mechanical energy of the rotating movement is done using wind turbines, which can have different structures. Modern wind turbines have wind generators with a horizontal shaft that has a system to rotate the shaft in a horizontal plane to monitor wind direction changes. Wind turbines may have a different number of blades, but, as for high powers, turbines with three blades are most commonly used. The diameter of wind turbines depends on their strength and ranges from 1 m with an output of 0.5 kW to 120 m with an output of 5 MW. The turbine drives the wind generator which may be of a different structure. The wind turbines and wind generator, together with the pillar they are placed on, make the wind generator.

The range of wind speed during which the wind generator generates electrical energy is typically 3-25 m/s, and the maximum (nominal) power is achieved at the wind speed from 12-15 m/s. For wind speeds above 25 m/s, wind turbines stop due to mechanical reasons. It is not economically profitable to design wind turbines for active operation at wind speeds greater than 25 m/s, due to the fact that such winds are relatively rare. The construction of a modern wind generator is designed to withstand windsup to 280 km/h.

Wind generators are built on windy sites where the mean annual wind speed is greater than 6 m/s (at a height of 50 m above ground). One wind generator with a capacity of 1 MW can generate approximately 2000 MWh of electricity annually, enough to meet the needs of 500 average four-member households. Most of the time, wind generators are grouped together on suitable sites called wind farms. Wind farms can have hundreds of wind generators and power over 300 MW. They are constructed on land, but also in shallow coastal areas where strong and steady winds blow (see Figure 7.15).

The price of one wind generator with a 1 MW capacity is about one million euros, which is equal with the cost of the installation of 1 MW in an average power plant using coal. Unlike thermal and hydropowers, wind farms are built

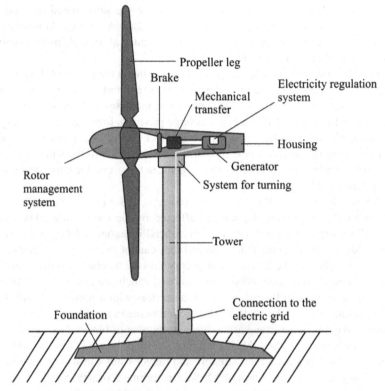

FIGURE 7.15 Construction of a wind turbine

very quickly, in just a few months. The life of a wind farm is about 25 years. However, the price of electricity generated from the average wind generator is still higher than the price of kWh from conventional power plants. The average wind generator with 1MW of installed capacity in electricity generates annually twice less energy than 1 MW of installed capacity in hydroelectric power, or about three times less energy than the average thermal power and about 3.5 times less energy than the same installed capacity in nuclear power. The production of a wind turbine is dictated by wind, whereas production management in a storage power plant is based on market demand. For example, reservoirs are filled at night and production is kept to a minimum, and during the day when 1kWh is the most expensive, the power plant operates at full capacity and produces the most expensive "current." On very windy sites, with an average annual rate greater than 7 m/s, wind farms can be more economical, but those locations are relatively rare.

It is considered that, according to the cost of produced kWh, wind farms would be fully competitive with conventional sources when the impact on the environment is included in the cost of electricity production. To produce 1 kWh in an average power plant using lignite, about 1.5-2 kg of coal is used, with the release of about 1 kg of carbon dioxide into the atmosphere and release of about 2 kWh of thermal energy that is dissipated into the environment and locally warms the river and atmosphere. Besides the water vapor, CO_2 is a leading cause of global warming (greenhouse effect). According to the Kyoto agreement, every kilogram of CO_2 that is emitted into the atmosphere has its price and these so-called external costs can increase the cost of producing 1 kWh of electricity in power stations even up to 200%.

Technology development of wind farms is intense and it includes improvements in all aspects of their use. First, it is aimed at further increasing the power granted by the individual wind generators. Special attention is paid to the further development of wind generators for work in mountainous locations with harsh climatic conditions and turbulent winds. Some manufacturers have already successfully installed commercial turbines for extreme mountain conditions.

Construction of Wind Turbines

The conversion of kinetic energy into electricity is achieved using wind turbines of different structures, some of which are shown in Figures 7.16 to 7.18

According to the mutual position of rotary axes of the rotor and wind direction that runsit, wind turbines are divided into axial (axis is generally parallel to the rotor surface, i.e., the wind direction is along the axis) and radial (the rotor axis is generally perpendicular to the surface, i.e., the wind direction is perpendicular to the axis).

FIGURE 7.16 "MAG"wind turbine

FIGURE 7.17 Spiral wind turbine

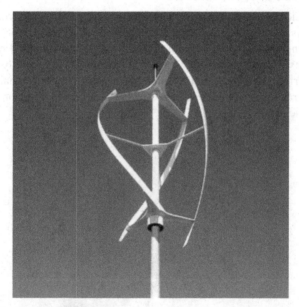

FIGURE 7.18 Silent wind turbine

Axial Wind Turbines

Today, axial wind turbines are the most common and usually have two or three wings (Figure 7.19). The first three-bladed axial wind turbines, designed for launching electro-generators, were constructed in the middle of the last century in Denmark, and based on this design, most modern wind turbines of this

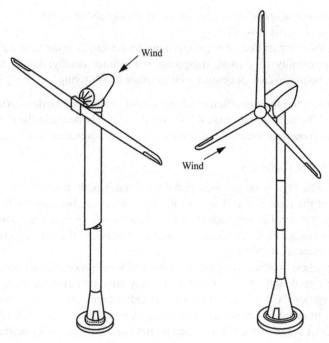

Wind

Wind

FIGURE 7.19 Axial wind turbines

application have been developed. Over time, airfoil wing profiles, rotor material technology, and transfer mechanisms, and today, an electro-generator with electronic controllers, have been specialized. Electric power industries of developed countries have already disposed of axial turbines in the form of wind plantations. A single turbine already achieves the price of KWh generated electricity from 0.06 Euros.

The length of the rotor wing, which is proportional to the torque at the output shaft, in particular contributes to more effective axial length of the wind turbines, and the risk of breaking them at wind speeds above 15 m/s. The problem is to some extent overcome by the development of the wings from strong and light but expensive material (e.g., carbon fiber), or cheaper, but heavier materials. While the first approach can make the cost of creating them unacceptable, the second reduces work efficiency since the increase in the rotor weight slows down the adjustment of the rotor wings' position toward the change of the wind direction, and those that change direction and blow short bursts are the most unfavorable ones.

The price of active adjustment and protection (rotation of wing around its own axis and/or breaking the rotor in a horizontal plane) is only justified in these high-power wind turbines. Some specific shortcomings of axial turbines include:

Suspension of operation in case of winds mainly above 15 m/s;
A large area that covers rotor;
High tower for its assembly (proportional to the length of the wing); and
The possibility of causing dangerous (sometimes deadly) bodily harm to living beings in case of contact with the rotor in full swing.

However, the high coefficient of utilization of wind of modern axial wind turbines (50% peak and average 45%) give them an advantage in the use for the projected strength in the range of several tens of kilowatts to several megawatts.

Radial Wind Turbines

In contrast to axial wind turbines, radial wind turbines do not have to adapt the position of the rotor to the change of the wind direction, but their coefficient of efficiency is lower. The best known types of these turbines were designed by author Sigrid Savonius (Finland), Georges Darrieus (France) (H-rotor (Figure 7.20), (French patent nr. 2659391).

The highest coefficient of the utilization of wind power has a Darrieus turbine (40% peak and 35% average), which may have two or more thin, slightly twisted, ribbon "C"-shaped wings. The disadvantages are that because of its balanced moment, it cannot independently start and it occupies a large space when it is of bigger dimensions, since in that case, the shaft is strengthened by

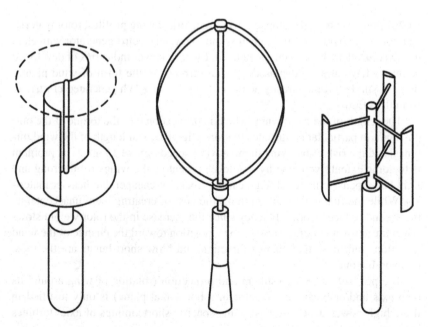

FIGURE 7.20   Radial wind turbines

tightening ropes. This type is only found in commercial applications (the last known is 4200 KW with a two-wing rotor diameter of 100 m from Quebec, Canada, and is no longer functioning). The H-rotor turbine is similar to the Darrieus turbine, while the Riviere turbine, with its plate-like wings that revolve around its own axis. (Figure 7.21).

The Savonius turbine, whose rotor consists of two cylindrical surfaces, partially folded and faced by a concave part, has the lowest coefficient of utilization of wind power (peak 15%), but has the advantage that it does not require a large area and cannot harm living beings that come into direct contact with the rotor working in full swing. While the axial wind turbine is close to reaching the maximum of its technical features, what has been observed in recent years is the increasing interest of specialists for underexplored possibilities of their radial cousins, and therefore new variants are frequently reported and patented.

Wind Farms

Usually, on suitable sites, many wind generators are grouped and create wind farms, which may have several hundred wind turbines and power of over 300 MW. The advantage of connecting wind turbines in one location lies in the lower cost of location development, simpler interconnection with the power system,

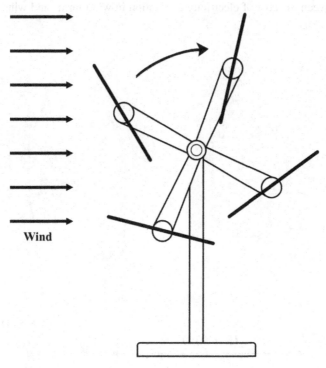

Wind

FIGURE 7.21   Riviere turbine

and a centralized approach to its management and maintenance. The problem that arises in connection with grouping wind generators refers to the optimal deployment of wind generators with the aim of maximizing their effects. The optimal distances are presented in Figure 7.22, where D is the diameter of the wind turbine.

The annual utilization factor of a wind farm is 10-15%. About 25% of the time during a year, plants cannot work because the wind speed is lower than the minimum and about 5% of the time they cannot work due to the fact that it is greater than the maximum. In terms of energy, wind generators are used to fill in the base of the load diagram, and due to low power, they are usually included only in the energy balance of the local (regional) consumption.

Economics of Wind Farms

As the starting speed threshold of the useful work of wind generators is relatively high, it is obvious that their location, given the intensity of the wind and the likelihood of windy days, is a basic factor when it comes to cost effectiveness of their use.

Windy areas are often very remote from populated areas and from adequately developed distribution networks, and thus, the costs of their integration into the energy system influence the economy of wind powers. Figure 7.23 shows the ratio between the cost of electricity generation in wind farms and wind turbine power.

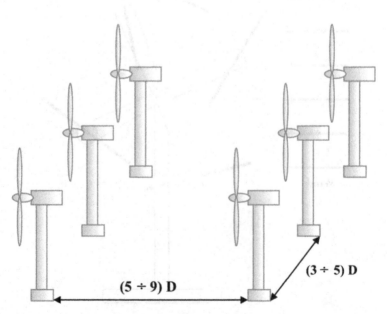

FIGURE 7.22 Optimal distance between wind generators

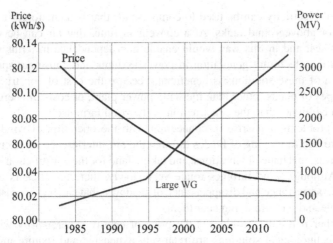

FIGURE 7.23 Ratio between the cost of electricity generation in wind farms and wind turbine power

The running costs of wind generators are negligible; they participate in the final price of energy with only 10% (no fuel costs, there are only maintenance costs, personnel, taxes, insurance, taxes, administrative costs, etc.). Depending on the number of windy days and the wind speed above the wind generator's useful work threshold (between 4-5 m/s or 14.4-18 km/h), their production price (of the order 6 c/kWh during the period of 15-20 years of exploitation) is in some countries competitive to the price of energy from fossil fuels, whereas specific investments, except for gas turbine power plants, are slightly lower (about 1000-1500 $/kW).

Advantages of Wind Farms

Power generation using wind power has numerous advantages. First, the technology for generating power using wind energy has been known since ancient times and considerably improved since. Modern technology offers solutions that can be installed in virtually all places – in deserts, at sea, on mountaintops, in populated areas, and so on. Power generation from wind does not create any environmental pollution. This energy is produced over a system of transmission lines and thus can be transferred to places where energy is needed.

Disadvantages of Wind Farms

One of the most unfavorable aspects of wind farms is that their production is variable and unpredictable. This is exactly why a share of wind farms in total production should not be greater than 10%. The share of these plants could increase if accumulation of energy is ensured. Some of the possible solutions imply combining wind farms with pump-storage systems or solar power plants.

Excess of electricity can be used to compress air that is then stored in under-ground or aboveground tanks. At a convenient time, that air can be used to run turbines, and in this way would enable non-dependence in terms of time between the electricity generation in power systems and consumer spending. But none of these solutions is beneficial, because the cost of construction of other types of plants, other than the wind power plant, or costs of energy stor-age, can be higher than the construction of the wind farm itself.

The problems that arise in connection with the operation of wind power farms are due to moving of blades through the air, interference of electromag-netic waves, mechanical vibrations, and taking land for the construction of wind farms. Although wind turbines rotate very slowly, there are flickers of solar light, i.e., appearance and disappearance of shadows, that can be very disturbing for people and wildlife living near them.

Another particular problem is the price of energy generated by wind tur-bines. Technological solutions are relatively expensive and require multiyear exploitation in order to make the initial investment profitable.

7.4 GEOTHERMAL ENERGY

Potential of Geothermal Energy

The word geothermal comes from the Greek words geo (earth) and Therme (heat) and means heat of the earth. The heat in the earth's interior is the result of the formation of planets from dust and gases more than four billion years ago, and the decay of radioactive elements in rocks continuously regenerates the heat, and thus geothermal energy is a renewable source of energy. The main medium that transfers heat from the interior to the surface is water or steam, and this component is renewed in the way that water from rains penetrates deeply through crevices; it is heated there and circulates back to the surface, where it appears in the form of geysers and hot springs.

The outer rigid crust of the earth is deep, from 5-50 miles, and is made of rocks. The mass of the inner layer is constantly coming to the surface through volcanic vents and cracks on the ocean floor. Below the crust is a layer that extends to a depth of 1800 miles, and is made from compounds rich in iron and magnesium. Below all this are two layers of the core – the liquid layer and solid layer. The radius of the earth is about 4000 miles, and no one really knows what exactly is in the interior. These are all scientific assumptions about the inside of the planet and are based on experiments in conditions of high pressure and high temperature.

Down through the earth's outer layer, the crust, the temperature rises approx-imately 17-30°C per kilometer of depth. Below the crust, there is a layer that is composed of partially molten rock, and the temperature of this is between 650 and 1250°C. In the earth's core, by some estimates, temperatures could be between 4000 and 7000°C. Since heat always moves from warmer to colder parts, the heat from the earth's interior is transferred to the surface and the heat transfer is the main driving force of tectonic plates (Tectonic plates are pieces of the Earth's crust and uppermost mantle, together referred to as the lithosphere). In places where tectonic plates merge together, there could be a failure of magma in the upper layers, whereby the magma cools and creates a new layer of the earth's crust. When the magma reaches the surface, it can create volcanoes, but mostly it remains below the surface of the vast mass of the pool where it begins to cool, a process that takes 5000 to one million years. Areas that are located below these magma pools have a high temperature gradient, i.e., the temperature rises rapidly by the increase in depth, and these areas are very suitable for geothermal energy.

The geothermal energy potential is enormous; there is 50,000 times more energy than all the energy that can be derived from oil and gas worldwide. Geothermal resources are found in a wide range of depths, from shallow surfaces to several-kilometers-deep reservoirs of hot water and steam can be brought to the surface and used. In nature, geothermal energy is found most often in the form of volcanoes, hot springs, and geysers of water. In some countries, geothermal energy has been used for centuries for recreational and medicinal bathing. The development of science was not limited to the area of medical exploitation of geothermal energy, but geothermal energy use is directed toward the process aimed at obtaining electricity, heating households, and industrial plants. Heating of buildings and use of geothermal energy in the process of generating power are considered the main but not the only way to use this energy. Geothermal energy can also be used for other purposes such as the production of paper, pasteurizing milk, the process of drying wood and wool, and many other purposes (see Figure 7.24).

The main disadvantage of geothermal energy is that there are not many places in the world that are very suitable for exploitation. The most suitable areas are on the edges of tectonic plates, i.e., areas of high volcanic and tectonic activity. Figure 7.25 shows a tectonic map of the world and areas suitable for geothermal energy.

Electricity Generation from Geothermal Sources

One of the most interesting forms of exploitation of geothermal energy is for the production of electricity. Hot water and steam from the earth are used to run generators, and therefore there is no burning of fossil fuels and as a result, there are no harmful emissions into the atmosphere, only water vapor is emitted. An additional advantage is that such power can be implemented even in various environments, from farms and sensitive desert areas all the way to forest and recreational areas.

The beginnings of using the heat of the earth for generating electricity are connected with a little Italian place called Landerello in the year 1904. It was in that place and that year that experimenting with this form of electricity generation

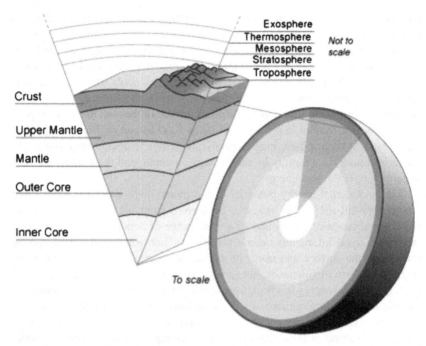

FIGURE 7.24 Layers of the earth

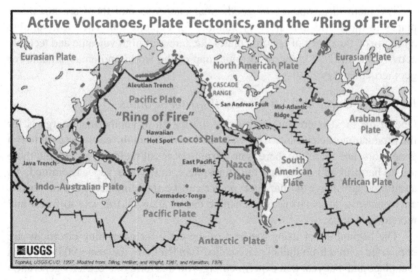

FIGURE 7.25 Areas suitable for utilization of geothermal energy

started, with steam used to run a small turbine that fed five lightbulbs. This experiment is considered the first use of geothermal energy for electricity generation. There, in 1911, construction of the first geothermal power plant started, which was completed in 1913, and its power was 250 kW. This was the only geothermal power plant in the world for almost half a century. The working principle is simple: cold water is pumped to the hot granite rocks located near the surface, hot steam comes out at over 200°C under high pressure, and the vapor then drives a generator. Although the plant in Landerello was destroyed in World War II, it was rebuilt and extended and is still in use today. This plant is still powered by electricity that supplies about one million households, i.e., produces almost 5000 GWh per year, which is about 10% of total world production of electricity from geothermal sources (see Figure 7.26).

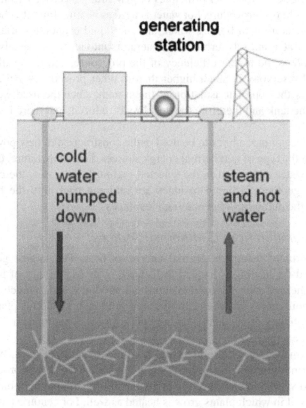

FIGURE 7.26   **Principle of generating electricity from geothermal sources**

Types of Geothermal Power Principles

The principle of dry steam – only hot steam is used, typically above 235°C. This steam is used for direct setting of turbine generators into motion. This is the simplest and oldest principle and is still in use today since it is by far the cheapest process of generating electricity from geothermal sources. The Landerello geothermal power plant just discussed used this principle. Currently, the largest power plant using the "dry steam" principle is in Northern California and is called The Geysers. It has generated electricity since 1960, and the amount of electricity produced from this plant is still sufficient to supply a city of San Francisco's size.

Flash principle (flash steam) – the use of hot water from geothermal reservoirs which is under great pressure and at temperatures above 182°C is used. Pumping water from these reservoirs toward power on the surface reduces the pressure and the hot water is turned into steam that sets turbines into motion. The water that did not turn to steam comes back to the reservoir for reuse. The majority of modern geothermal power plants use this principle.

Binary principle (binary cycle) – water used in the binary principle is colder than the water used for other principles of generating electricity from geothermal sources. In this procedure, hot water is used as heating fluid that has a much lower temperature than boiling water, and the liquid evaporates at a temperature of hot water and sets turbines of generators into motion. The advantage of this principle is the higher efficiency of the procedure, and the availability of geothermal reservoirs is much higher than in other procedures. An additional advantage is the complete isolation of the system, since the used water flows back into the tank and therefore the heat loss is reduced, and there is almost no water loss.

The choice of procedures to be used in the construction of new power plants depends on the type of geothermal energy sources, i.e., temperature, depth, and quality of water and steam in the selected region. In all cases, the condensed steam and geothermal liquid residues are returned back into the bore, thus increasing the durability of geothermal resources.

Heat Generation from Geothermal Sources

Another form of using geothermal energy is heat. The largest geothermal system used for heating is located in Iceland, in its capital city of Reykjavik, where geothermal energy is used in almost all buildings, and as much as 89% of Icelandic households are heated in this way. Although Iceland is plausibly the largest beneficiary of geothermal energy per capita it is not alone in the field of geothermal energy use.

Geothermal energy is also used in agriculture to increase yields. Water from the geothermal reservoir is used for heating greenhouses for the production of flowers and vegetables. The heating of greenhouses does not imply only heating of air; the soil in which plants grow is heated as well. For centuries, it has been used in central Italy and Hungary, and currently it covers 80% of energy needs for greenhouses heating.

Heat pumps are another way of using geothermal energy. Heat pumps use electricity to circulate the geothermal fluid, a fluid that is later used for heating, cooling, cooking, and hot water, thus significantly reducing the need for electricity.

Since the estimated total quantity of geothermal energy that could be used is much higher than the overall amount of energy sources based on oil, coal, and natural gas together, geothermal energy should be given more importance because it presents a cheap, renewable source of energy that is more environmentally friendly. Since geothermal energy is not readily available everywhere, at the very least it should be used in the places where the energy is readily available (edges of tectonic plates) to reduce at least a little pressure on fossil fuels and thus help the earth recover from harmful greenhouse gases.

7.5 HYDROPOWER ENERGY

Ocean Energy

Oceans cover more than 70% of the earth's surface and thus represent a very interesting source of energy that in the future could provide energy to households and industrial plants. Currently ocean energy is an energy source that is very rarely used since there are a small number of power plants using ocean energy and the plants are still small in size. But as a renewable sector it is becoming more important. There are three basic types of solutions that are used in the exploitation of ocean energy:

The energy of waves;
Ocean tidal energy; and
The thermal energy of the sea.

Wave Energy

Wave energy is a form of kinetic energy that exists in the motion of waves in the ocean. Wave motion causes the winds blowing over the surface of the ocean. This energy can be used to run turbines, and there are plenty of places where the winds are strong enough to produce a continuous wave motion. Huge amounts of energy lie in wave energy, offering enormous energy potential. Wave energy is captured directly beneath the surface of waves or from various fluctuations of pressure below the surface. This energy could drive a turbine: the wave rises in the chamber and the growing power of water forces air from the chamber and then the movable air drives the turbine, which then powers the generator.

The main problem with the energy of waves is the fact that this energy source cannot be used uniformly in all parts of the world; many research projects are dedicated to solving this problem of uniformity. But there are also many areas with very high levels of utilization, such as the west coast of Scotland, Northern Canada, Southern Africa, Australia, and the northwestern coast of North America. There are many technologies for the exploitation of wave energy, but only a few are actually commercially exploitable. Technologies for the use of energy waves are not only installed at the coast, but also far out on the open sea, with systems set installed in deep water, at depths exceeding 40 meters (see Figure 7.27).

Most technologies for the exploitation of wave energy are still focused on the closeness of the coast, or are on the coast, and the difference between them lies in their orientation toward the waves with which they interact, and in the working principle by which the wave energy is converted into the desired shape of energy. So-called "terminator devices, point absorbers, attenuators, and overtopping devices" are certainly among the most popular technologies. "Terminator devices" such as "oscillating water columns" are usually somewhere on the coast or near the coast, spreading perpendicularly with regard to the travel direction of a wave, whereby after the wave power is captured and reflected, the "oscillating water column" is then like a piston moving up and down, forcing air through a hole connected to the turbine. "Point absorbers" are different types of technology that include a floating structure with components that move in relation to each other due to wave energy and then create energy since the movement pushes electromechanical or hydraulic energy converters. Attenuators can also be floating structures that are oriented parallel to the direction of the wave, where the difference in the heights of waves varies according to the length of the device that causes bending in places where the parts merge together, and this bending is coupled with a hydraulic pump or other converters for further transformation into useful forms of energy.

"Overtopping devices" are actually reservoirs that are filled with the coming waves to levels above the average of those from the surrounding ocean; after the water drains, the gravity forces them back towards the ocean surface, and then that energy drives water turbines. Although the potential of wave energy is unquestionable, there are certain aspects that should be taken into consideration, especially problems concerning the surrounding environment.

 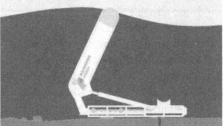

FIGURE 7.27   Use of wave energy

Tidal Energy

The amplitude of tides depends on the mutual position of the sun, moon, and earth, and the amplitude and frequency are different on some coasts. The amplitudes of the tides in the Mediterranean Sea are 10 cm, and in the Atlantic, Pacific, and Indian oceans, they are approximately 6-8 m. In some parts of the coast in western France and in the southwestern part of Great Britain, the amplitude reaches more than 12 m. On the West European Atlantic coast, the time gap between the two tides is 12 hours and 25 minutes and on the shores of Indochina there is only one tide in 24 hours.

For energy utilization of the tides, it is necessary to select a suitable place on the coast, with large amplitude of the tide, as well as with the possibility to isolate a part of the sea surface (constructing the partition) to create the storage pool. The simplest way to achieve this is by building turbines that operate in only one direction of water flow.

In order to prolong the time of operation, a turbine operating in both directions of water flow can be set up: from the pool and into the pool. There is a variant of a turbine that works like a pump, either to transfer water from the pool into the sea, or from the sea into the pool. In this way, the potential energy of the tide is better exploited.

However, regardless of the selected mode of exploiting tidal energy, neither uninterrupted drive nor constant power can be achieved. This shows that the production of electricity on this basis is impossible in isolation, without cooperation with other plants for generating electric energy. Such plants must be included in the power system in which other plants have a total force of several tens of times greater than the power of these plants. For the plant to be economically viable, it must fulfill two requirements: the length of the dam should not be too long and the depth of the dam site should not be too deep.

The total tidal energy is estimated at 26,000 TWh or 2230 Mtoe in a year. A third of this is lost in the shallow seas. The mean amplitude of oceanic tide widths is less than 1 m, and the energy utilization of tides is economically justified if the amplitude is greater than 2 m. Therefore, it is estimated that out of the total energy of tides, only 2% is usable, which is 520 TWh per year, about 3% of the current annual production of electricity (about 15,000 TWh). The actual electrical energy is even smaller, because even through the best sizing only 25% of theoretical production can be used.

A tidal energy plant was built in France, called the La Rance plant, which was set into motion in 1966. It has 24 turbines that can operate as pumps for both directions of water flow. The total power is 240 MW turbines. The annual production is 608 GWh (20% of the theoretical production), and 64 GWh is required for pumping (the energy is taken from the system). The usable pool capacity is 184 million m^3, the surface area is 22 km^2, and the length is about 12 miles. The length of the wall is about 720 m. In Russia, not far from Murmansk, an experimental plant was built (in operation since 1968) with a power of 800 kW.

This type of power requires high investment costs, and the total possible production of electricity on favorable locations is only a marginal amount of the energy needed.

Thermal Energy of the Sea

Ocean thermal energy conversion is a method used to create electricity that uses the temperature difference that exists between deep and shallow water, since the water is colder at greater depths. The greater the temperature difference, the greater the efficiency of the whole method, and the minimum temperature difference should be 38°C. This method has a long history of use and dates from the early nineteenth century. Most experts believe this method is a feasible investment and even with the existing technology, a very large amount of electricity could be produced. But this is not the case today as OTEC (ocean thermal energy conversion) requires huge, expensive large-diameter pipes that must be set at least a kilometer into the sea so that they can bring colder water from greater depths, which is of course is very expensive. The types of OTEC systems include:

Closed-cycle systems that use liquids with low boiling points, usually ammonia, and thus drive a turbine, which then generates electricity. Warm surface seawater is pumped through a heat exchanger and because of the low boiling point, it evaporates, and the generated vapor then runs the turbo generator. The colder deeper water is then pumped through a second heat exchanger where, due to condensation, it passes through the vapor back into liquid and the liquid is then recycled through the system. In 1979, the Natural Energy Laboratory, in collaboration with several partners, created a mini OTEC experiment, which was the first successful OTEC system of a closed sea that was constructed in the sea. The mini OTEC vessel was transported 1.5 miles from the Hawaiian coast, and it managed to produce enough energy to power lights on the vessel and to ship computers and TVs. Twenty years later, in 1999, the Natural Energy Laboratory tested a pilot plant of a closed system, with power of 250 KW, the largest plant of this type ever put into operation.

Open-cycle systems that use warm surfaces of tropical oceans for getting electricity due to the fact that hot water boils after being placed in a container with low pressure. After that, the expanding steam begins to force a low-pressure turbine connected to an electrical generator, and eventually it condenses back into liquid due to exposure to cold temperatures in the deep ocean.

Hybrid systems that are designed in a way that combines the positive effects of both of open and closed systems. Operation of hybrid systems includes warm seawater that enters the vacuum chamber where it is converted to steam (a process similar to the one in the case of open systems). After that the vaporization of steam into a low boiling liquid (as in closed systems) then drives a turbine and creates electricity (see Figure 7.28).

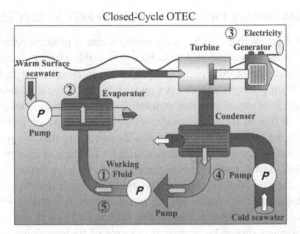

FIGURE 7.28   OTEC system

OTEC has a lot of potential to generate electricity, but electricity is not the only positive benefit of using an OTEC system. As a by-product, cooling of the air can be produced, and the used cold seawater from the OTEC plant can either chill freshwater in the heat exchanger or flow directly into a cooling system. Additionally, some types of fish, salmon, for example, can reproduce much better in deep water rich in nutrients, a product of the OTEC's work. But since OTEC plants require a large initial investment, create certain environmental issues, and require a lot of building space, these systems are not ideal. Another factor affecting the commercialization of OTEC projects is the fact that the world has only a few hundred sites suitable for construction, mostly in the tropics, where the deep ocean is rather near the coast, and thus the additional costs that would be incurred during the construction of an OTEC project away from the coast are avoided.

Ocean energy is a renewable source of energy that definitely requires more research, mainly in order to increase investment effectiveness and reduce the huge start-up costs, which is the biggest disadvantage of this renewable source. The oceans represent two-thirds of the earth's surface and as such, have huge potential worthy of further research. However, current technologies do not support the profitability of these projects. More research needs to be done. Problems related to the massiveness of these plants or return on investment are certainly highlighted, but they are not the only issues, since it is also necessary to meet certain environmental standards before entering into larger projects, as to preserve the environment to the greatest possible extent. Despite the fact that the ocean energy sector has not experienced a boom, projects such as construction of the OTEC power plant in Keahole Point in Hawaii inspire optimism and faith in this technology.

7.6 MANAGEMENT OF RENEWABLE ENERGY SOURCES

The impact of renewable energy on the environment is project specific for each site, but generalizations are still possible. Renewable energy sources tend to be more on the side of environmental protection in comparison to the energy that comes from conventional sources, especially in terms of emissions to air, and the lifecycle of emissions from renewable energy sources is insignificant in comparison to that of fossil fuel plants. Renewable energy projects should be considered in the framework of sustainable development that integrates energy demand reduction and energy efficiency, using a mixture of energy from various renewable sources to meet the increased scale of energy demand and, on the other hand, the protection of biodiversity and local communities.

Technologies for renewable energy sources vary in terms of efficacy and acceptability of environmental protection. Therefore, each individual project that is planned should be evaluated in terms of finding alternatives to find the most appropriate technology for each particular case. In the development phase of the project, it is necessary to give equal importance to social and environmental, as well as technical, economic, and financial aspects, whereby the assessment solutions should include all political, program, and project options. Strategic assessment of environmental impact and life cycle analysis should be integrated and done as an initial step in the process and should give equal importance to the options both on the demand and supply side. Renewable energy must be supported to enter the energy market, but this support must be limited to renewable energy produced in a sustainable manner [21,22].

Hydropower

Small and mini hydropower plants, which produce between 100 KW and 10 MW of electricity, often produce enough electricity to be integrated into the power grid. These plants do not require accumulations and do not disturb the flow of a river or stream, and they can be very effective in energy supply to networks in those areas where there are rivers or waterfalls.

Large hydroelectric power plants cause problems when it comes to regulating water courses, damaging landscape, impacting flora and fauna, the emission of greenhouse gases (methane which is released from the flooded area), affecting the quality of drinking water (changes in dissolved substances in water and oxygen level, presence of toxic substances, changes in temperature and acidity, turbidity, etc.), and creating noise and negative visual impact for residents in the area. Large dams can cause the relocation of hundreds of thousands of people, destroying their lives and communities. Ensuring adequate compensation, moving, and reconstruction can prove be very difficult and, in many cases, communities affected by such projects remain in a worse position than they were before the project.

The World Commission on Dams estimates that 1-28% of the total world emissions of greenhouse gases come from hydroaccumulations. In some cases, emissions from accumulations may be equal to or greater than those from power plants working on coal or gas, and emissions are highest in shallow, tropical accumulations.

The hydrological cycle is renewable, but large hydropower plants do not use only water stream as a drive, but hydroaccumulation, which are reduced in some cases very rapidly (accumulations around the world are losing capacity upon the sediments increased at an average rate of 0.5-1% per year).

The criteria for the realization of getting energy from water courses include:

Project is less than 10 MW;
Project does not include a dam, accumulation, and resettlement;
Project in no way affects the regime of watercourses and movement of wild animals;
Project does not affect biodiversity or human needs for water; and
Project does not affect the possible investment in the renewal and increase in efficiency of existing energy units in the project area.

Solar Energy

Solar energy, including concentrated systems using reflective materials such as mirrors to concentrate solar energy and then convert heat into electricity, is becoming increasingly cost effective for the supply of electrical networks, although it is still far more expensive than alternative technologies. Solar PV systems, which directly convert sunlight into electricity, are particularly suitable for rural electrification in areas that do not have the conditions for micro-hydropower plants. Such systems can be used for power generation, water pumping and maintenance, health systems, and communications. PV systems have few negative effects during use, but the production of their cells requires careful control of the use of potentially toxic and hazardous materials.

The use of solar collectors for generating heat is a far more cost-effective option. It is estimated that the costs of heating and water heating are more than 60% higher than the total cost of energy for a single building. It is possible to cover 50-65% of annual hot water needs with solar energy and the appropriate-sized system. The overall need for hot water in summer in most cases can be met with solar heating systems. Then the conventional system can be completely excluded. This is particularly advantageous because in summer, the system operates with a low level of capacity utilization due to lower demand for heating.

So far, significant impacts of solar technologies on the environment or society have not been identified. However, land-related issues can be raised – when selecting locations for plants, agricultural land must be avoided. The potential impact on nature should be taken into consideration.

Wind Power

Wind power is generally a cheaper option than solar energy in locations where the average wind speed is greater than 4 m/s during the weakest windy weather.

Although there are doubts about constancy of wind, they can be solved by combining wind with other renewable sources such as solar energy and hydropower. Wind power plants have very low emissions throughout their life cycle, but have a number of consequences for the environment that may reduce their potential. The most important consequences are:

Visual effect – wind turbines are too conspicuous, since they must be installed in prominent places. This (reasonably) can be considered a violation of the landscape, and dissatisfaction on this point grows along with the increasing dimension of the new generation of turbines.

Noise – wind turbines generate aerodynamic noise from wind passing over the blades and mechanical noise from the moving parts of turbines, particularly from the housing of the generator. However, improvements in design to some extent reduce the noise. Turbine farms built away from densely populated areas are generally less irritating.

Electromagnetic interference – wind turbines can dissipate electromagnetic signals causing interference to communication systems. Placing in the proper location (avoidance of military training grounds or the airport) can reduce this impact.

Security of birds – birds can be hurt by contact with rotating turbine blades. Migratory birds are more vulnerable than species that are not moving. Placing turbines outside the path of migratory birds reduces this impact.

The criteria for the realization of getting energy from wind power include:

Project is not running in protected nature areas;

Project is not along the migratory paths of migratory birds;

Project has no impact on the population of bats (in addition to problems of contact and disruption of habitat, also the issue of ultrasound emissions needs to be addressed);

Projects of wind turbines to be based on the studies of biodiversity and to be carried out along with impact assessment on the environment, like any other industrial project;

Projects on wind energy production should be supervised to ensure they do not create a negative impact on human communities and wildlife;

Project should use the latest equipment in order to eliminate noise, vibration, and electrical and magnetic fields, and international financial institutions should not support the usage of old, used installations; and

Projects of coastal wind power plants should be based on accurate analysis of potential impacts on birds and mammals, including their habitat and feeding areas and resources.

Geothermal Energy

Geothermal energy is a clean, renewable energy source from which the world can get heat and electricity. It is considered a renewable source after the heat is released inside the earth and is essentially unlimited.

Geothermal energy can be used to produce electricity for immediate use and for home heating (through geothermal heat pumps). Geothermal energy relies on an existing, permanent source of heat for power generation and therefore is considered a base, constant supply. Since some renewable energy sources can be used only under favorable weather conditions, they are considered to have limited availability to meet the constantly growing needs of the twenty-first century. However, geothermal energy has the potential to provide reliable sources of electricity with significantly lower levels of emissions from fossil fuels and it eliminates the problem of radioactive waste disposal. The factor availability of geothermal energy is about 95%, which means that geothermal power plants can be used during 95% of any period of time, based on many decades of observation of these plants. The capacity factor of geothermal energy ranges from 89-97%, depending on the type of system that is in use.

Geothermal energy during operation may discharge emissions into the atmosphere. As for the gases, they are primarily carbon dioxide and hydrogen sulfide, along with traces of ammonia, hydrogen, nitrogen, methane, and radon, and volatile metals such as boron, arsenic, and mercury. Emissions should be controlled by strict regulations and control methods of the geothermal industry itself used to check the compliance with the requirements of this regulation. Systems of decreasing hydrogen sulfide reduce damage to the environment but are expensive to install.

The criteria for projects in this area include:
Project relies on the return of the used geothermal water back into the earth, so there are no possibility of contamination of river or lake systems with hot water; and
Process equipment is used for the elimination of hazardous emissions of greenhouse gases, hydrogen sulfide, and other gases from the thermal waters.

Biomass

Biomass is a renewable source of energy that occurs in various ways, both by people and nature. It is obtained from many sources, including from by-products of the wood industry and agricultural crops. Biomass does not release carbon dioxide into the atmosphere after it absorbs the same amount of carbon during the growth that is released during use (but the energy consumed during transportation and processing should be taken into account). The advantage of biomass is that it can use the same equipment to produce electricity in existing power plants that burn fossil fuels. Biomass is an important source of energy and the most important fuel worldwide after coal, oil, and natural gas.

Production of biomass for energy purposes implies the use of a large area, which, when connected with the usual method of land, creates a significant impact on biodiversity and the manner of its production. Therefore, the use of crop residues to produce electricity, heat, and biodiesel, whether from sugar beet, rice chaff, or straw and other waste from crops, forestry activities, or vegetable oil, etc., is one of the best ways of sustainable energy production, while it does not preclude other important ways to use agricultural waste such as soil conservation, for example. Obtaining methane for energy production from the decomposition of municipal solid waste (for sanitary landfills) or agricultural waste (e.g., pork and poultry manure) may be a viable alternative, depending on the method of treatment of these wastes. Burning – incineration – of municipal and industrial waste is not an acceptable solution, because it creates a wide variety of polluting emissions, which are very dangerous to human health and the environment.

Decisions about acceptance of burnable waste are less sustainable and represents a bigger threat to the society compared to other alternatives for solid waste management.

- Solid biomass: Organic, non-fossil material of biological origin that can be used as fuel for heat or electricity.
- Wood, wood waste, other solid waste: Dedicated crops grown for energy (poplar, willow, etc.), a number of materials from wood obtained in industrial processes (especially in the wood and paper industry) or indirectly in the forestry and agriculture (burnt wood, waste branches, wood chips, sawdust, wood chips, resin, etc.), and other wastes, such as grass, rice chaff, shells of nuts, husk, poultry waste, and others.
- Dedicated crops: In theory, any plant material can be used for bioenergy production, but the material that is grown specifically for this purpose is characterized by large amounts of biomass and high energy potential. Bioenergy can take many forms, including bioelectrical energy and biofuels. Biogas and biodiesel fuels are derived from plants, which can be used in much the same way as natural gas or petroleum. When the biofuel is produced from domestic plants or plants grown without fertilization and without irrigation, it creates significantly less emissions than conventional petroleum. If they are produced and used locally, biofuels can help the local economy.

Biofuels are a comparatively clean alternative to oil as a fuel source and can be particularly useful for traffic. They have the potential to provide a fuel that emits small amounts of carbon compared to the amount of conventional fossil fuels. Also, biofuels generally produce less polluting emissions than other fossil fuels. However, one must carefully consider the impact of production and the use of biofuels on the environment in a broad sense, including the impact of biofuel production on local ecosystems. For example, the planting of forests as plantations for biofuel is problematic, since such plantations require enormous amounts of water or the use of dangerous supplements in fertilizers and pesticides. The amount of conventional fuel needed for the production of certain biofuels cancels or nearly cancels the usefulness of biofuels in reducing the amount of greenhouse gases. Biofuels must be accepted, with caution, as an alternative fuel source since they are one of the few alternatives to conventional oil use.

Biodiesel is a fuel whose properties are the same as those of mineral diesel, but it is derived from renewable energy sources. In Great Britain, the most attention has been focused on the production of biodiesel from rapeseed oil. The technical term for the fuel from seed is methyl ester. Any cooking oil, including used (fried) oil, generally has the potential to produce biodiesel. Crude oil from rapeseed goes through the esterification process that removes glycerin, allowing oil to have properties similar to mineral diesel. Vehicles do not require any modification to use biodiesel, since it can be mixed with normal diesel. There are no significant differences found in engine performance upon the comparisons conducted between biodiesel and conventional diesel. Glycerol is a valuable byproduct from the reaction and is used in over 1500 products, including pharmaceuticals, polymers, paints, and canvas. The sale of glycerol may recover the production costs. Biodiesel is significantly safer than diesel generated from oil since it has a lower flash point, it is more difficult to ignite, it does not produce explosive gases, and it even has a lower degree of toxicity for humans and animals if swallowed. Biodiesel is biodegradable, and in case of spillage less damage is created to the environment. Usefulness of the program includes reduced levels of carbon monoxide and total hydrocarbons, and mild odor. Contrary to conventional diesel, it is completely free of aromatic compounds and sulfur, which are poisonous and subjects to legislation.

The criteria for biomass energy projects includes:

Planning and planting promote the protection, restoration, and conservation of natural forests and do not increase pressures on natural forests and protected nature areas;

Plantations do not have a negative impact on natural habitats;

Use of genetically modified organisms in crops is excluded;

Native species are preferable to exotic species in plantations and restoration of degraded ecosystems. Exotic species, which should be used only when their properties are better than native species, should be carefully monitored to detect unusual mortality, disease, or excessive multiplication of insects and other adverse impacts;

Project improves the composition of soil, fertility, and biological activity;

Project does not involve the use of dangerous fertilizers and insecticides;

Project does not have negative impacts on water availability and its quality or impact on river and lake systems in terms of water;

A single plant species is not planted on large areas, unless tests and/or experience demonstrate that it is ecologically suitable for it, that it is not invasive, and it does not have significant negative ecological impacts on other ecosystems;

Project does not raise issues of land ownership, use, or access;

Project does not jeopardize the safety of food at any level (plantations for energy dramatically reduce/eliminate food crops in the area where they are present);

Project does not include the increase in greenhouse gas emissions;

Biomass as an energy source should be of domestic origin (no import of bio-mass from third-world countries);
Project cannot give rise to social conflicts; and
Production of biomass must have an obvious positive balance/energy balance (the produced energy against the all energies spent in the process).

Biogas is a gas that comes from the decomposition of biological (organic) waste regardless of whether the decomposition takes place at the landfill, in a closed equipment for the anaerobic decomposition of plants, or in wastewater treatment facilities. Utilization of this atypical source of waste gas is not only beneficial for the environment, but also for owners of milk farms. Although it requires the use of digesters, animal waste can be converted into valuable liquid natural gas, fiber, and fertilizer. Biogas, whether generated as a byproduct of the digester processing brewery hops or more often from the decomposition of animal waste, is an excellent source of energy after the predominant methane gas is purged. Liquefied natural gas obtained from biogas can be a locally produced, clean, sustainable, and cheap source of renewable energy, which makes it a very attractive option for growers on farms for milk production and end-users of gas in the local community. A direct negative impact has not been identified. The recommendations are possible in terms of decentralization – local use of biogas, for example.

Fuel Cells

Fuel cells are electrochemical devices that convert fuel energy directly into electricity. They function like regular batteries when they are fed with the fuel (hydrogen) through the anode and an oxidant (e.g., air) through the cathode. Fuel cells bypass the traditional production of thermal energy in the form of combustion, the conversion of heat into mechanical energy (as with the turbine), and ultimately convert mechanical energy into electricity (e.g., using the generator). In contrast, fuel cells chemically combine the molecules of fuel and oxidant without burning, thus freeing inefficiencies and pollution of traditional combustion. Hydrogen can be produced by electrolysis using energy generated from renewable sources, or hydrogen-rich fuel transformation such as oil and natural gas. The process of reshaping transmits relatively small amounts of CO_2, but it does not produce emissions of other pollutants generated by burning fossil fuels. Technology is at a relatively early stage of development and not yet cost-effective compared with the price of conventional energy sources. Possibilities of future use of fuel cells are, however, very broad and include transportation (private vehicles, public transport) as well as domestic and industrial needs.

Non-financial Effects of Renewable Energy

Improvements in energy management are the consequence of long unplanned exploitation of natural resources and pollution of the atmosphere as a consequence of their exploitation and use. The energy problem is globally recognized as one of the priorities of sustainable development that can and must be solved by applying different activities at all levels. For all projects related to quality management of sustainable energy development, there are large groups of interested entities including:

Companies – as places of production or consumption, through which the application of the appropriate management methods acquire a number of advantages that are reflected in the positive image acquisition, financial gain, lower costs, and the provision of certain subsidies;

Users – as consumers of the goods or services of companies who wish to support the energy efficient companies and therefore produce very strong pressure on the responsible institutions;

Citizens – as consumers of energy, who are encouraged to save energy and to independently invest in renewable energy sources;

Legislation – as necessary framework for the functioning of the system, has to regulate the area of sustainable energy management and to be adjusted depending on changes in the demands of global integration;

Investors – as a part which wants to direct their capital towards companies that do environmentally responsible business, investing in projects aimed at energy saving or production of energy from renewable sources;

International community – this group is interested in efficient energy management and energy stability after one of the key issues for the survival of the world community;

Current generations – as a party which has shown interest in the regulation of the complex issue of energy management, regardless of the fact that full effects of the implementation of projects in this field cannot be expected in the short term; and

Future generations – as the biggest beneficiaries of changes in the safe area of production and consumption of energy, who are entitled to have the same amount of resources that the generations before them disposed of, which is a key principle of sustainable development and the principle of intergenerational justice.

7.7 ECONOMIC EFFECTS OF SUSTAINABLE ENERGY MANAGEMENT

Sustainable energy management is one of the key demands of managing natural resources and sustainable development in general. Sustainable development, in essence, represents a system of our integrated subsystems (economic, environmental, social, and institutional), and any change in one subsystem causes effects in the remaining three. Therefore, no changes or activities should be observed independently and separately from the others.

Special problems exist in balancing economic and environmental development, as they are largely interconnected and often directly opposed. Economic development has always been the only indicator of development and is measured by specific economic indicators. The development measured in this way did not take into account the impact that economic development had on the environment, as measured primarily by resource depletion and pollution production. Economic development has had its price, which was not considered until sustainable development concept set the criteria and caused a reconsideration of the impact of economic development of the environment.

The modern concept of sustainable development questions the justification of economic development at any cost and, in the process of measuring the development level, it introduces new indicators that are directly opposed to them. Indicators of environmental quality and quality of life are becoming equally important parameters. In this way, economic development ceases to be the dominant goal.

Economic development is essentially based on the exploitation of natural resources and environmental pollution, which becomes the subject of a separate study. Modern business shows a particularly strong connection between energy consumption and development in general. The modern world consumes large amounts of energy, and the intensity of energy production and consumption is different in different parts of the world. Energy production and consumption have a direct impact on environmental quality, but they are a necessary precondition for economic progress. To determine the balance between the consequences of energy production and consumption on the one hand, and welfare resulting from economic development on the other is a very sensitive issue. There are numerous efforts to create a balance between energy resources, consumption, and satisfactory development, which are linked with a great number of global problems and disagreements on all levels.

When planning development, each company should take into account all the consequences that will happen after introducing particular changes. A sustainable approach to business and energy management presents a big change that brings about certain consequences, whereby the changes in the economic sphere are of special importance. While the sustainable approach is desirable, it does not lead to economic collapse of the company. Thus, it is necessary to consider and plan the company development with respect to the rules of sustainability, but that should not threaten a company's survival.

Globally, sustainable energy management has an economic effect that can be viewed on two grounds [90]:

> Sustainable energy management in a company produces a range of positive effects, whereby positive economic effects arise primarily as a result of savings arising from the proper implementation of sustainable energy management; and
>
> Realization of projects from sustainable energy management and/or exploitation of alternative energy sources are a means to achieve certain incomes, where these investments are treated as separate business ventures.

Both these approaches offer the possibility of considering sustainable energy management as a change in the company conditioned by the need to operate in a socially responsible manner while also having strong economic justification.

These basic characteristics of energy production from alternative sources point to the fact that obtaining energy in this way is certainly desirable from the aspect of sustainable energy management and should be considered in each situation. Therefore, an overview of the costs of obtaining electricity in this way are given in Table 7.4, so that companies can compare them with investment costs and thus assess the justifiability of investment and time of the investment return [92].

These amounts must be considered along with other indicators in each country. Government subsidies given to producers of alternative energy should also be taken into account. In many countries (the EU region and all other European countries), producers of alternative energy have a privileged status, they receive certain subsidies, and the state is obliged to purchase all the energy they produce as a pre-emptive right [25].

SECOND STRATEGIC OBJECTIVE – ENERGY EFFICIENCY

7.8 ENERGY EFFICIENCY

Definition

Defining energy efficiency involves consideration of several different approaches but covers a wide range of energy consumption indicators.

TABLE 7.4 Economic assumptions of energy production from renewable sources

Renewable power generation costs 2010		
Power generator	Typical characteristics	Typical electricity costs (U.S. cents/kWh)
Large hydro	Plant size: 10-18,000 MW	3-5
Small hydro	Plant size: 1-10 MW	5-12
Onshore wind	Turbine size: 1.5-3.5 MW	5-9
Offshore wind	Turbine size: 1.5-5 MW	10-14
Biomass power	Plant size: 1-20 MW	5-12
Geothermal power	Plant size: 1-100 MW	4-7
Rooftop solar PV	Peak capacity: 2-5 kilowatts-peak	20-50
Utility-scale solar PV	Peak capacity: 200-100 MW	15-30
Concentrating solar thermal power (CSP)	50-500 MW trough	14-18

Energy efficiency encompasses all changes that result in a reduction in the energy used for a given energy service or level of activity. This reduction in energy consumption is not necessarily associated with technical changes, since it can also result from better organization and management or improved economic efficiency in the sector (e.g., overall gains of productivity). This definition of energy efficiency is provided by the World Energy Council (WEC) [34].

Energy efficiency improvements refer to a reduction in the energy used for a given service or level of activity. The reduction in energy consumption is usually associated with technological changes, but not always since it can also result from better organization and management or improved economic conditions in the sector ("non-technical factors").

In some cases, because of financial constraints imposed by high energy prices, consumers may decrease their energy consumption through a reduction in their energy services (e.g., reduction of home temperature or car mileage). Such reductions do not necessarily result in increased overall energy efficiency of the economy, and are easily reversible. They should not be associated with energy efficiency.

To economists, energy efficiency has a broader meaning: it encompasses all changes that result in decreasing the amount of energy used to produce one unit of economic activity (e.g., the energy used per unit of GDP or value added). Energy efficiency is associated with economic efficiency and includes technological, behavioral, and economic changes.

Energy efficiency is first a matter of individual behavior and reflects the rationale of energy consumers. Avoiding unnecessary consumption of energy or choosing the most appropriate equipment to reduce the cost of energy helps to decrease individual energy consumption without decreasing individual welfare.

Avoiding unnecessary consumption is certainly a matter of individual behavior, but it is also often a matter of appropriate equipment: thermal regulation of room temperature or automatic de-activation of lights in unoccupied hotel rooms serve as good examples of how equipment can reduce the influence of individual behavior.

Energy efficiency is a concept that differs from the concept of energy management. Energy efficiency essentially means consumption of less energy to obtain the same level of energy service, whereby energy management means the planning, control, and optimization of the energy use, in particular toward achieving energy goals.

The Main Elements of an Energy Efficiency Monitoring Process

Energy efficiency monitoring is in essence a very complex process, consisting of several elements (input, output, and activities).

Inputs that characterize energy efficiency can be defined by many criteria separately and/or simultaneously. In addition, the measurement can be performed using conventional or modern, more complex methods of measurement, using various metrics. The energy is obtained from various sources in various forms, the distribution and transformation of energy from one form are frequent and linked with losses, and contemporary needs are based on the use of energy in different forms and quantities.

Defining the output elements of an energy efficiency process is also complex and subject to a great number of factors and needs. In the majority of cases, output indicators, used to determine the level of energy efficiency, are determined by the energy that is transformed and incorporated into the final product or service, which is added to the waste energy and energy that is lost in the form of externalities. Definition of waste energy and the energy lost in the form of externalities is linked with many problems. The methodology for its determination is undefined and incomplete, and the need for more precise work in this area is evident. Only a precise determination of the total output power can ultimately determine the level of energy efficiency and its changes.

All activities that occur in the process of changing energy efficiency are the subject of particular interest and constant improvement. The improvement of energy efficiency is an activity defined as one of the strategic objectives of improvement in the energy management process, both on global and national levels, and at the level of each company individually. Changes in the level of energy efficiency may result from the implementation of various activities (of technical and/or organizational nature), and opportunities for improvement are found in virtually every sphere of activity associated with energy consumption.

The Main Principles of Global Energy Efficiency Policy

Reconciliation of a company's plan with global and national energy efficiency policies means that the company's policy will be compiled in a way that will be fully adjusted to the basic regulatory documents existing in the particular area of the country, region, and world. Globality of the energy-related problem has created the need for cooperation at all levels, and compliance with global policies in this field is imposed as one of the imperatives. All strategic documents existing at the global level recognize the problem of energy efficiency as a key one and accordingly, resolving and improving is placed on the list of long-term priorities (in the energy policies of the United States, Canada, Japan, European Union, and Russia). In all these countries and regions, the improvement of energy efficiency, along with the increase in the volume of energy generated from renewable sources and reduction of greenhouse gas emissions, is one of the top three priorities.

A special aspect of global energy efficiency policy includes the development of clean technologies that will contribute to the implementation of energy efficiency improvement. There are specific standards and methods that support and promote energy efficient technologies and products. Energy efficiency is being studied and the solutions for use in individual households are being improved. In addition to technological improvements, the mechanisms and instruments that accompany the economic aspects of energy efficiency are being developed, implementing changes in the way of determining GDP and other conventional methods for determining the economic indicators of the condition.

One of the goals of global energy policy is to increase the amount of energy obtained from renewable sources. The U.S. and EU strategic documents define that the amount of energy produced from renewable sources shall be 20% by 2020. The technologies of energy production from renewable sources are being improved. There are special procedures that support investing in this area.

Reduction of greenhouse gas emissions and other emissions occurring in the production, distribution, and consumption of energy (primarily CO_2, NO_2, and SO_x) are also defined as a strategic objective at the global level. Technological solutions that contribute to reducing these emissions, as well as methods of measurement, monitoring, and trading carbon credits, are being developed.

Energy efficiency management is a process based on several assumptions. First of all, it is designed in a way to ultimately enable improvement of energy efficiency parameters both in physical and in economic terms. In addition, the energy efficiency management process must be adequately integrated into all other management processes within the company as well as the process of strategic management in general.

Management in this area is aligned to the general starting points that observe the problem of increasing energy efficiency as one of the priorities of sustainable development in general. Therefore, any change in a positive direction can be considered acceptable and fully justified. The management process must be designed to ensure positive progress, no matter how small. Management based on monitoring incremental changes is imperative for energy efficiency management.

The energy efficiency management process can be viewed in two ways. On the one hand, it can be seen as part of a sustainable energy management process, and on the other hand, as an independent process. Sustainable energy management is a complex process based on implementation of management processes that comply with specific production and consumption, and goals are aligned with the basic objectives of sustainable development. Both observation methods are considered necessary. Energy efficiency is just one of the parameters in the complex process of sustainable energy management and it must be regarded as one of its main goals. Energy efficiency is the goal and energy management is a way to achieve this goal (see Figure 7.29).

7.9 ENERGY EFFICIENCY AUDITS

The Importance of Energy Efficiency Audits

Modern society, which is turning to operations based on the principles of sustainability, largely puts the problem of energy consumption at the forefront. Development plans incorporate issues associated with energy into strategic objectives and undertake extensive activities that enable the improvement of the

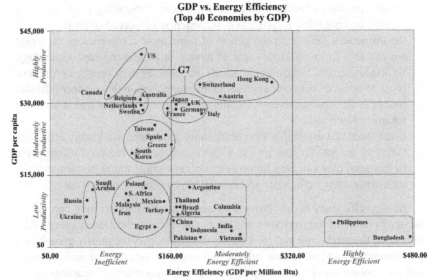

FIGURE 7.29 GDP vs. energy efficiency

most important indicators of sustainable energy development, such as reduction of greenhouse gas emissions, increase in the energy production from renewable sources, and increase in energy efficiency.

Given the nature of activities, the increase in energy efficiency can be regarded as potentially the most important indicator of success of the company in the contemporary and future long-term period. Energy efficiency is the amount of energy that is used for obtaining a certain amount of products and is expressed in physical size. Much greater importance is attached to the indicator of energy efficiency as measure of the amount of money spent to cover energy costs for performing these activities. Energy efficiency as an economic size is a particularly important indicator in the following cases:

Companies that perform activities that are, by their nature, great energy consumers, which in most of the cases means energy generated from non-renewable sources. Globally, most companies today consume energy in a quantity equal to or greater than the amount that the company spent in the previous period. In the developed world, companies pay increasing attention to reducing energy consumption, but generally speaking the average consumption is still high;

Companies spending imported energy-generating products, which greatly increase the level of import dependence and threaten the security of both the company and the country in which it operates;

Companies that use outdated technology are particularly large consumers of energy, and they exist in most of the world. Only a small number of countries are in a position to make significant investment in installing modern energy friendly technologies. Regardless of the extensive evidence of profitability of the investment after a certain period of time, most companies in the world are not able to invest large sums of money in purchasing new technology. By exploiting old technologies, which are, besides high energy consumers, prone to disorder, the company uses more energy, and in that manner it further invests in the losses, thereby reducing the amount of income that is available for investment; and

Companies that are highly environmentally aware also see energy efficiency audits as an activity that provides them with certain required information. In this way, they can plan future activities and adjust future performance for improving energy efficiency and sustainable business in every sense.

Energy efficiency audits provide companies with data so they can get an accurate picture of the degree of sustainable business in the area of energy consumption, the environmental load which thereby occurs (primarily greenhouse gas emissions), and economic effects that occur at the same time. Measuring and monitoring energy consumption using only quantitative measures is not considered sufficient for the purpose of obtaining complete information. It is necessary to monitor indicators of energy consumption per unit of activity per certain working step, process, and unit of product. The full picture is obtained in the end by measuring indicators of external costs, which represent the economic value of environmental pollution caused by energy exploitation.

Energy Efficiency Audit Objectives

Depending on the type of activity, size of company, availability of energy resources, and other parameters, different objectives for performing energy efficiency audits can be defined, but in most cases the objectives include:

Identifying the type, quality, quantity, and price of energy resources used for activities of the company;

Special monitoring of energy consumption from renewable sources;

Definition of consumption units (profit centers – work processes, operations, technological lines) on which the monitoring of energy efficiency indicators will be performed;

Determining the energy cost of previously defined profit centers, thus providing insight into the energy efficiency of individual profit centers;

Defining all parameters of energy inputs and outputs;

Identifying internal energy transportation routes and potential losses;

Identifying the place of origin, quantity, and characteristics of energy waste, along with special monitoring of pollution that is the subject of legislation and internationally accepted agreements;

Monitoring all of the previously listed indicators by hours, days, months, seasons, and years, identifying regularities and irregularities in certain form, and defining the cause; and

Determining savings in some profit centers, whereby the potential savings, investment required, the list of priorities, and expected results are particularly defined.

An energy efficiency audit, through its objectives, obtains large amounts of data, which lead to one, final result. The main objective of an energy efficiency audit is to get a broader picture of the level of sustainable energy company operations, including energy, environmental, and economic indicators. In this way, companies can assess their current position and plan future activities.

Energy Efficiency Audit Levels

Level I Audit

The Level I energy audit includes an overview of business' annual energy consumption and a walkthrough to evaluate the performance of existing facilities. This audit will encourage businesses to reduce energy use by 10%. It includes the inspection of a facility to identify specific issues and needs. The audit provides a description of the facility, including:

Brief description of what type of activity occurs in the facility;
Square footage;
Year built;
Building envelope and insulation;
A summary of business' annual gas consumption by month;
Operation and maintenance recommendations; and
A brief overview of equipment replacement and projected annual energy savings.

Level II Audit

The second phase is a more in-depth study that includes capital cost items such as boiler or burner replacements, energy management upgrades, etc. If applicable, this audit will work in cooperation with an electric utility. The Level II audit provides a comprehensive report with detailed energy efficiency recommendations. This report includes the following:

On-site inspection of the facility to identify specific issues and needs, including building envelope, windows/doors, sidewall and attic, and insulation levels;
Inspection of HVAC equipment and current operation of this equipment;
Evaluation of HVAC system efficiency;
Operations and maintenance procedures;
A description of the facility, including description of what type of activity occurs in the facility, square footage, year built, building envelope and insulation, HVAC system, domestic hot water, commercial cooking and age;
Major equipment list including hours of operation;

A summary of important findings along with any supporting data, measurements, drawings, etc.;

A summary of business' annual gas consumption by month;

Gas usage pie chart with calculated annual heating costs for windows/doors, attic/roof, sidewalls, fresh air ventilation including domestic hot water, laundry, kitchen appliances, and other major equipment; Operation and maintenance recommendations, including ways to optimize equipment operation, update energy management protocol, and improve maintenance practices and operational changes;

Major equipment investment recommendations, including

replacement of HVAC equipment size calculation based upon heat loss and ventilation analysis;

Calculated annual energy savings based upon building space; and

Economic analysis (payback, ROI, etc.).

The proposed model of energy efficiency audit is based on the average needs of the organization and should be changed and adjusted for specific needs. However, all audit techniques should simultaneously observe both physical and economic indicators of energy flow, and thus provide information relating to the energy and economic viability of the company.

7.10 Improvement of Energy Efficiency

The changes in energy prices are the best and a natural driver of activities aimed at increasing energy efficiency in all sectors and their establishment, along with the necessary social programs of the state for those categories of energy users that need to be subsidized.

The practice of developed countries has shown that the state must play a leading role in systemic actions aimed at increasing energy efficiency. This is especially necessary in economies where the cost of energy and energy-generated products is controlled by the state. In such conditions, end users see energy as a necessary cost, the reduction of which, as a rule, is not a priority. This whole area of activity aimed at increasing energy efficiency discourages interest for investing in such projects. When energy and energy-generated products become goods fully subject to all rules of the supply and demand law, the leading role of the state will no longer be necessary. Free investment funds due to their nature "look for" projects where the realization will increase their value, by increasing energy efficiency.

Until these barriers are removed, the leading role of the state in terms of increasing energy efficiency is reflected in the creation of adequate legal framework (laws, bylaws, regulations, etc.), in creating incentives for the implementation of a number of technical measures that lead to its increase, and in influencing the behavior of end users by raising their awareness in this area.

In addition, the establishment of liability in respect to energy management (introduction of energy management) and a constant concern about increasing energy efficiency in all sectors of energy consumption and production are a prerequisite for the implementation of systematic measures to increase energy efficiency in the country as a whole.

Bylaws govern the rational use of energy, among other things, so it is important to regulate the obligation of energy entities (energy users and consumers) to establish appropriate units for management of energy, so-called energy management, which should follow energy parameters and periodically report to the relevant government institutions on the current parameters of energy, promote energy efficiency by taking measures for its increase, and report to relevant government institutions achieved results, observing the prescribed procedure.

Encouraging energy efficiency can be enhanced by applying specific measures, and the following can be considered the most important:

Providing a variety of tax and other incentives for companies that implement projects in order to improve energy efficiency;
Introducing the principles of energy efficiency in public procurement and in the criteria for the allocation of other funds and loans;
Ability to support projects aimed at increasing energy efficiency;
Cooperation with international financial institutions in order to realize favorable credit arrangements;
Promoting the interest and building capacity of local banks to lend to projects in the field of rational use of energy;
Creation of legal conditions for a special type of financing by "third parties" or for the operations of companies that implement saving measures (in the companies or municipalities) and pay their for services from the achieved energy savings; and
Ratification of the Kyoto Protocol and the implementation of other necessary activities in order to open opportunities for the use of Clean Development Mechanism under the Kyoto Protocol for the realization of energy efficiency projects. The Clean Development Mechanism (CDM) is one of the flexibility mechanisms defined in the Kyoto Protocol. It is defined in Article 12 of the Protocol, and is intended to meet two objectives: (1) to assist parties not included in Annex I in achieving sustainable development and in contributing to the ultimate objective of the United Nations Framework Convention on Climate Change (UNFCCC), which is to prevent dangerous climate change; and (2) to assist parties included in Annex I in achieving compliance with their quantified emission limitation and reduction commitments (greenhouse gas (GHG) emission caps).

In addition to these measures, it is also necessary to work on raising awareness of end users, the implementation of education in this area, and the implementation of demonstration projects to identify and promote good practices.

Technical and Organizational Measures

The basic organizational measure aimed at the rational use of energies implies the introduction of managing energy (energy management) for all consumers whose total installed capacity is more than 1 MW. This measure involves the monitoring of energy consumption and constant concern about increasing energy efficiency. The measure is introduced gradually and does not require special investment funds (it is realized from current maintenance), and it brings the minimum annual savings in industry, in case of power engineering and PUC, to 3% of the final energy consumption, i.e., annual savings of 23.7 million Euro. For the implementation of measures it is necessary to adopt the Law on the Rational Use of Energy as well as develop and implement training of needed staff.

As for the sectors of industry and building construction, energy audits aimed at recording the current status of spending and identification of measures to improve energy efficiency is of great importance. In this sector of industry a 15% reduction of energy consumption may be achieved between 15 and 20%.

Significant potential is found in improving the combustion process along with regular auditing of this process. This measure can reduce energy consumption by 2-3%, equal to 0.0809 million ton (0.9402 TWh) or 28.2 million Euro decrease in fuel costs. For the implementation of these measures, additional funds are not required, since existing maintenance funds can be used. However, it is necessary to adopt a regulation and for energy entities to implement it, whereby the inspection will strictly control the execution of this obligation.

By adding condensing areas on boilers in industry and in heating plants designed and built for heavy oil where natural gas is currently used, an increase in the energy efficiency of the boiler units of 6-7% can be achieved. This investment should be considered since as a rule, it is repaid in less than one year if the boiler works 4,500 hours per year. This measure can be achieved through partial stimulus by the country or through the company's own funds.

Using waste heat from power plants and production processes has the potential to increase energy efficiency that can reach up to 20% of the total heat production needs. This is typical for the drying, processes in the chemical and food industry, industry of building materials, and the like.

The energy integration of the production process, especially in the chemical industry, presents a potential that can be used to increase energy efficiency of a factory's heating systems up to 5% with a relatively short repayment period (usually less than one year and not longer than three years).

Replacement of the existing engines. This measure can be explained in the following example. 4000 MW of electric motors were installed. Replacing them with engines having greater efficiency of class EF1 and EF2 reduces consumption of electricity of 188 GWh per year or 7.5 million Euro. Investment for the implementation of these measures is 200 million Euro. This measure provides substantial energy effects, but given the current electricity prices its implementation is economically unrealistic, since due to the present conditions of factories and parity pricing to end users, it brings effects only after 15 years.

Transport Sector

Measures to be implemented in the transport sector include the following:

Defining, identifying, and adopting a national strategy on the development of transport systems. The national strategy on the development of transport is necessary, since the current situation is alarming in all sectors.

Rejuvenation of the fleet in all sectors. Besides the other issues and considering the energy efficiency aspect, the age of the fleet is among the key issues. It is therefore necessary to adopt measures to stimulate the purchase of new cars and discourage the use of vehicles older than 15 years. Each country has its own regulations regarding car import.

Building Sector

The following activities have been defined in this sector:

Switching from heating using electrical energy to other forms of energy. It is estimated that it is possible to achieve a reduction of the current electric power system load for heating using electrical energy, which currently amounts to 3062 MW, by 1222 MW by replacing it with the systems of remote and central heating or natural gas, which would result in annual savings in electrical energy of 1500 GWh. The value of that energy is, at the current purchase price of electricity in the market, 60 million Euros a year. This requires a planned construction of these systems in areas dominated by heating with electricity. It is likely that these zones are suitable for the development of natural gas for easier construction and less expensive installation. However, due to the possible amount of investment, this measure requires detailed individual analysis of each site separately.

Replacement of lightbulbs in homes and commercial buildings. By replacing only two lightbulbs of 100 W with corresponding fluorescents of 20 W, an electricity savings of 701 GWh can be achieved and the source capacity of 480 MW released, i.e., 28 million Euros. Investing in the implementation of this measure (purchase of bulbs) amounts to 18 million Euros with a repayment period of 8 months. Implementation of these measures is possible in various ways.

Adoption of new regulations on the outdoor design temperature. The aim of this measure is the reduction of the projected installed power of home installations and heating sources for buildings, improving energy efficiency in the use of heating systems, and reduction of capital costs for these plants. Reduction of the middle design temperature at the level of the state by 40°C would result in 10% energy savings for heating (the implementation of this measure would eliminate the currently evident heating above the projected values). Implementation of the measure does not require investment funds.

Consistent application of ISO 14000 and other standards. Consistent application of basic and other related standards for the design of new buildings and their thermal protection can reduce the design installed capacity of heating by 30-40% and achieve approximately such an energy savings for heating. Implementation of measures requires only strict enforcement and compliance with existing regulations, as well as the achievement of real price of apartments with the energy quality control of the facility on the handover of the buildings and heating systems in them. It is necessary to adopt a maximum allowed value of the final energy for heating buildings of kWh/m2 year and to expect the designer to meet the stated value in terms of heating physical characteristics. Data on annual consumption of energy should be an integral part of the documentation of each building. Licensed individuals would be expected to determine consumption based on the project.

Establishment of incentive funds aimed at improving the thermal protection of existing residential buildings. It is necessary to establish incentive funds for improving the thermal protection of existing residential and non-residential buildings at national, province, and local levels. For that purpose, the budget, dedicated foreign and domestic credit, and donor funds are to be used, with the mandatory participation of owners in cofinancing of additional thermal protection of buildings.

To replace the windows in all multifamily residential buildings connected to central heating systems, in 60% of multifamily residential buildings that are now using electricity in all single family buildings with central heating, and 40% of single family residential buildings using electricity or natural gas for heating, the amount of 936 million Euros would be necessary for multifamily residential buildings, or 1326 million Euros for single residential buildings. The overall effect of the final energy savings for heating upon the realization of this investment amounts to about 7.4 TWh per year (4.05 TWh for multifamily and 3.35 TWh for single family residential buildings) and if the money is invested now, the total repayment period would amount to 8.9 years (for multifamily 7.1 years with an annual saving of 132 million Euros and for single family 10.8 years with annual savings of 122.7 million Euros).

For the purpose of improving the insulation of the walls of the facilities it is necessary to invest 341 million Euros for multifamily residential buildings, and 497.9 million Euros for single-family residential buildings. The total effect of the realization of energy saving activities indicated above is the savings in the final heating of about 3.5 TWh per year (1.9 TWh for multifamily residential buildings and 1.6 TWh for single family residential buildings). In the period from 2007 to 2012, it is envisioned that the investments would achieve 10% savings of the stated potential.

Adoption of regulations for the constructors of new buildings and owners of the existing building to obtain a certificate of energy efficiency for each building.

For new buildings, which will be heated from the system of remote or central heating, it is necessary to use TPV in the substations and boilers of these systems.

Only basic measures to increase long-term energy efficiency have been proposed. Given the fact that saving of energy and its rational use is one of the imperatives of sustainable energy management, determination of new measures, specific to a particular activity, is under development.

THIRD STRATEGIC OBJECTIVE – RISK MANAGEMENT IN ENERGY FACILITIES

7.11 ENERGY MANAGEMENT RISKS

Energy management is a complex, long-lasting process conditioned by numerous different factors. As such, it is implemented in an environment that is more or less unstable, where the environment influencing the processes of getting, distributing, and consuming energy in the world is particularly variable and unpredictable. Certainly, there exist and are applied numerous possibilities to predict fluctuation of the global energy market, but particularities and significance of energy for overall development inevitably lead to problems and events that are difficult to manage. Energy management represents the process attached to increased risk and thus must be planned and implemented with special care.

Sustainable energy management is a layered energy management process and thus it becomes an even more complex process with the risk assessment that accompanies and influences it. Risk, as a probability of a specific event, represents a complex phenomenon that can be studied and observed from many aspects. In the process of sustainable energy management risk can be regarded as an individual process that is integrated with the sustainable energy management process.

During risk assessment in the process of sustainable energy management it is necessary to take into account and analyze numerous internal and external factors. Apart from that, it is necessary to analyze economic and non-economic risk factors that can influence the success of sustainable energy management implementation in a particular business system.

Economic risk factors include a large group of events that can occur in an economic environment, which above all includes:

Supply market – where it is necessary to estimate all factors of energy source supply, their availability, purchase price, distribution terms, quality, and specificities with specific analysis of energy supply capacity produced by alternative resources. It is also necessary to analyze capacity to produce energy from domestic resources;

Sales market – a company that decides to implement process of sustainable energy management at the market discloses its product enhanced with the quality in the form of responsible behavior and as such it must find its place in the range of all other products on the market. In certain cases, products of such companies may cost a bit more since introducing the process of sustainable energy management demands high initial investment. Thus it is necessary to analyze market competence to react positively to emersion of such products; Competition – every company deals with existing competition. It is necessary to consider the strengths and quality of the competition, its geographic distribution, and the market segments that it covers. Contemporary companies, among other things, fight for their customers by practicing environmentally responsible business, so that gaining strong competitive position in great measure depends on behavior toward the use of energy resources; Legislation – every country has certain law regulations defining business regulations and every change can result in major problems for the company. Modern legislation in great measure supports sustainable energy management, but risks of changes in this sphere are still possible; Technology innovations – are of particular importance in the process of sustainable energy development. All changes in technology must be specifically anticipated. It is also necessary to take into account quality, price, availability, delivery, and installation of certain technology as well as other factors of this kind; and Quality product regulation – modern quality parameters in great measure demand products of particular quality regarding the energy exploited for their production, as well as energy characteristics of products where energy saving and energy efficiency are set as top priority.

Non-economic factors are a sequence of various factors that in great measure can have positive or negative influence on the implementing process of sustainable energy management. Regarding the specificity of global distribution and significance of energy sources for human race development, it is necessary to analyze these factors as precisely as possible;

Availability of energy resources – uneven distribution of energy sources puts a great number of world economies into unequal positions, since many countries are bound to import energy sources. Thus every country strives to provide a sufficient amount of energy for its own functioning, but this greatly depends on the regulations dictated by the countries disposing of energy resources; Conditions of natural disaster occurrence – energy resources are the products of countries that are greatly vulnerable to the possibility of a natural disaster, which can greatly endanger their capacity to supply the world market; Conditions of disaster occurrence due to human activities – facilities for producing and processing energy, as well as objects and means of distribution, are in their essence complex and as such they can be particularly subject to occurrence of minor or larger industrial disasters; Political factors – are the factors that determine functioning of the world today in great measure, especially in the energy distribution sphere, which is a limited resource. All political changes must be considered and special attention should be given to political predictions in order to minimize the risk caused by their occurrence; and

Warfare – risk of military action is always present in regions disposing of large energy source supplies, which represents continuous risk for energy source availability to all, depending on energy import.

These are only fundamental economic and non-economic factors that need to be taken into account when estimating the risk of performing a sustainable energy management process. Their analysis is complex but necessary since transition from the ways of traditional to sustainable energy management represents a kind of strategic change that inevitably leads toward big and long-lasting changes.

7.12 BASIC STEPS OF RISK MANAGEMENT

Risk assessment includes performing whole sequences of assessments that are extremely complex for activities that are associated with energy production and consumption since they might lead to environmental accidents with huge consequences. Facilities dealing with energy production and transport in a conventional way are the ones carrying the high environmental risk, while obtaining energy from renewable sources may be regarded as environmentally quite acceptable. Obtaining energy from nuclear power plants represents a particular problem in risk assessment. The risk management process consists of three basic steps, as shown in Figure 7.30.

Identifying the risk is the first step in the process of estimating the risk for the environment. Estimating the risk means observing company technology process from the aspect of environmental protection. When observing technology this way, it is most important to realistically examine all ongoing events (Figure 7.31).

When identifying the risk it is necessary to analyze the production process in detail and to note all pollution points that arise as a consequence of energy consumption that represent potential risks. It is quite possible that during any phase of the production process hazardous emissions are affecting the environment that can be easily and quickly eliminated with a minor investment of money and labor, but there are also events that are difficult to manage. When identifying risk points, it is priority to note events that are highly dangerous that primarily affect closest work environment. Modern technologies are constructed so that

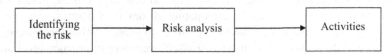

FIGURE 7.30 Risk management process

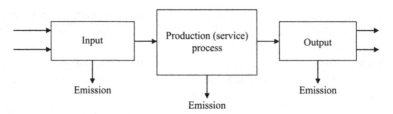

FIGURE 7.31   Basic steps in identifying the risk

a number of risk points are minimized as much as possible, but risk estimating process remains the necessary part of the risk management.

Risk analysis, as a part of the risk management process, implies analyzing every observed critical point. In the first phase it is necessary to note all parameters of observed risk, including:

Risk origin, where it should be determined why the critical point in the technology process has occurred in a specific place, i.e., what caused the pollution. Proper evaluation of pollution origin is of particular significance since it can provide valuable information that can instantly offer adequate solutions. It is a different situation if it turns out that the cause of the harmful effect is a human factor, faulty installation of the facility, a bad technological solution, etc.;

Type of pollution means detecting pollution type, i.e., its physical, chemical, and biological nature;

Activity direction, the various waste substances directed toward different parts of nature; some pollute the air, some water, some directly affect health of people;

Effect intensity, defining the strength of the polluting emission effect; and

Hazard, the final analysis, giving the real danger of pollution to the environment.

Upon evaluating characteristics of the risk event itself, a list of priorities can be determined. Energy consumption is often associated with risks of various intensities, but in its essence it is the cause of a great number of risks that exist in performing certain activities. The energy process is the process expected to cause pollution and thus requires particular attention.

Risk evaluation, where all previously determined parameters are perceived together and a decision (evaluation) is made where individual critical points are labeled as a high, medium, or low risk. Further actions depend on this evaluation. High-risk influences should be eliminated as soon as possible, while low-risk influences in most cases can be eliminated in a simple way. Creating a priority list is important in every management process since it is inefficient and unproductive to invest efforts in eliminating all observed harmful influences emerging due to energy consumption in different places within the company. Experience shows that goals must be ranked according to urgency level and solution possibility, and afterwards they should be approached individually.

Only when one problem caused by energy consumption is solved can the second problem can be approached.

Activities include sequences of activities (applying protection measures, employee training, strict and continuous control) performed in order to adequately detect critical points detected by previous procedures. As an example, it is nuclear power plants or industry with particularly dangerous chemical compounds carrying high environmental risk, but risks are controlled and nothing is left to chance.

It is possible to establish that a certain technology has a great number of critical points whose rehabilitation is quite expensive and thus unprofitable. In such cases it is indicative to consider changing the technological process or changing the whole technology. As a rule technologies are energy efficient and energy consumption is controlled. Nevertheless, modern technologies are not available to all of the world and the problem of transport is still present.

Finally, it is necessary to emphasize that environmental risk assessment is a necessary parameter when evaluating future investment activities, i.e., making decisions about beginning new activities. Every business or investment project must provide information on the possible environmental risk that a new business carries as well as provide information on the method by which the company will keep these risks under control.

7.13 RISK ASSESSMENT

Risk is an event that can be of a different nature or intensity. Risk assessment is in its essence a complex and multifactor analysis associated with all business aspects and is particularly analyzed for the purpose of business economic adequacy assessment. In fact, certain risks may bring into question a company's survival or demand investments that are so large they impose strategic shifts in doing business with far-reaching consequences. Modern business rules and the concept of sustainable development demand operation risk assessment regarding energy consumption of existing companies as well as of investment analysis, i.e., investment in new activities. There are three fundamental questions that should be considered prior to discussing the ranking of the project and difficulty criteria for managing project finance.

First, it is necessary to estimate the strategic scope of the analyzed project or venture, i.e., to estimate to what extent the project should be strategic and opportunistic. Projects with distinct strategic character should be subjected to detailed analysis. Most projects relating to energy consumption can be regarded as projects of this type since the energy is a strategic resource and every activity in this field is of particular significance.

Second, it is necessary to analyze existing types of risk. In conventional portfolio management it is understood that all investments, regardless of their potential profitability, may carry significant failure risk. This is especially true in the case of an environmental investment whose potential profitability may depend on uncontrolled risks unevenly distributed in the observed region.

Third, it is necessary to take into account the economic implications the project entails. Regardless of the priority, sustainable energy development has methods to achieve equity and equivalence by balancing means according to the demand of achieving the biggest environmental profit per money unit spent and should be explored. It is necessary to estimate whether indicators should be set so that they justify forced projects in order to achieve goals that are not economy oriented. Energy production and energy resource management represent an issue of strategic importance to every country, so projects of this kind cannot be regarded solely as economic ventures.

Assessment of potential risk that a certain project entails is extremely complex and the assessment of the energy risk makes it even more complex. Global problems boosting the risk in this sphere further complicate the problem. Namely, problems of uneven distribution of energy resources, conditioning the supply of energy resources, unstable markets, and price fluctuations are only some of the risk factors that should be taken into account. General factors that should be considered include:

Exposure – it is necessary to estimate how many and what kind of uncontrolled risk factors may influence potential results. Consultants' help can be quite efficiently used for exposure analysis;

Timelines – it is necessary to examine whether there is a trend where major risk factors become more or less favorable. If the assessment of time is wrong, this can lead to project failure due to premature investment or it can even become profitable after a long time. Most requests from the sphere of sustainable energy management are strongly expressed, and it is expected that in the future this will be even more true. Thus it can be considered that the timeline factor for energy management will have a stimulating influence on the projects of this sphere;

Volatility – it is necessary to estimate the probability of the anticipated result if all uncontrolled risk factors act together. In the case of analyzing this factor it is really important to have a good plan for the investment since this factor can have a devastating impact if all kinds of risk factors occur simultaneously or within a short period. For projects dealing with energy a high level of these influences can be expected, which is a consequence of instability in the global energy market and thus this analysis should be performed carefully;

Safety – regardless of the number of risks it is necessary to consider whether there is a way to minimize or to decrease the influence of uncontrolled risk factors. This mostly refers to the phase in the risk management process where the reaction decision is made. Safety is mostly achieved by risk transfer or by insuring the project outcome; and

Persistence – projects in the energy industry are mostly long-term projects, so it is necessary to estimate how long the uncontrolled risk factors will influence the success of project. In this way the possibility of activity termination of the noted risks during the process can be perceived. It is also necessary to predict whether the activity of uncontrolled risk factors will continue after the project is finished.

The above includes only the fundamental factors that should be taken into account when planning the investment necessary for the implementation of sustainable energy management. In case that process predicts major investments (which often is the case), risk analysis must be performed precisely. Investments in the process of sustainable energy management are quite often initiated as a result of a company's orientation toward socially acceptable business, so projects are long lasting, demand major initial investment, and positive economic effects take place a few years later. Because of this, the process of investing in sustainable energy management, although environmentally justified, must be performed in a way that will not endanger a company's long-term survival.

7.14 METHODS OF ENVIRONMENTAL INVESTMENTS RISK ASSESSMENT

In practice there are two basic groups of methods to estimate and measure environmental investment risk: non-monetary and monetary methods. Depending on the need, these methods are used in order to perform risk assessment to provide information about the extent to which the risk of investing in sustainable energy management is a risky venture, achieved by risk quantification.

Non-monetary methods are methods that are in the development phase and are featured by specific advantages and disadvantages. The main advantages of non-monetary methods are that they can focus on necessary terms to provide services, can be based on conventional tools, reflect factors that determine aggregate payment capacity, reflect risks of service process failure, can be used as a benefit-transfer method, can be used to create lists of environmental priorities, and can reflect correspondent preferences for the product mix. The main disadvantages are that they cannot be used in conventional cost-benefit analysis, they are not particularly useful to justify increase in costs, are generally not popular with economists and scientists, demand certain decisions that must be additionally estimated, and demand initial investing in data acquisition; and

Monetary methods are methods based on the assessment of the currency value and thus have certain advantages and disadvantages. The main advantages of risk assessment based on the monetary method are that they are a generally accepted economic benefit measurement method, are easy to interpret and use for the purpose of comparison, practical to use in conventional cost-benefit analysis, and for certain individuals they are the only acceptable measurement method. The main disadvantages of monetary methods lie in the fact that evaluation methods of non-market products are not sufficiently developed, and they are expensive to implement. Results of non-market evaluations are often conservative, and in most cases specific for a certain region. There are difficulties in determining the connection between benefits and decision and utility difficulties in order to create environment priority list. Monetary methods are mostly used for investment risk assessments in investments with sustainable energy management as a goal. Basic monetary methods have been developed for practical applications:

Market Price Method – this method estimates the economic value of the ecosystem product or service that can be bought or sold on the market. This method can be used to estimate changes in quality and quantity of the goods and services. It uses standard economic techniques to measure the economic benefit of market goods based on quality of the supply by different prices. The standard method for the measurement of utility values that can be found on the market is the assessment of consumer and producer surplus. The economic benefit total or economic surplus represents the sum of consumer and producer surplus. Assessment of the energy amount used for obtaining each product is in the essence of this method.

Productivity Method – this method estimates the economic value of the ecosystem product or service that supports production of commercial market goods. The productivity method is also regarded as the derived value method used to estimate the economic values of an ecosystem product or service involved in production of market goods and services. It can be applied when products or services are used together with other inputs for the production of market goods. For example, the use of alternative energy resources in some production processes improves the value of the final product and thus becomes environmentally acceptable. The economic benefit obtained by the production of the energy from alternative resources can be measured by the increase of revenue resulting from the consumption of energy from traditional sources.

Hedonic Pricing Method – this method estimates the economic value of environmental services that directly influence the market price of other goods. This method is used for the assessment of an economic value of environmental service. It is mostly applied to determining house prices that also reflect the value of environmental attributes. It takes into account environmental qualities (air pollution, water pollution, noise, etc.) as well as environmental comforts (nice view, aesthetic values, etc.). The basic premise of the hedonic pricing method lies in the fact that the price of market goods is connected to their characteristics or to the services they provide. For example, the price of a car reflects some of its characteristics–its ride, comfort, luxury, cost effectiveness, and so on. Thus we can evaluate certain characteristics of a car or some other market good based on considering how much people are willing to pay for the adjustment of certain characteristics. The stated method is most commonly used to determine value of the products whose energy characteristics have been improved, making them environmentally acceptable, such as cars and technology facilities that use alternative fuels. In this way it will be estimated whether consumers are willing to pay a specific (in most cases) higher price for the product of technology based on sustainable energy management.

7.15 CASE STUDY – SUB-SAHARAN AFRICA

Analysis of Opportunities for Biofuel Production in Sub-Saharan Africa

1 FUEL USE IN SUB-SAHARAN AFRICA

As a continent, Africa has the lowest per capita energy consumption. Biomass in the form of charcoal and fuelwood continues to be the main fuel source for most Sub-Saharan countries. Low consumption levels reflect low economic development and also hinder economic development. Biofuels are a relatively new concept and although bioethanol was used as a petroleum supplement in the past in countries including Malawi, South Africa, and Zimbabwe, it is only in the last few years that large-scale biofuel production has been seriously considered.

2 BIOFUEL EXPANSION DRIVERS IN AFRICA

The European Union (EU) has proposed mandatory biofuel blends with petroleum of 5% by 2015, and 10% by 2020 (to be reviewed in 2014) as a mechanism to reduce carbon emissions. Since the EU is unable to meet its biofuel target due to insufficient agricultural land available in Europe, this has created an international market for biofuels. European and other investors are placing strong pressure on African countries to make more land available for biofuel feedstock to meet European biofuel and global carbon reduction targets. Africa's interest in this market is driven predominantly by the need for economic growth, especially among rural communities. Biofuel production represents an opportunity to boost rural economies by supplying international markets with fuel crop products and, in turn, opening markets for agricultural surpluses. The concern is that the economic benefits of biofuels to African countries may be minimal, especially if raw feedstock is exported for processing elsewhere. Of the 40 countries within Sub-Saharan Africa, 14 are landlocked. Further, nearly 40% of Africa's population lives in landlocked countries where transport costs on average are 50% higher than in coastal countries. Biofuel production offers African countries importing petroleum a means to achieve energy security and the possibility of reducing the foreign currency demands for importing oil.

3 TYPES OF BIOFUEL PROJECTS IN SUB-SAHARAN AFRICA

It is impossible to obtain an accurate figure for the status of biofuel projects across Africa. The situation is very dynamic, with new investments being announced monthly, often to simply disappear. A large number of projects are in the planning phase, but few are fully operational. Current projects in ten West African countries represent an investment of $126 million, with plans to install processing capacity for 70 million liters of biodiesel and 165 million liters of bioethanol per year. But, as far as can be ascertained, none of these projects are operational. On an Africa-wide scale, the proposed biofuel expansion equates to tens of millions of acres. African biofuel projects can be divided into four basic types based on the scale of production and intended use of the biofuels:

Type 1 – Mali Folkecentre, dumpong biofuel sin Ghana; fuel used by producers at the village or farm level, small-scale growers on 1-99 acres, and outgrowers;

Type 2 – Commercial farmers in South Africa and Zambia, mines in Zambia; fuel used by producers at the village or farm level; large industrial farms of 100 acres or more;

Type 3 – Outgrowers linked to commercial plantations; small-scale farmers linked to commercial biofuels fuel-processing plants; national and international liquid fuel blends; small-scale growers on 1–99 acres and outgrowers; and

Type 4 – Large-scale commercial plantations in Tanzania, Mozambique, and Madagascar growing for EU markets; national and international liquid fuel blends; large industrial farms of 100 acres or more.

4 COMPETING FEEDSTOCKS

Sugarcane, *Jatrophacurcas*, and palm oil are the three main feedstocks being promoted and grown in Africa, but other feedstocks are also being considered, such as sweet sorghum, cassava, and cashew apples. Sugarcane is grown in many African countries and is a well-understood and established crop. Sugarcane produces more biofuel per acre than any other currently used biofuel crop. Areas in Africa have high sugarcane potential: Zambia has reported production of up to 200 tonnes per acre (t/acre), which is three times the international norm of 65 t/acre. Malawi and Ethiopia already produce bioethanol from sugarcane and many other countries are considering the option. In addition there is the potential for cogenerating electricity from sugarcane bagasse (and presumably sweet sorghum) as an added benefit.

Jatropha has sparked major interest throughout Africa, with projects currently being implemented in many African countries with suitable climates. South Africa does not support growing jatropha as the country fears that the plant may become a noxious weed. Enthusiastic claims for jatropha's drought hardiness and yields are being tempered by the realization that jatropha is only likely to yield more than 1 t/acre of oil in areas with more than 800 mm of rain and where plantations are well managed. Despite the large-scale plantings, and high expectations, many uncertainties still surround the long-term viability of jatropha.

African palm oil can produce up to 20 t/ha of fruit that gives about 4.5 t/acre of oil. It has been extensively grown for cooking oil, especially in West Africa. Palm oil plants require high rainfall and humidity; this limits plantations to tropical rainforest areas and raises concerns about deforestation. Sweet sorghum, though not currently produced for biofuels, is generating widespread interest as it approaches sugarcane production levels, but in areas of lower rainfall and possibly with less fertilization. Extensive agronomic trials are being undertaken by the International Crops Research Institute for the Semi-Arid Tropics

(ICRISAT), for instance, in Zambia and Botswana. A further advantage is that it may be possible to produce both sorghum grain for food and sugar for ethanol concurrently from the same field. Maize is the key bioethanol crop in America, but although farmers are promoting it as a biofuel crop for South Africa, the government has discouraged its use because of concerns about food security. Most African maize is a staple food crop and too important as a food to be seriously considered for first-generation biofuel (see Figure 7.32).

5 PRODUCTION POTENTIAL

With the exception of the extensive Sahelian region in North Africa and the Kalahari/Namib area in southern Africa, the rest of the continent has sufficient rainfall to support biofuel production. West Africa has the world's second biggest block of tropical rainforest, and there is also a thin belt of rainforest along the east coast. Although rainforest areas would be climatically suited to crops such as sugarcane and palm oil, only already deforested areas should be considered to avoid a net negative carbon balance that would be unacceptable under international biofuel and biodiversity policy and future climate agreements. Attempts to predict the potential scale of biofuel expansion are hampered by continuing uncertainty over the extent of available deforested and degraded land that is not already currently used for other purposes.

FIGURE 7.32 A small screw press for extracting jatropha oil stands idle due to low jatropha yields in a project just outside of Gorongoza Game Reserve, Mozambique

6 LAND NEEDED TO MEET FUEL SECURITY

On a global scale, biofuel production using first-generation technologies can only realistically replace a small percentage of fossil fuel. For many African countries, the situation is very different and even first-generation biofuels can provide full fuel self-sufficiency from very limited land areas. Land availability varies widely: countries such as Malawi, Rwanda, Burundi, and South Africa have limited land, while countries such as Mozambique, Angola, and Zambia have extensive land available. One must use caution in interpreting "availability," however, because much of this land is forested and is subject to customary claims. Reliable statistics on land availability are scant and most land is currently being used in some way. Agricultural productivity per acre in Africa currently falls way below international norms. It is therefore possible that agricultural intensification could generate an extensive agricultural surplus that could be diverted into biofuels without affecting local food security. However, caution must be used in ensuring families negatively affected by biofuel expansion are able to benefit in meaningful ways. Second-generation biofuels would have even greater benefits as they achieve significant "well-to-wheel" reductions in greenhouse gas emissions and require dramatically less land when compared with first-generation biofuels. This is because most biomass, including many organic wastes, can be used as feedstock.

Additionally, second-generation biofuels perform better as internal combustion engine fuels, as they do not have any of the technical problems of degradation and material incompatibility associated with first-generation biofuels.

7 KEY SOCIAL CONSIDERATIONS

Globally, the most serious concerns about biofuel expansion focus on the potential impact on global food prices and thereby poverty. At the global level, the immediate net effect of higher food prices on food security is likely to be negative. Although Sub-Saharan African countries are feeling the pinch from rising food prices, biofuel production at regional and national levels need not diminish regional food security. Large-scale biofuel plantations may displace current land users. Where land is governed by a chief or traditional authority, economic benefits to the traditional authority or community as a whole may well overrule existing resource use rights enjoyed by individual members. Caution must be used, however, given increasing evidence that local or customary leaders do not always act in the community's best interest when there are personal benefits to be gained. Conversion of land for new uses must be based on adequate prior information as well as representative consultation.

Women make up most of Africa's agricultural workforce as they are responsible primarily for growing food crops in rural areas, whereas men are responsible for cash-generating crops such as cotton and tobacco. Even though women play the prominent role in agriculture, there is much inequality. From an African perspective this inequality stems from traditional sociocultural roles and results in women being denied equal access to means of production such as land, credit, appropriate technology, and extension services. It is anticipated that in small-scale production, the emerging biofuel industry will have the greatest labor impacts on rural women. Converting land used by women predominantly for food crops to growing energy cash crops might cause the partial or total displacement of women's food-growing, suggesting that biofuel production may help stabilize agricultural food production by giving farmers assured markets for their surpluses, stimulating changes in food aid policies, and stabilizing agricultural production.

A Food and Agricultural Organization report (The State of Food and Agriculture, FAO, Rome, 2008) that while higher agricultural prices could revitalize the role of agriculture as an engine of economic growth over the medium- to long-term, urban residents and the large number of net food buyers in rural areas are likely to be negatively affected, with the poorest households the most affected. Other sources provide evidence for localized food insecurity where land has been converted from small-holder agriculture to commercially produced biofuels.

Depending on the global market prices of biofuels, and the feedstock and production model (and therefore the net area of agricultural land diverted from food to fuel), such impacts could play out at the national level through reduced food self-sufficiency at household or national levels.

Dispossession of land or resources is another key potential concern in Sub-Saharan Africa. Most available land is communal and governed by customary law. Communal land users do not have secured individual owners to the activities toward increasingly marginal land or divert their labor towards activities over which they have less control. This would undermine the ability of women to ensure a secure food supply to feed their families. In addition, if land traditionally used by women is switched to energy crops, women may then be marginalized in household decision making about agricultural activities because they control less land.

8 FINANCIAL VIABILITY

The factors and criteria affecting biofuels' economic/financial viability are national and local in their scope and specifics. Factors include: (a) the cost of biomass materials, which varies depending on land availability, agricultural productivity, labor costs, etc.; (b) biofuel production costs, which depend on the plant location, size, and technology; (c) fossil fuel costs in individual countries, which depend on fluctuating global petroleum prices and domestic refining characteristics; and (d) the strategic benefit of substituting imported petroleum with domestic resources. The economics of biofuel production and use will therefore depend upon the specific country and project situation.

Biofuel production is often a high up-front cost venture, and many programs require government support in the initial start-up phases. Access to affordable financing is a major constraint. Traditional banks are unwilling to provide financing due to market uncertainties and perceived high risks. Investors and financiers have limited data and information on which to base sound judgments and decisions. Biofuels require a ready market both locally and internationally to guarantee economic viability. Reliable and competitive markets are not yet fully developed in Africa, and the continent has limited access to international biofuel markets. Market prices for feedstock and fossil fuels largely determine biofuel competitiveness. Given that these prices are highly volatile, investing in biofuel requires closer examination of the long-term market potential and other determinants to minimize the risks. This is particularly important for small holders, who have limited capacity to weather failed investments. Investors need the security of markets and if the market is to be national, then they may need the assurance of mandatory blending with petroleum products. Economies of scale are also crucial, as is having the knowledge and capacity to select the appropriate feedstock and technology.

9 CONCLUSIONS

Africa has land available to support biofuel production, but availability varies widely from one region and country to another and competing uses need to be considered. Where land is available, it is important to ascertain that biofuels are the most appropriate land use and will provide greater benefits to the current land users and owners. The land rights and resource rights of indigenous people and disadvantaged groups need to be protected. No land should be allocated without adequate provisions for ensuring existing land users capture benefits from biofuels and without free, prior and informed consent. Such practices have proven extremely difficult to operationalize in practice.

Africa has huge potential for agricultural intensification. A key concern is why this is not occurring with food crops, which almost always are more valuable than fuel crops and should be a first priority. The Mapfura-Makhura Incubator assists small-scale farmers in the Groblersdaal area, South Africa, to grow oilseed crops for biodiesel. The growers are planning a cooperative scheme to run a processing plant.

Biofuels in Africa must be for Africa's benefit. Africa must not be used to meet global biofuel demand unless the development has social and economic benefits for Africa. For instance, African countries should be fuel self-sufficient before they export excess feedstock for international use. Policies should also support production models with greater gains for small-holder producers. Biofuel projects must balance local and national benefits. Economic or production efficiency might have to be forfeited to maximize local benefit, for instance through small local processing rather than large central processing.

Deforestation and loss of biodiversity remain key concerns. Checks and balances are needed to protect against both social and environmental bad practices. A national cap on land available, a set of land allocation criteria for biofuels, and monitoring systems to ensure these standards are respected need to be developed in each country to limit food-fuel conflicts, ensure social sustainability, and keep biodiversity loss within acceptable limits. The implications of second-generation biofuel technologies need to be considered as they may affect the economics of first-generation projects in the future.

Chapter 8

Sustainable Energy Development Monitoring

Planning and implementation of a sustainable energy development strategy is a complex process that represents a big strategic change, demands significant resource allocation, and has a great impact on all business indicators. Since energy is one of the fundamental actuating forces of modern society, the process of sustainable energy management is an extremely complex procedure that must be flexible and compliant with changes within an organization and in its environment.

Implementation of sustainable energy management is not a final process and its success cannot be evaluated after completion because there is no end date since it is a gradual process that is ongoing. Reaching a certain proficiency level is certainly an imperative of the process, but its continuous improvement is always a part of the overall process.

Thus the efficiency of a sustainable energy management strategy can and must be evaluated solely by monitoring specific indicators. Developing evaluations exclusively by economic indicators is highly desirable, but a sustainable energy development strategy cannot be evaluated by standard methods. It has to be done by monitoring specific indicators that take into account the concept of sustainable development, confrontation of economic and environmental subsystems, and energy significance for worldwide development.

The implementation of efficiency control of a sustainable energy development strategy must be evaluated in a specific manner, taking into consideration all the particulars of this type of strategy.

Monitoring the level of the achieved sustainable energy development is a way of control which should regard the fact that sustainable development is a multidimensional phenomenon and that every change in a subsystem influences the state of other subsystems.

Due to the particularity of an observed issue, regular annual control using monitoring and measurement of energy-related indicators can be considered as well-acceptable [44].

8.1 INDICATORS OF SUSTAINABLE DEVELOPMENT

Indicators of sustainable energy development represent a special group of indicators that can define the achieved level of sustainable development. In order to define strategic goals and operational goals of sustainable development derived from them, it is necessary to define indicators of sustainable development. As a multidimensional phenomenon, sustainable development has an obligation to integrate economic, environmental, social, and institutional subsystems into one entity while taking into account their interaction. Qualitative indicators, which enable evaluation of states and changes in environmental, economic, social, and institutional subsystems, are the indicators of sustainable development. Indicators are not simple "value measures," they are composed of more simple parameters and show complex dynamic values. Indicators are a term without which a strategy of sustainable development cannot be established and at the same time they are a basis by which evaluation of the strategy is conducted. Indicators detect problems early and warn of economic, social, and environmental losses.

Indicators of sustainable development differ from traditionally accepted economic, social, and environmental indicators and are expressed in different ways. They are distributed within all four existing subsystems of sustainable development, which makes studying them quite complex and demanding. Simply defining and studying indicator values is not enough. The evaluation of sustainability based on energy indicators is the integral part. Since the whole human race is based on consumption of certain energy resources, and exploitation of energy resources and usage of energy sources are some of the greatest sources of nature pollution today, sustainable energy development is an important part of sustainable development as a whole.

On the other hand, unresolved problems in the field of energy development may leave far-reaching consequences and endanger implementation of the whole sustainable development concept. Thus, particular attention must be given to studying and evaluating sustainable energy development indicators, especially if we take into account that there is no complete scientific consensus regarding methodology and interpreting acquired results. Nevertheless, defining sustainable energy development by evaluating its indicators is the only possible way to perform adequate strategic analysis for all further planning in this field.

In order to use indicators of sustainable energy development in analysis, they have to be particularly efficient. The characteristics of efficient indicators of energy sustainability include:

Relevance regarding the goal: they should show significant characteristics of the monitored energy subsystem;

Comprehensiveness: they should be understandable to the public, not only to the experts of the monitored field;

Reliability: they should show information incorporated into an indicator precisely; and
Data availability: they should comply with national statistical systems of data and information processing.

The significance of indicators was perceived in 1992 at the *Earth Summit* and was institutionalized in Agenda 21, Chapter 40, inviting all countries to develop indicators on an international, national, governmental, and non-governmental level. Moreover, the Agenda pointed out the significance of effort harmonization, suggesting incorporation of indicators into official reports and databases. In 1995, as a response and reaction to these suggestions the Commission on Sustainable Development (CSD) was established and an indicator draft program was initiated within the UN.

In order to achieve goal targets, various types of indicators are used. On this occasion, the European Spatial Planning Observatory Network, or ESPON, developed a large number of indicators for the purpose of monitoring and measurement. The number of indicators used for monitoring of the achieved level of sustainable development varies depending on the need of the country in question. The U.S. Interagency Working Group on Sustainable Development Indicators originally identified up to 400 possible indicators, of sustainable development. Currently, there are 30 indicators related to energy development in use.

After including the attitudes and opinions of the public and working groups, 10 more indicators were added. Thus the final number of sustainable development indicators for the United States is 40. About 70 indicators were systematized in Germany. The number of indicators varies, depending on size of the organization: for larger organizations indicators should be within range from 50 to 100, while 10 to 20 indicators satisfy the needs of public application.

Defining networks and organizations is a key factor for development of indicators. Although every single network (schema) is imperfect, it is essential that it is observed in the context of users' needs. According to the subsystems of Agenda 21, indicators were classified as social, economic, environmental, and infrastructure. Later, further categorization as *influence indicators* ("*pressure*"), *state indicators* ("*state*"), and *response indicators* ("*response*") was used. An overview of conceptual "pressure-state-response" frame based on division of indicators to these three groups is explained:

State indicators showing environment and development states ("*state*");
Pressure indicators showing activity load on to environment, i.e., level of energy resource exploitation ("*pressure*"); and
Response indicators pointing out the success of environmental policy, energy resource protection, new solutions, and economy development ("*response*").

The featured starting point for monitoring sustainable development through evaluating its pressure, state, and response indicators can be regarded as one of the most efficient methods for sustainable development control since indicators are exact measurable values. An efficient control system is based on the ability to control its implementation to achieve target goals within a specified period of time.

In order to maintain efficient control and make necessary adjustments, goals must be clearly defined and measurable. Indicators of sustainable development should be selected in a way to be measurable, which enables defining their level and thus evaluating efficiency of activity performance. If it appears (measures) that indicator values do not match previously planned values, there is way to determine to what extent they deviate from the goal, what caused the deviations, and what is needed to meet the target. It is also possible to reach a conclusion about the functionality of the indicator itself. If it turns out that a certain indicator is unsuitable for measuring and monitoring, it can be altered or adjusted until it is defined so that it is applicable in the control process.

8.2 INDICATORS OF SUSTAINABLE ENERGY DEVELOPMENT

Indicators of sustainable energy development belong to the group of indicators which are very significant, since problems of providing energy represent specific load to the environmental subsystem and at the same time it is an important premise of economy development. Economy and environment development give specific significance to studying indicators related to energy as a whole.

When determining the level of achieved sustainable development as a whole, development energy indicators bear the largest specific weight, which means that slight changes in an indicator's state can cause big changes in the final result. Regular value monitoring of these indicators is highly significant for precise evaluation of achieved sustainable development. Correct and real evaluation of the energy-related indicators is necessary in performing all other control activities.

The priority is to determine the set of indicators that can be regarded as the most suitable in a time when GDP, as a traditional development indicator, began to lose its significance. In order to achieve a certain GDP, a country must exhaust a certain amount of resources and in the process create pollution, which up until now was not subject to specific evaluation.

Gradual elimination of development indicators based on GDP can be performed only by development and testing of the new set of indicators, and by comparison between the results obtained using GDP and the results obtained by using other indicators. This is possible only by monitoring development in two ways: evaluating development level only on the basis of traditional economic indicators (which include the GDP) and by evaluating development based on monitoring other indicators (which replace the traditional GDP). So far, the following indicators have been widely used in determining sustainable development level (see Table 8.1).

All of these indicators are economic and monitor development solely as an economic category. Modern approaches involve evaluating sustainable

TABLE 8.1 List of economic indicators

Indicator No. i_n	Indicator	Measure
	GDP/pc Gross domestic product per capita – A measure of the total output of a country that takes the gross domestic product (GDP) and divides it by the number of people in the country [82].	$
	Debt	% GDP
	Road infrastructure	1.000 km
	Inflation	%
	Gini coefficient GINI – A measurement of the income distribution of a country's residents. This number, which ranges between 0 and 1 and is based on residents' net income, helps define the gap between the rich and the poor, with 0 representing perfect equality and 1 representing perfect inequality [83]	Index
	Growth of GDP	% GDP
	Investments as part of GDP	% GDP
	Industrial growth	%
	External debt	Billion - The number equivalent to the product of a thousand and a million; 1,000,000,000 of $
	Export	Billion of $

Source: World Bank, Annual report, 2011.

development based on indicators that are not exclusively economic and measure specific GDP results in exhaustion of resources and reaching certain levels of pollution. Energy resource exploitation is a primary indicator of economic development cost (see Table 8.2).

TABLE 8.2 List of indicators of sustainable development (beyond GDP)

1.	GDP load by CO_2 emission	metric tons CO_2/GDP
2.	GDP load by methane emission	metric tons methane/GDP
3.	GDP load by energy consumption	eq. tons/GDP
4.	GDP load by pesticide usage	kg/ha/yr/GDP
5.	GDP load by water consumption	metric tons/yr/GDP
6.	GDP load by waste production	metric tons/GDP
7.	Arable ground	%
8.	Irrigation	km^2
9.	Usage of fertilizers	kg/ha/yr
10.	Organic agriculture/ploughed ground	%
11.	Protected areas	%
12.	Usage of pesticides	kg/ha/yr
13.	Emission of methane	1.000 metric tons
14.	Emission of carbon dioxide	metric tons
15.	Forestation	km^2
16.	Usage of energy	eq. tons
17.	Renewable sources of energy	%
18.	Usage of energy per capita	eq. tons/pc
19.	Poverty	% under the poverty limit
20.	Population number	1.000.000
21.	Literacy	%
22.	Urban population	%
23.	Unemployment	%
24.	Birth rate	% per 1.000
25.	Mortality rate	Number per 1.000
26.	Phone network	Users per 1.000

TABLE 8.2 Continued

27.	Internet network	Users per 1.000
28.	Efficiency of government	Index
29.	Index of political freedom	Index
30.	Corruption	Index
31.	Investments on education	% GDP
32.	Municipalities that accept Agenda 21	Number
33.	Investments on health care	% GDP
34.	Index of democracy	Index
35.	Women in the parliament	%

All of these indicators show the need to thoroughly study production and exploitation influence on overall development. The importance of state monitoring in the energy sector is especially important. Thus sustainable energy management and balancing the interaction of economic, environment, and energy are one absolute priority in this field.

8.3 DETERMINATION OF SUSTAINABLE DEVELOPMENT LEVEL

Indicators of sustainable development can be divided into four categories and four levels according to their scope in order to present indicators of sustainability in more detail. On the first level are *sector indicators* that show aspects of sustainable development at the level of separate sectors and industries. Sector indicators are associated with *subnational* (territory) *indicators* that cover all subsystems but within geographic (special) entities of a monitored country.

The second level deals with *resource indicators* that cover exploitation and accumulation of capital forms (manufactured, natural, humane, and human capital). On the next level of the "indicator pyramid" there are indicators of results in all of these four subsystems of sustainability. On the last level there are cumulative indicators that are particularly significant since they sublimate information. General evaluation of progress toward sustainable development can be given based on such presented and condensed information. *Indexes* can be found in this category.

Indexes of sustainable development are constructed in order to get one indicator that shows achieved sustainability through monitoring of indicators systematized into subsystems (environmental, economic, social, and institutional). The value of a created index as a significance measure of the observed indicator group also depends on the estimated significance value of some indicator groups.

By displaying indicator values we can note "critical values" ("*hot spots*") and the severity of the situation in the monitored field. In this respect a number of models have been developed, which graphically represent the value of each indicator, and calculation process and quantify the index of sustainability.

Separate values of monitored indicators are mostly quantified by histograms. If information for a long time period is obtained, the logical sequence of drafting a histogram would present changes of the observed phenomena through the trend. Individually calculated subsystem values may be displayed through various polygons. The number of breaking points that will be defined and the shape of the polygon depend on the number of components observed.

Visual models point out indicators of non-sustainability to people in charge. In *"cockpit terms"* the decision maker is regarded as a *"pilot flying a large aircraft."* During the "flight" cockpit instruments give important information in order to perform corrective activities in due course.

According to this analogy, early aircrafts had simple measuring instruments for monitoring fuel and oil, engine pressure, and direction, but modern cockpits are equipped with various measuring instruments, warning indicators, and integrated computers. Indicators are mostly aggregated so they do not burden the pilot with information, but each problem can be monitored by a separate instrument, identifying a certain problem.

Within the *Consultative Group of Sustainable Development Indicators* (CGDSI), four highly integrated visual models of sustainable development have been developed. Wide-ranging application of visual models is a consequence of their "optical" affinity since they enable easy detection of critical values. Some of the models established by the CGSDI include:

Four-sided Pyramid model;
Elliptical Indicator cluster;
Compass of Sustainability; and
Dashboard of Sustainability.

The first three models are based on the four-clustered approach to sustainable development which uses the Internet to access detailed information and data. The latest model – the Dashboard model – uses only three clusters of indicators and information respecting a multidisciplinary approach and application of various information sources (media).

8.4 DETERMINATION OF SUSTAINABLE DEVELOPMENT LEVEL BY INDICATORS OF DEVELOPMENT

Sustainable development represents extremely complex phenomenon and as such it is subjected to evaluation and criticism of experts and public. Apart from that, in its essence, sustainable development in great measure uses traditional ways of determining development course and goals, which in the past have only regarded economic development. Conventional economics was the only way to predict, define, and describe past, present, and future development, and in this process it was exclusively directed by traditional economic indicators without considering other aspects of society and environment.

But since the development of the sustainable development concept, economics is no longer the only priority in all societies and countries.

Principally, economic development was completely based on exploitation of natural resources, which inevitably led to worldwide degradation of the environment, and exploiting natural resources of a certain country did not necessarily lead to its economic development. This statement complicates the issue of sustainable development as a global concept that can be implemented globally.

Economic Indicators of Sustainable Development

Economic indicators of sustainable development are certainly the first group of development indicators to be discussed in the process of determining a sustainable development level in general. Traditional approach to development of the humankind are principally based on economic criteria. In the past, a country was considered developed if certain desired values of economic indicators were met. Long-lasting reliance exclusively on economic development indicators brought into question the adequacy of such a manner to determine a level of (non) development since contemporary development science points to the fact that development does not exclusively represent economic development.

Numerous indicators are used in order to determine a level of achieved economic development, but practice has shown that certain indicators can be regarded as more efficient and relevant to the goal. Thus, in order to calculate the level of achieved sustainable development it is best to use available, scalable, and easily compared indicators.

All these mentioned indicators of economic subsystem can be shown graphically by using the following formula, which takes into consideration the importance (weight) coefficient of every single indicator:

$$S_1 = i_1 \cdot 25 - i_2 \cdot 5 + i_3 \cdot 5 - i_4 \cdot 10 + i_5 \cdot 10 + i_6 \cdot 5$$
$$+ i_7 \cdot 5 + i_8 \cdot 5 - i_9 \cdot 10/i_{23} + i_{10} \cdot 10/i_{23}$$

After calculating the mutual value, the current state of the economic subsystem in a chosen country is determined.

When determining the level of achieved economic development it is also necessary to study environment load as a consequence of such development, which primarily it refers to the use of energy and emissions into the atmosphere generated as a result of energy source exploitation.

Environmental Indicators of Sustainable Development

Environmental indicators of sustainable development represent the state of an environmental subsystem of sustainable development and they represent a measure of environmental devastation in order to achieve particular economic growth. In fact, the development of the human race is inevitably connected with exploitation of natural resources and various types of pollution that result from the deterioration of environmental indicators.

There are numerous environmental indicators that can define a state in an environmental subsystem of sustainable development. Only indicators that have the greatest individual importance and whose values are of a particular interest to the observed ones must be taken into consideration.

All the indicators of an environmental subsystem can be shown graphically by using the following formula, which takes into consideration the importance (weight) coefficient of every single indicator:

$$S_2 = i_{11} \cdot 5 + i_{12} \cdot 10 + i_{13} \cdot 5 - i_{14} \cdot 5 + i_{15} \cdot 5 - i_{16} \cdot 5 - i_{17} \cdot 20 - i_{18} \cdot 25 + i_{19} \cdot 10 - i_{20} \cdot 10$$

Indicators of energy resource exploitation and atmosphere pollution generated as a result of using energy sources are significant and defined by the high importance (weight) coefficient and thus greatly influence the final result.

Social Indicators of Sustainable Development

Economic development must inevitably be followed by development of a certain level of social security and quality of life in general.

All the indicators of a social subsystem can be shown graphically by using the following formula, which takes into consideration the importance (weight) coefficient of every single indicator:

$$S_3 = i_{21} \cdot 10 - i_{22} \cdot 18 + i_{23} \cdot 2 + i_{24} \cdot 5 - i_{25} \cdot 5 - i_{26} \cdot 20 + i_{27} \cdot 10 - i_{28} \cdot 10 + i_{29} \cdot 10/i_{23} + i_{30} \cdot 10/i_{23}$$

Institutional Indicators of Sustainable Development

The development of certain institutions and social regulations are quite significant for sustainable development implementation. System institutions are in great measure carriers of changes, new legislative solutions, and rules that change traditional ways of development.

All the indicators of an institutional subsystem can be shown graphically by using the following formula, which takes into consideration the importance (weight) coefficient of every single indicator:

$$S_4 = i_{31} \cdot 15 - i_{32} \cdot 10 - i_{33} \cdot 10 + i_{34} \cdot 16 + i_{35} \cdot 2 + i_{36} \cdot 22 + i_{37} \cdot 15 + i_{38} \cdot 10$$

Overall Indicators of Sustainable Development

The definition of the development state in certain measure by observation of the indicators of sustainable development also presents important differences in relation to the traditional economic way of measuring. Contemporary movement in the area of development planning indicates the need to include non-economic development indicators – environmental and social.

In the following formula the total results can be seen in all four observed subsystems of sustainable development, together with their final value for every country (C) that was covered by this research. Given that this position was accepted, i.e., that all four subsystems of sustainable development have an equal influence on the all-inclusive state of sustainable development, the authors used uniform importance (weight) coefficients for every subsystem itself by using the following formula for calculation:

$$C = S_1 + S_2 + S_3 + S_4$$

Determining the overall level of sustainable development by applying the indicator method is an acceptable and useful method. It regards the need to observe all four subsystems equally. In this way, economic indicators that were dominant in past decades are slowly losing their significance. Current science is more oriented to determination of the relations that exist between the economic and environmental subsystems. Defining a new set of indicators that will be able to define the interaction of changes between subsystems represents an absolute priority. Energy consumption (the environmental subsystem indicator) is closely connected to the total development level (using GDP as a typical indicator of an economic subsystem). Increasing GDP, favorable change in economic subsystem, is a consequence of energy consumption that is extremely unfavorable change in environmental subsystem. Sustainable energy management offers a frame and possible solution models of this noted paradox and also the possibility to plan, predict, observe, measure, and correct future changes in accordance with the need for development and better quality of life at all levels.

8.5 DETERMINATION OF SUSTAINABLE ENERGY DEVELOPMENT LEVEL

Providing sufficient amounts of energy for the needs of the human race undoubtedly represents one of the greatest problems today. Providing sufficient energy was and continues to be a priority of all world economies, and awareness of energy resource capacity limitations and of pollution levels generated by energy source exploitation imposes the need for new approaches to energy management processes. The fundamental goal of energy management is to provide sufficient energy for all the needs of a certain country.

Energy development based only on satisfying short-term needs without regarding environmental costs of such development and without considering the rights of future generations, has led to limited energy sources, unevenly divided resources, and various pollutions. The concept of sustainable energy development is being developed and the level of its success can be determined by defining the level of achieved sustainable energy development.

Sustainable energy development is a subsystem of sustainable development and as such it has its characteristics as well as its methods of observing and measuring. Sustainable energy development can be defined as the development of obtaining, distributing, and exploiting the energy sector and is based on sustainability principles. In other words, today energy management is modified in accordance with sustainability development and thus the contemporary goal of energy management can be defined as providing and exploiting energy sources in a way that will enable adequate economic and environmental development without jeopardizing the possibility for future generations to enjoy the same.

There are numerous indicators in the sphere of energy production and exploitation. However, with sustainable energy management and sustainable development as a goal, new indicators are created by combining existing indicators to provide adequate insight into the economic and environmental implications of this process.

8.6 CASE STUDY – SOUTHEASTERN EUROPEAN REGION

Assessment of the Effectiveness of Policy Implementation for Sustainable Energy Development in Southeastern Europe

Abstract

Using solely economic indicators as a measure of development as a whole is no longer justified and acceptable. This study covered 11 countries in the region of Southeastern Europe – the region for which indicators of sustainable energy development were defined. The authors suggest a specific way of expressing their needs for treatment using the method of weight coefficients. The results indicate that the efficiency of energy policy in the region varies widely. Countries at higher levels of economic development record a high dependence on energy imports. Countries in the region at a similar stage of development show different energy consumption per capita and different intensity of energy production from renewable resources. Research shows that energy efficiency policies in the region, apart from the available energy resources, largely depend on the level of economic development that a country wants to achieve. Countries at lower levels of development show higher levels of sustainability in terms of energy management and vice versa. The most favorable situation was observed in Bulgaria. Negative trends were observed in Greece and Hungary, while in Slovenia the indicators were on the border of sustainability. Data used for research refer to the year 2010.

Keywords: Energy policy, Efficiency, SE Europe

1 INTRODUCTION

Countries of Southeastern Europe form a natural, historical, and economic entity. Regardless of mutual differences and historical circumstances, all countries of the region have adopted the basic regulatory documents that are directed toward the joint policy for sustainable development of Europe. Considering the specific impact of energy production and consumption on the level of sustainable development, countries in the region have defined and ratified the following basic documents: [48]
Kyoto Protocol;
National strategies for sustainable development; and
Strategies for energy sector development.

Contemporary conditions of production in highly developed countries are characterized by a compromise *(trade-off)* between the quality of the environment and the economic development of the country. The economic system that does not value natural resources and stimulates economic growth regardless of the consequences on limited resources is not sustainable in the long run. Basically, the entire economic system is to act in accordance with the mature environmental system. Both systems are characterized by cyclicity. An optimal economic system needs to be more productive but also remove unwanted residuals – waste materials and a surplus of used or emitted energy. As a matter of fact, the Kyoto Protocol mechanisms allow their control and gradual elimination.

Energy efficiency involves rational and effective usage of natural resources, replacement of imported fuel with domestic energy sources, use of renewable and alternative energy sources, and increase of energy efficiency in the production and final consumption of energy. The direct link between the implementation of the Protocol mechanisms and characteristics of high energy efficiency is evident. The relationship between consumed energy and economic results is considered to be one of the main indicators of energy intensity of the economy. After examining key settings contained in these documents, taking into account the basic characteristics of the region, we can conclude the following:

In all countries of the region unique commitment to sustainable development exists;
All countries in the region largely depend on the import of energy;
All countries in the region accepted increasing the share of energy produced from renewable sources to a minimum of 20% of the total energy produced;
In all countries of the region some subventions for producers of energy from renewable resources exist; and
In all countries strong commitment to implement activities that increase the energy efficiency level exists.

These basic parameters of the general situation in the energy sector in the region of Southeastern Europe indicate that a shared starting point exists there and that the expectation, that implementation of energy policy would be relatively coordinated in these countries was a realistic one. However, previous studies of individual indicators of sustainable development (that include indicators of sustainable development in the energy sector) indicate significant differences between countries in the region. Taking this into account, it is necessary to study indicators that are exclusively related to the energy sector in the region of Southeastern Europe and to determine on one hand, the energy policy effectiveness level, and on the other hand, to define, quantify, and explain the reasons for individual differences and the diversity among the countries in the sample.

2 METHODOLOGICAL PRINCIPLES OF THE RESEARCH

The study was planned and carried out in accordance with the basic recommendations for the scientific research work, which include previous definition of objects and goals of the research and refinement of the methodology to be used for data processing.

2.1 Subject of the Research

The subject of the research is the effectiveness of energy policy in the region of Southeastern Europe. A sustainable energy policy represents a set of countries' legal, organizational, management, technical, and economic measures. Energy policy in its essence is multidimensional, comprehensive, and long-term and therefore represents a phenomenon that is particularly complex for monitoring. That is why research as a subject has to create a methodology that will be acceptable for evaluation of the implementation energy policy. Only on the basis of accurate monitoring and defining the degree of the achieved level of sustainable energy development is it possible to come to certain conclusions and, based on them, to implement corrective measures.

All the countries that make up the research sample have the same initial legal framework that regulates this area, and all have shifted to a model of sustainable energy development. In addition, there is a significant degree of similarity in terms of available energy resources. All countries in the sample have adopted their own strategies for sustainable energy development and took over the obligations of the Kyoto Protocol almost simultaneously [37].

However, countries of this region are largely distinguished by the efficiency of the implementation of energy development policies. The essence of research is precisely to try to define efficiency levels of energy development in individual countries in the region as a whole. In addition, an important aspect of the research is defining differences between countries in the sample, their quantification, and an explanation of the reasons for emergence of the observed differences.

Given the same or similar starting conditions, i.e., the legal and natural conditions, the research will point to the fact that differences in the efficiency of the energy policy implementation largely depend on the influence of other factors. In this way, the research will emphasize the importance of an effective management process for sustainable energy development, as well as the need to modify those parameters, which turns out to be of paramount importance and has the greatest impact in sustainable energy development. The study will define the specific factors that in some countries of the region affect effective or ineffective energy policy implementation.

2.2 Aim of the Research

Research of energy policy effectiveness in the region of Southeastern Europe was designed to provide results that are precisely defined, comparable, and usable. Any national strategy for sustainable energy development in the region has to define specific goals, but precisely defined measurement parameters do not exist. In fact, only the measurable goals can indicate the degrees of achievement of goals. In accordance with the need to obtain these results, fundamental and derived research goals were defined.

Fundamental Goal of the Research

Determining the general efficiency of energy policy implementation in the Southeastern Europe region.

Derived Research Goals

Determine the effectiveness of energy policies in individual countries of the region;

Determine the average efficiency level of policy implementation of sustainable energy development of the region as a whole;

Compare the achieved level of sustainable energy development with an average of the region and mutually;

Define the values of certain energy parameters in the region of Southeastern Europe;

Define the specific indicators of energy policy implementation in individual countries of the region;

Determine the achieved level of energy independence;

Define the amount of energy supplied from renewable sources;

Determine the differences in the implementation of energy policy between individual countries, and the interpretation of observed differences and reasons; and

Propose measures and activities, based on the results, that may have positive impact on improving the efficiency of energy policy implementation in individual countries of the Southeastern Europe region as a whole.

On the basis of defined basic and derived research objectives, the study provides a concise review of energy policy objectives and measures determined by the study, which will serve as one of the parameters for determining the final result, the level of overall efficiency of energy policy implementation. The main selected performance indicators of sustainable energy policy implementation proposed by the authors of the study are given in Table 8.3.

Realization of basic and derived research objectives enables obtaining precise insight into the basic parameters of efficiency of energy policy implementation in individual countries and in the Southeastern Europe region as a whole. This approach provides the basis for realistic consideration of developments in this field, the current situation, and prediction of future values based on which the necessary adjustments of national strategic plans for sustainable development of the energy sector can be made.

2.3 Input Parameters

In defining the basic indicators of energy policy implementation in the Southeastern Europe region the authors started from the fact that there are numerous indicators that are directly or indirectly related to a given issue. The selection of indicators to be used for the purpose of research has been made on the basis of basic starting points that are scientifically defined and used to select indicators of sustainable development. For an indicator to be acceptable for research, it should:

Reflect the essential features of a system or process – all the indicators for processing in the study show the most important parameters of the situation in the energy sector; and
Be expressed in a measurable and comprehensive way – all the indicators used for the purpose of research are expressed in internationally recognized and comparable measurement units.

The authors proceeded from the fact that an insufficient number of indicators can not provide reliable and credible results, but on the other hand, too many

TABLE 8.3 The basic measures of efficiency of energy policy implementation

The goal of energy policy	Measure of efficiency	Mark
Energy independence	Energy output/total energy consumption	*EI*
Increasing the share of energy produced from RES – renewable energy sources	% of energy from RSE in relation to total energy production	*PRES*
Increasing of energy production	Energy production per capita	*EPP*
Reduction of energy consumption per capita	Energy consumption per capita	*ECP*

indicators can lead to obtaining polysemous and therefore less useful results. In accordance with the scientific attitude and experience of the authors in similar studies, a set of five indicators, given in Table 8.4, was selected.

For the purpose of research the most important indicators of energy development were chosen. Auxiliary indicators were selected in accordance with the fact that the production and consumption of energy to a large extent is determined by the achieved level of economic development, expressed in the GDP. In addition, energy consumption is largely dependent on the number of inhabitants in the country. If the data can be compared, it is necessary to express them through the GDP, or by the number of inhabitants.

To survey the research results, individual values of these indicators as well as the new indicators derived from their combinations were used.

2.4 Data Processing Procedure

Starting from the research objectives and the need to define the effectiveness of energy policy implementation as a means of processing, the statistical method of weight coefficients was chosen. This method was used and modified for the purpose of improvements in previous research, and it can be considered acceptable for this type of research.

The weight coefficients method is designed so that it allows treatment of data that are expressed in different measuring units and the final result, is the index – untitled number that allows easy and simple comparison between various countries in the sample.

The ultimate objective, i.e., the measure of the effectiveness of energy policy implementation (EEP), was obtained by applying the following formula:

$$EEP = EI \cdot wc_1 + PRES \cdot wc_2 + EPP \cdot wc_3 + ECP \cdot wc_4$$

TABLE 8.4 Review of indicators of energy sector development

Basic indicators of sustainable energy development	
Name of the indicator	Unit
Total energy production	GWh
Total energy consumption	GWh
Total production of energy from renewable sources	GWh
Auxiliary indicators of sustainable energy development	
Gross domestic product	$ per capita
Population	000

where:

EI – energy independence
PRES – energy production from renewable sources
EPP – energy production per capita
ECP – energy consumption per capita
wc_i - the selected weight factors

To obtain the values of each parameter chosen weight coefficients were used. Specifically, the authors recognized the fact that each indicator had its own impact on the final result. At the time of research, certainly the greatest influence on the level of effectiveness of energy policy implementation were the main parameters of energy production and consumption. The indicator of dependence on imports was less important because it indicated that countries in the sample could not have a significant impact as they had limited amounts of energy resources available to them. The values of the weight coefficients, i.e., the significance level of each indicator used for the purpose of research, are given in Table 8.5.

Depending on the level of the influence on the final result, the values of weight coefficients were expressed in a range from 1-100. The authors believe that at the time of the research the parameter that had the greatest impact on the efficiency of energy policy was the percent ratio between the total production and consumption of energy – an indicator of energy independence for which the weight coefficient was 30. The strategic sustainable development of any country of crucial importance is that it relies on energy imports as little as possible.

The level of energy efficiency, i.e., the indicator of energy production per capita, had less influence. The indicator of energy consumption per capita had a proportionally large impact on the result, with the coefficient 25. In light of efforts to increase energy efficiency and increase the beneficial use of energy, consumption habits and attitude towards energy were exactly expressed by this indicator.

TABLE 8.5 Weight coefficients of energy policy indicators

Indicator	Weight coefficients (*wc*)
EI – energy independence	30
PRES – energy production from renewable sources	20
EPP – energy production per capita	15
ECP – energy consumption per capita	25

Energy production from renewable sources certainly had some impact on the final result, but the authors consider that this influence was much smaller than that of previously mentioned ones, because in the majority of the observed countries the production of energy from renewable sources is at initial stage. In order to increase energy production from renewable sources it is necessary to realize a number of factors, and a certain time period, that would enable more precise conclusions [39].

In last place by significance the authors used an indicator of energy production per capita, with 15 as the weight coefficient. This situation was caused by the fact that in these countries significant deposits of oil or liquefied gas do not exist, so that energy production was not much of a choice in these countries, but a necessity that must be taken into account and therefore quantified as shown.

3 RESULTS

Research on the effectiveness of energy policy in the Southeastern Europe region was carried out based on data for 2010. The sample consisted of 11 countries in the region: Albania, Bosnia and Herzegovina, Bulgaria, Croatia, Greece, Hungary, Macedonia, Montenegro, Serbia, Slovenia, and Romania. First, individual indicators of sustainable energy development in each country in the sample were defined, and data processing and deriving of the results were performed afterwards.

3.1 Individual Inputs

For the purpose of the research, the definitions of the input indicators for 2010 were previously performed and the values are given in Table 8.6.

The presented data provide a general picture of energy production and consumption in the region of Southeastern Europe. The most obvious regularity in this data is that in all countries of the region energy consumption far exceeds energy production. In this respect, it is possible to define two groups of countries.

The first group consists of countries in the region where the energy production was two or two and a half times lower than its consumption. This group includes Bulgaria, Croatia, and Slovenia. This situation can be to some extent justified by the fact that these are countries that have strong economic development but at the same time tend to improve their energy policies in different ways. In these countries there is a lack of natural resources and thus this situation was as expected.

The second group consists of countries that spend three times more energy than they are able to produce. This group includes most of the countries: Albania, Greece, Macedonia, Hungary, and Montenegro. This situation was to some extent expected, with respect to the lack of natural energy resources available in these countries. However, a large span of GDP per capita suggests that the intensity of economic development clearly does not depend directly on

TABLE 8.6 Values of indicators of sustainable energy development in region of Southeastern Europe

Country	GDP ($ pc)	Population (000)	Total energy production (GWh)	Total energy consumption (GWh)	Production of energy from RES (GWh)
Albania	7381	2986	12939.09	31003.99	5373
Bosnia	7751	3843	56955.43	76588.26	5455
Bulgaria	14226	7577	113421.43	242613.02	4258
Croatia	17607	4487	50434.60	128769.57	6526
Greece	28833	11305	108327.86	429410.66	8100
Hungary	18815	9986	112504.12	326686.32	212
Macedonia	9350	2114	13358.18	34582.39	1492
Montenegro	10432	667	586.14	1872.72	41
Serbia	10808	7307	16227.76	54950.83	110
Slovenia	27899	2849	41129.76	93272.80	3461
Romania	11766	21959	340021.06	505011.28	18400

energy production in the country. Albania had GDP which is twice smaller than in Greece, but produced ten times less energy. On the other hand, Slovenia and Greece had similar values of GDP, but Slovenia produced nearly three times less energy than Greece, which is not particularly rich in energy resources.

Romania and Bosnia and Herzegovina are the countries with the most advantageous situation where the question of energy independence is concerned. In these countries, energy consumption was greater than production, and the level of energy independence was still lower than 100%, which means that these countries were to some extent closest to the ideal situation – i.e., producing the energy they need. Certainly, achieving this situation is unrealistic or hard to attain. It should be kept in mind that Bosnia and Herzegovina was the country closest to the ideal situation, but it is at the same time a country that has a very low level of economic development and therefore does not spend a lot of energy. The worst situation in terms of energy independence was found in Serbia, since it consumes more than four times the amount of energy it produces [40].

Correlation between energy production and consumption and of GDP was significantly disproportional. Specifically, countries with a higher GDP at the same time recorded higher energy production and consumption, but there was no regularity observed. Namely, Albania and Bosnia are countries with very similar levels of GDP per capita, but the production of energy in Albania was almost four times lower than in Bosnia. Similar GDPs were recorded in Croatia and Hungary, but energy production in Hungary was twice as high.

An even more drastic difference was recorded in Serbia and Romania. These are countries at a similar level of development measured in the GDP, but at the same time Romania produces 20 times more energy, which is primarily due to wealthier energy resources in Romania. The worst situation was observed in Montenegro, which shows a GDP similar to Serbia but produces 27 times less energy. All this indicates that Montenegro creates its GDP with high and inefficient use of energy.

At this point it is necessary to note that in four countries of Southeastern Europe there are working nuclear power plants, i.e., one plant in Slovenia, one in Romania, four power plants in Hungary, and two in Bulgaria. An overview of energy production in these plants is given in Table 8.7.

It is evident that the production of energy from nuclear sources in the region is small, but it certainly must be considered as an element of strategy for sustainable energy development in the region. Acceptance or rejection of nuclear power is defined by separate legislation in each country in the region.

3.2 Processing of Input Data

In order to determine the efficiency of the implementation of sustainable energy development policy, it was necessary, according to predetermined indicators, to establish the selected criteria of efficiency, an overview of which is given in Table 8.8.

TABLE 8.7 Production of energy from nuclear power plants in Southeastern Europe

Country	Number of nuclear power plants	Total power of nuclear power plants (MW)
Slovenia	1	666
Hungary	4	1829
Romania	1	1310
Bulgaria	2	1906

Source: Survey of policy implementation efficiency of sustainable energy development, Educons University, 2011 (in Serbian)

Energy Independence – EI

All regional countries in Southeastern Europe are highly dependent on imported energy. The least amount of energy import was in Bosnia and Romania. Energy production in Bosnia covers about 77% of the country needs, while the production of energy in Romania meets 67% of its needs. All other countries in the region form a group of countries whose own production of energy can meet 25-45% of the country's needs. Greece and Serbia have similar parameters of energy independence, but Greece has a considerably larger GDP. This leads to the conclusion that the level of economic development of a country does not directly affect its ability to produce enough energy for its own purposes, and vice versa.

Energy Production from Renewable Sources – EPRS

All regional countries in Southeastern Europe have defined the production of energy from renewable sources as one of the strategic objectives of sustainable energy policy. In this sense, the research points to a few specific results. By far the largest share of energy produced from renewable sources was registered in Albania, which produces large amounts of energy from the energy of water flows, while the country is on the low level of economic development, so that production and consumption of energy are not high. The second group consists of Croatia and Macedonia, which produce slightly more than 10% of energy from renewable sources, an excellent result. The largest group consists of countries that produce between 5 and 10% of energy from renewable sources, such as Bosnia, Greece, Montenegro, Slovenia, and Romania. The most unfavorable situation was in Serbia and Hungary, which recorded minimal energy production from renewable sources.

TABLE 8.8 Criteria of implementation efficiency of sustainable energy development policy in the region of Southeastern Europe

Country	EI	PRES	EPP	ECP
Weight coefficient	30	20	15	25
Albania	0.417	41.53	4.33	10.38
Bosnia and Herzegovina	0.744	9.58	14.82	19.93
Bulgaria	0.467	3.75	14.97	32.01
Croatia	0.392	12.94	11.24	28.70
Greece	0.252	7.48	9.58	37.98
Hungary	0.344	0.19	11.27	32.71
Macedonia	0.386	11.17	6.32	16.36
Montenegro	0.313	6.99	0.88	2.81
Serbia	0.295	1.77	0.85	7.52
Slovenia	0.441	8.41	20.07	45.52
Romania	0.673	5.41	15.48	23.00
Average	0.429	9.93	9.98	23.36

Source: Survey of policy implementation efficiency of sustainable energy development, Educons University, 2011 (in Serbian)

Energy Production per Capita – EPP

The greatest energy production per capita was recorded in Slovenia with 20 GWh, followed by Romania, Bosnia, and Bulgaria, which produced about 14 GWh of energy per year per capita. Less than 5 GWh of energy per capita was produced in Albania and in Macedonia. The poorest results were recorded in Serbia and Montenegro, where annual per capita production of energy was below 1 GWh. All of the above points to the fact that the amount of GDP and intensity of economic development have no direct connection with the production of energy, as expected. Greece and Slovenia have similar GDPs, but Slovenia produced two times more energy than Greece per capita. This difference can be interpreted in different ways, but it seems that the crucial factor is just (in) adequate management of energy development.

Energy Consumption per Capita – ECP

The largest energy consumption per capita in the Southeastern region was noted in Slovenia. Lower consumption, an average of 35 GWh per year, was recorded in Bulgaria, Greece, and Hungary. About 20 GWh of energy per capita was spent in Bosnia and Romania, and 10 GWh in Albania. A low level of energy consumption was recorded in Serbia, and even lower in Montenegro. Regardless of the similar level of GDP, Montenegro spent almost four times less energy per capita. These data indicate that energy consumption per capita does not depend directly on the level of economic development and logical assumption that countries at a higher level of development are spending more energy.

Average Values of Indicators

On the basis of the determined values for each country in the sample, the average values of the selected indicators are defined. The countries of Southeastern Europe must import slightly more than half of the energy needed to meet their own needs. This result indicates a high level of energy dependence. In the region 9.93% of energy was produced from renewable sources, which can be considered a good indicator, if keeping in mind the fact that half of the region consists of countries at the lower level of development in relation to the European average, and investment in energy production from alternative sources requires a certain economic strength in a country.

Such a high result is necessarily affected by the fact that the group of alternative sources of energy includes energy derived from water courses, which are very numerous in this part of Europe. The last important indicator refers to the ratio of produced and consumed energy per capita in the region of Southeastern Europe. Inhabitants of the region consume energy in a quantity that is on average two times higher than the amount of energy they can produce.

3.3 Result of the Research

The final result of the research is the assessment of the implementation efficiency of sustainable energy development in individual countries of Southeastern Europe, determined by applying the following formula, which integrates the values of the selected weight coefficients:

$$EEP = EI \cdot 30 + PRES \cdot 20 + EPP \cdot 15 - ECP \cdot 25$$

This formula was applied to individual indicators for each country in the sample, and finally the value of the implementation efficiency level of sustainable energy policy was determined for the region of Southeastern Europe as a whole. In order to enable the addition of all parameters that were defined in different computational ranges, all indicators were transformed into a common one and on that basis their individual share was determined, so the formula was adjusted as follows:

$$EEP = EI \cdot 18.354 + PRES \cdot 0.22 + EPP \cdot 0.075 - ECP \cdot 0.25$$

With the application of this formula it was possible to determine deviations of individual countries from the regional average. The final level of efficiency of sustainable energy development policy in the countries of Southeastern Europe is given in Table 8.9.

The results of the implementation efficiency of sustainable energy development policy in the region indicate considerable diversity between the respective countries. In this sense four groups of countries can be distinguished in the observed sample.

The first group includes Bosnia and Herzegovina, Albania, and Bulgaria. These are countries where satisfactory levels of sustainable development have been achieved, and the main reason for this is the inability of these countries to produce the amount of energy they need for domestic needs, but compared to other countries they show the largest positive value. However, the degree of energy development should not be considered separately from the achieved level of economic development, expressed through the GDP. Albania and Bosnia are two countries in the region with the lowest values of GDPs and that is why in these countries the need for greater energy consumption does not exist. An extremely positive result, also recorded in Albania, is the consequence of energy

TABLE 8.9 Estimated level of implementation efficiency of sustainable energy development policy in the region of Southeastern Europe

Country	Degree of sustainable energy development	GDP [89]
Albania	14.521	7.381
Bosnia & Herzegovina	11.892	7.751
Bulgaria	10.519	14.226
Croatia	3.711	17.607
Greece	−2.505	28.833
Hungary	−0.977	18.815
Macedonia	1.289	9.350
Montenegro	6.646	10.432
Serbia	3.987	10.808
Slovenia	0.069	27.899
Romania	8.953	11.766
OVERALL	4.967	14.988

production from water sources, which is considered as energy from renewable sources. The result for Bulgaria was largely determined by demand that exceeds production, as in all other countries but not as pronounced.

In the second group of countries are Montenegro and Romania, where the positive effects of energy policy were definitely recorded. These are countries with completely different economic characteristics, but with similar levels of economic development. Energy policy in this case can be evaluated positively. When it comes to Montenegro, its result can be explained by the structure of the economy, which is oriented to tourism, and there is no need for high energy consumption. Montenegro is a country with by far the lowest energy consumption per capita in the region, and it spends ten times less energy than Romania. On the other hand, the great advantage of Romania is extremely high energy independence.

In the third group of countries, which includes Croatia, Macedonia, and Serbia, the efficiency of energy policy was positive, but certainly not as much as would be acceptable in the long-term. This results for Macedonia and Serbia can be explained by the low level of economic development. Croatia is a country of this group, which records a higher GDP and leads in the use of renewable energy sources in the region. In addition, Croatia is at the very top when it comes to energy production per capita. Croatia did not record a better end result and is extremely dependent on energy imports.

The fourth group consists of countries that are located on the border or have a negative degree of sustainable energy development. This group consists of the EU countries with high GDPs: Greece, Hungary, and Slovenia. Energy sustainability is at the border in Slovenia. Slovenia records such an adverse result primarily because of the vast differences that exist between production and consumption of energy per capita. Import of energy is needed to assure the strong swing of economic development, which is expressed through the GDP in Slovenia, and it clearly has a negative effect on the energy stability of the country. Hungary showed a negative result, while it consumed three times more energy than it produced, and production of energy from renewable sources was at the lowest level in the region. This result can certainly be justified by the lowest level of available energy resources in the region and the need to provide energy-intensive economic development in the country (Golušin M, Munitlak Ivanovic O, Domazet S, Dodic S, Vucurovic D. Assessment of the effectiveness of policy implementation for sustainable energy development in Southeast Europe. J Renew Sust Energy 2011). The worst situation was observed in Greece, which is also the country with the highest value of GDP in the region. The reason for this lies primarily in the excessive energy consumption per capita and lack of natural resources. Greece is a country whose economic development, like that of Montenegro, largely depends on tourism, but Greece showed significantly worse results because of the irrational consumption of energy per capita. A graphical representation of the research results is given in Figure 8.1.

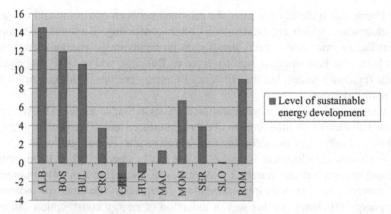

FIGURE 8.1 Level of efficiency of sustainable energy development policy in individual countries of Southeastern Europe

Overall, in the eleven countries in the region countries of the region more efficient energy policy was when compared with the established average.

4 CONCLUSION

The research of energy policy efficiency in the Southeastern Europe region was intended and implemented to provide adequate insight into the current situation and to provide useful, measurable, and usable data. Based on the results set forth in the 11 countries in the region, the respective conclusions can be clearly defined.

First, all countries in Southeastern Europe are heavily dependent on imports of energy. This is supported by the fact that all countries in the region are poor in oil and gas and import of energy sources is therefore inevitable. Future reorientation to other energy sources is the only way to increase the level of energy independence of the region.

Countries in the region showed different levels of economic development, expressed in the GDP. Research clearly points to the fact that a particular GDP can be achieved, among other things, more or less, with a successful policy of sustainable energy development. In this respect, the research clearly shows the negative situation with regard to the most developed countries in the region – Greece, Slovenia, and Hungary. With the aim of rapid economic development in these countries, energy is imported in large quantities, and the rationality of energy consumption could certainly be questioned.

On the other hand, the positive effects of energy policy implementation were recorded in the countries with the lowest development levels, Albania and Bosnia and Herzegovina. It is clear that the above result is just a consequence of reduced energy needs in economies that are poorly developed and not the result of successfully applied strategies of energy development.

The research clearly indicates the relations between economic and energy developments, which are bilateral and often conflicting. Economic development has its price and reflects directly on problems in the energy sector. On this basis, the best situation was observed in Bulgaria, whose GDP was equal to the regional average, but the efficiency of energy politics in this country was far above average.

Energy production from nuclear sources exists in four countries in the region. Its contribution to the total amount of energy produced and to the general energy stability of the countries, according to research results, is not significant.

Economic development is based on the use of energy. When taking into account the natural constraints, the countries of the region can only improve the current situation with implementation of rigorous measures for increasing energy efficiency. In this way, a reduction of energy consumption can be expected, which is, on the basis of the research results, currently the largest problem of sustainable energy development.

All of this points to the fact that monitoring and management of sustainable development is a complex process, where measurement of the achieved effects of energy development is one of the phases of the energy management process. In order to design more efficient measurement methodology it is necessary to conduct further research to achieve certain consensus in theory and practice.

It is helpful to observe sustainable energy management as a multivalued phenomenon, process, and operation method that complies with the sustainable development concept. In sustainable energy management, it is necessary to conduct major and multivalued changes in all society sectors and levels so that the process of sustainable energy management takes place gradually. Initializing and beginning the implementation of sustainable energy management represents numerous analyses, assessments, changes, and adjustments, and implementation of the process is linked with the implementation of numerous and various measures and activities which to a larger or lesser extent are completely different than the traditional method of energy management.

The need for universal energy management emerged, in part, after the big energy crisis of the early 1970s, and initial concepts attempted to solve the greatest problem – uneven distribution of energy resources and monopolistic position of countries in the disposal of energy resources. The concept of sustainable energy management appeared in the 1990s as a result of defined sustainable development in general.

Energy in great measure represents the backbone of the development of societies. But regarding energy solely as a resource that contributes to producing goods and services for economic development has contributed to creating the complex energy problem we face today. Measuring society's development by traditional indicators based on the GDP created the need to exploit, produce, and consume energy irrationally without taking care of environmental damage occurring at the same time.

Today, business development generally promotes and encourages application of sustainable energy management in all spheres of life. Certain approaches to implementation appropriate for various circumstances have been developed. The traditional approach as well as (to some extent) the approach of applying the quality model are acceptable for application in companies that are at the beginning of introducing sustainable energy management into practice or performing activities with minor energy demands. The systematic approach is more acceptable in companies where arranged control structures and principles are already in place since it requires application of more complex tools and methods. Life cycle analysis and gap analysis are regarded as universal since they can be successfully applied within companies that are at different levels of organizational structure and culture, where the starting point is introducing changes that do not require major financial investments. Apart from the criterion of the company's general attitude, the optimal application model greatly depends on the type of activity, economic and non-economic effects, and general attitude and politics of the company and its wider working environment.

Sustainable energy management is most conveniently conducted in practice by using tools and techniques that arrange and observe the process, in its essence complex, as a sum of operations that are to be performed at certain time and in a specific way. Analyzing, planning, directing, implementing, and control phases of sustainable energy management implementation are performed according to goals that are above all focused on providing sustainability as well as preserving positive economic business effects. A combination of reduced energy consumption from non-renewable resources, increased energy consumption from renewable resources, and increased energy efficiency on one side and creating positive economic and non-economic effects on the other, create the definition of modern sustainable energy management.

Complex application effects of sustainable energy management concepts are basically threefold. Sustainable energy management produces numerous positive and negative effects, and the need to work with them. Secondly, there are problems in all implementation phases and certain deviations from the planned goals imposing the need for correction. Finally, every operation segment in this field opens numerous question and fields to numerous kinds of scientific and practical work.

The positive effects of applying sustainable energy management are numerous and multivalued, internal and external. By applying the concept of sustainable energy management, a company can above all begin (or continue) its goal to increase the quality of all business performances. Energy supply costs are reduced, available energy is consumed more rationally and efficiently, and waste energy is reduced, and every decrease in costs creates bigger profits that can be used to improve the business further. Additionally, reducing energy costs allows reducing the price of final products, positively influencing the market. By applying the concept of sustainable energy management, companies can greatly reduce environmental pollution (at local, regional, national, and global levels), reducing the risk of penalties while creating a positive public image that encourages and promotes socially responsible business.

The negative effects of applying sustainable energy management are manifested in certain segments and are registered as such since the consequences of this energy management method are monitored in all society spheres. Application of sustainable energy management reduces the need to hire employees, which can imperil social welfare. Consistent application of Kyoto Protocol recommendations, which in great measure relate to changes in the energy management sphere, creates problems in the field of development planning. Reducing energy consumption from non-renewable sources (where energy production from renewable sources is still not big enough to compensate for it) inevitably leads to slowing of industrial and economic growth. Negative effects are particularly expressed in examples of economically underdeveloped countries.

Problems in implementing the concept of sustainable energy management are anticipated, are mostly predictable, and thus are the subject of specific research and solution models. Some of the most common problems that complicate the application of sustainable energy management include insufficient knowledge and information, insufficiently developed consciousness of the necessity to respect intergenerational justice, insufficient and/or inadequate system support, high initial costs, difficulties in applying observation, insufficient application tools and techniques, and undeveloped methodology for measuring effects of sustainable energy management.

Further directions of development and improvement of sustainable energy management include research in rapid and efficient energy production from renewable resources, adjusting distribution networks, optimizing energy consumption of operation processes, improving existing and developing new techniques and tools, and the development of globally accepted methodology solutions to estimate all effects resulting from applying sustainable energy management.

Sustainable energy management generally creates a number of challenges whose solutions require a combination of strategic planning, efficient operation work, technical and technological improvements, training, and global cooperation. Socially responsible business and respecting the rights of future generations are the starting points and the final results of all complex efforts that lead toward successful implementation of sustainable energy management as well as all other sustainable development trends, in general.

Glossary

Analysis – detailed examination of the elements or structure of something, typically as a basis for discussion or interpretation.

Audit – an evaluation of a person, organization, system, process, enterprise, project, or product.

Biofuel – a type of fuel whose energy is derived from biological carbon fixation. Biofuels include fuels derived from biomass conversion, as well as solid biomass, liquid fuels, and various biogases

Biomass – the biodegradable fraction of products, waste, and residues from agriculture (including vegetal and animal substances), forestry, and related industries, as well as the biodegradable fraction of industrial and municipal waste.

Biorefinery – a facility for achieving large-scale integrated production of fuels, power, and chemicals from biomass.

Biotechnology – any technological application that uses biological systems, living organisms, or derivatives thereof, to make or modify products or processes for specific use.

Concept – an abstract idea; a general notion.

Control – determine the behavior or supervise the running of.

Eco-Management and Audit Scheme (EMAS) – a voluntary environmental management instrument developed in 1993 by the European Commission. It enables organizations to assess, manage, and continuously improve their environmental performance.

Effective – adequate to accomplish a purpose; producing the intended or expected result.

Efficiency – accomplishment of or ability to accomplish a job with a minimum expenditure of time and effort.

Energy – a measure of the capacity, expressed as the work that it does, changing to some specified references state. It is measured in jules.

Energy balance – the arithmetic balancing of energy inputs versus outputs for an object, reactor, or other processing system.

Externalities – a side effect or consequence of an industrial or commercial activity that affects other parties without it being reflected in the cost of the goods or services involved.

Fuel cells – electrochemical devices that convert fuel energy directly into electricity.

Goals – the results or achievements toward which effort is directed.

Implementation –putting into effect according to or by means of a definite plan or procedure.

Indicator – a thing, especially a trend or fact, which indicates the state or level of something.

Management – the process of reaching organizational goals by working with and through people and other organizational resources.

Matrix – an environment or material in which something develops.

Method – a particular procedure for accomplishing or approaching something, especially something systematic or established.

Methodology – a system of methods used in a particular area of study.

Monetization – the process of converting or establishing something into legal tender.

Objective – an end that can be reasonably achieved within an expected timeframe and with available resources.

Optimization – to find the best compromise among several often conflicting requirements.

Organizational culture – a pattern of shared basic assumptions invented, discovered, or developed by a given group as it learns to cope with its problems of external adaptation and internal integration that have worked well enough to be considered valid and therefore able to be taught to new members as the correct way to perceive, think, and feel in relation to that problem.

Organizational structure – Explicit and implicit institutional rules and policies designed to provide a structure where various work roles and responsibilities are delegated, controlled, and coordinated. Organizational structure also determines how information flows from level to level within the company. In a centralized structure, decisions flow top down. In a decentralized structure, the decisions are made at various levels.

Planning – a scheme or a method of acting, doing, proceeding, etc., developed in advance.

Procedure – a series of actions conducted in a certain order or manner.

Process – a series of actions or steps taken to achieve an end.

Quality – the ongoing process of building and sustaining relationships by assessing, anticipating, and fulfilling a stated or implied need.

Resources – any property that can be converted into money.

Risk – the potential that a chosen action or activity (including the choice of inaction) will lead to a loss (an undesirable outcome).

Strategy – a plan of action or policy designed to achieve a major or overall aim.

Sustainability – the capacity to endure.

Sustainable development – development that meets the needs of the present without compromising the ability of future generations to meet their own needs.

System – a set of connected things or parts forming a complex whole, in particular.

Technique – a way of carrying out a particular task, especially the execution or performance of an artistic work or a scientific procedure.

Technology – the application of scientific knowledge for practical purposes, especially in industry: "computer technology"; "recycling technologies." Machinery and equipment developed from such scientific knowledge.

Tool – a device or an implement, especially one held in the hand, used to carry out a particular function.

Bibliography

[1] Adizes I. Managing Corporate Lifecycles. revised edition. Prentice Hall Press; 1999.

[2] Akashi O, Hanaoka T, Matsuoka Y, Kainuma M. A projection for global CO_2 emissions from the industrial sector through 2030 based on activity level and technology changes. Energy 2011;36(4):1855–67.

[3] Allcott H. Rethinking real-time electricity pricing. Resour Energy Econ 2011;33(4):820–42.

[4] Andrews K. Concept of Corporate Strategy. 3 subedition. Richard D. Irwin; 1986.

[5] Ansoff HI, McDonnell A. Implanting Strategic Management. 2nd edition. Prentice Hall; 1991.

[6] Apergis N, Payne J. Renewable energy consumption and economic growth: evidence from a panel of OECD countries. Energy Econ 2010;38(1):656–60.

[7] Belke A, Dobnik F, Dreger C. Energy consumption and economic growth: new insights into the cointegration relationship. Energy Econ 2011;33(5):782–9.

[8] Bhide A, Rodriguez C. Energy poverty: a special focus on energy poverty in India and renewable energy technologies. Renew Sust Energy Rev 2011;15:1057–66.

[9] Certo SC, Peter P. Strategic management: concepts and applications. McGraw-Hill; 1991.

[10] Daly HE. For The Common Good: Redirecting the Economy toward Community, the Environment, and a Sustainable Future. updated and expanded edition. Beacon Press; 1994.

[11] Dale A. At the edge: Sustainable Development in the 21st Century. Ontario, Canada: UBC Press; 2011.

[12] Dalgard CJ, Strulik H. Energy distribution and economic growth. Resour Energy Econ 2011;33:4.

[13] Dodić J, Ranković J, Jokić A, Dodić S, Popov S. Optimization of ethanol production from thin juice as intermediate product of sugar beet processing. XXI congress of chemists and technologists of Macedonia, Ohrid 2010;9:23–6.

[14] Dodić J, Vućurović D, Dodić S, Grahovac J, Popov S, Nedeljković N. Kinetic modelling of batch ethanol production from sugar beet raw juice. Appl Energy 2012; doi:10.1016/j.apen-ergy.2012.05.016.

[15] Dodić S, Ranković J, Zavargo Z, Golušin M. An overview of biomass energy utilization in Vojvodina. Renew Sust Energy Rev 2010;14(1):550–3.

[16] Dodić S, Popov S, Dodić J, Ranković J, Zavargo Z, Jevtic Mućibabic R. Bioethanol production from thick juice as intermediate of sugar beet processing. Biomass Bioenerg 2009;33(5):822–7.

[17] Dodić S, Popov S, Dodić J, Ranković J, Zavargo Z. Biomass energy in Vojvodina: market conditions, environment and food security. Renew Sust Energy Rev 2010;14(2):862–7.

[18] Dodić S, Popov S, Dodić J, Ranković J, Zavargo Z. Potential contribution of bioethanol fuel to the transport sector in Vojvodina. Renew Sust Energy Rev 2009;13(8):2197–200.

[19] Dodić S, Popov S, Dodić J, Ranković J, Zavargo Z. Potential development of bioethanol production in Vojvodina. Renew Sust Energy Rev 2009;13(9):2722–7.

[20] Dodić S, Popov S, Dodić, J, Ranković, J. Effect of different strains *Saccharomyces cerevisiae* on bioethanol production on thin juice from sugar factory. I International congress "engineering, materials and management". Books of Abstracts, Jahorina 2009;10:14–16.

[21] Dodić S, Vućurović D, Popov S, Dodić J, Ranković J. Cleaner bioprocesses for promoting zero-emission biofuels production in Vojvodina. Renew Sust Energy Rev 2010;14:3242–6.

[22] Dodić S, Vućurović D, Popov S, Dodić J, Zavargo Z. Concept of cleaner production in Vojvodina. Renew Sust Energy Rev 2010;14(6):1629–34.

[23] Dodić S, Zekić V, Rodic V, Tica N, Dodić J, Popov S. Situation and perspectives of waste biomass application as energy source in Serbia. Renew Sust Energy Rev 2010;14:3171–7.

[24] Dodić S, Zekić V, Rodic V, Tica N, Dodić J, Popov S. Analysis of energetic exploitation of straw in Vojvodina. Renew Sust Energy Rev 2011;15:1147–51.

[25] Dodić S, Zekić V, Rodic V, Tica N, Dodić J, Popov S. The economic effects of energetic exploitation of straw in Vojvodina. Renew Sust Energy Rev 2012;16(1):2355–60.

[26] Dodić S, Zelenović Vasiljević T, Marić R, Radukin Kosanović A, Dodić J, Popov S. Possibilities of application of waste wood biomass as an energy source in Vojvodina. Renew Sust Energy Rev 2012;16:2355–60.

[27] Drucker P. The Classic Drucker Collection. Routledge; 2011.

[28] Duić N, Guzović Z, Lund H. Sustainable development of energy, water and environment system. Energy 2011;36(4):1839–2314.

[29] Đorćević B. Strategijski menadžment. Kruševac: Fakultet za industrijski menadžment; 2005.

[30] Đuran J, Golušin M, Munitlak Ivanović O, Jovanović L, Andrejević A. Renewable energy – economic development in the EU. Problemy Ekorozwoju/Prob Sust Dev, 2012, in press.

[31] EU STRATEGY 2020 – strategy on sustainable and inclusive growth. European Commission; 2008.

[32] European Environment Agency, Annual reports, 2000–2011.

[33] Eurostat, Annual Reports 2000–2010.

[34] Energy Efficiency Policies around the World: Review and Evaluation, World Energy Council, London, UK, 2008.

[35] Fayol H. General and Industrial Management. New York: Pitman Publishing; 1949.

[36] Giampietro M, Sorman AH. Are energy statistics useful for making energy scenarios? Energy 2012;37:1.

[37] Golušin M, Dodić S, Vućurović D, Ostojić A, Jovanović L. Exploitation of biogas power plant – clean development mechanism project, Vizelj, Serbia. J Renew Sust Energy 2011;3:5. doi:10.1063/1.3631820.

[38] Golušin M, Mihić S, Mihić M. Policy and promotion of sustainable inland waterway transport in Europe – Danube River. Renew Sust Energy Rev 2011;15(4):1801–9.

[39] Golušin M, Munitlak Ivanović O, Bagarić I, Vranješ S. Exploitation of geothermal energy as a priority of sustainable energetic development in Serbia. Renew Sust Energy Rev 2010;14(2):868–71.

[40] Golušin, M., Munitlak Ivanović, O, Redžepagić, S.: Transition from traditional to sustainable energy development in the region of Western Balkans – current level and requirements, Applied Energy, in press.

[41] Golušin M, Munitlak Ivanović O, Dodić S, Vućurović D. Implementation of sustainable energy management in companies – gap anaysis technique. 6th conference on sustainable development of energy water and environment systems, Dubrovnik, Croatia 2011;9:25–9.

[42] Golušin M, Munitlak Ivanović O, Domazet S, Dodić S, Vućurović D. Assessment of the effectiveness of policy implementation for sustainable energy development in Southeast Europe. J Renew Sust Energy 2011;3. doi:10.1063/l.3663953.

[43] Golušin M, Munitlak Ivanović O, Jovanović L, Domazet S. Determination of ecological-economic degree of development in countries of SE Europe – weight coefficients technique. Problemy Ekorozwoju/Probl Sust Dev 2012;1:87–93.

[44] Golušin M, Munitlak Ivanović O, Teodorović N. The review of the achieved degree of sustainable development in Southeastern Europe – the use of linear regression method. Renew Sust Energy Rev 2010;15(1):772–6.

[45] Golušin M, Munitlak Ivanović O. Definition, characteristics and state of indicators of sustainable development in countries of Southeastern Europe. Agr Ecosyst Environ 2009;130(1–2):67–74.

[46] Golušin M, Munitlak Ivanović O. Kyoto protocol implementation in Serbia as precognition of sustainable energetic and economic development. Energy Policy 2011;39(5):2800–7. ISSN 0301-4215.

[47] Golušin M, Ostojić A, Latinović S, Jandrić M, Munitlak Ivanović O. Review of the economic viability of investing and exploiting biogas electricity plant – case study Vizelj, Serbia. Renew Sust Energy Rev 2012;16:1127–34.

[48] Golušin M, Tešic Z, Ostojić A. The analysis of the renewable energy production sector in Serbia. Renew Sust Energy Rev 2010;14(5):1477–83.

[49] Grahovac J, Dodić J, Dodić S, Popov S, Jokić A, Zavargo Z. Optimisation of bioethanol production from intermediates of sugar beet processing by response surface methodology. Biomass Bioenerg 2011;35(10):4290–6.

[50] Grahovac J, Dodić J, Dodić S, Popov S, Vućurović D, Jokić A. Future trends of bio-ethanol co-production in Serbian sugar plants. Renew Sust Energy Rev, doi: 10.1016/j.rser.2012.02.040.

[51] Grahovac J, Dodić J, Jokić A, Dodić S, Popov S. Optimization of ethanol production from thick juice: a response surface methodology approach. Fuel 2012;93:221–8.

[52] Grahovac J, Dodić J, Jokić A, Dodić S, Popov S. Optimization of bioethanol production from sugar beet thin juice. Alumni meeting of international summer schools Novi Sad 5, Tehnološki fakultet, Novi Sad, str. 15; 2011.

[53] Grahovac J, Dodić J, Popov S, Dodić S, Vućurović D, Tadijan I, et al. Bioethanol production technologies: current situation and perspectives in Vojvodina. Tractors Power Mach 2011;16(3):121–8.

[54] Home R, Grant T, Verghese K. Life Cycle Assessment, principles, practice and prospects. Csiro Publishing; 2009.

[55] Hicks JR, Simon H. Modes of economic behavior: variations on themes. USA: Massachusetts Institute of Technology, Department of Economics; 1979.

[56] Higgins JM, Vincze JW. Strategic management: text and cases. Harcourt Brace; 2000.

[57] International Energy Agency, Annual Reports 2000–2010, Paris.

[58] Jevtic Mucibabic R, Dodić J, Ranković J, Dodić S, Popov S, Zavargo Z. The valorization of the intermediates in the process of sugar beet as the alternative raw materials for the bioethanol production. Food Process Qual Safety 2008;35(2):71–6.

[59] Johanson G, Scholes K. Exploring corporate Strategy. Prentice Hall; 1998.

[60] Jokić A, Grahovac J, Dodić J, Dodić S, Popov S, Zavargo Z, et al. An artificial neural network approach to modelling of alcoholic fermentation of thick juice from sugar beet processing. Hemijska Industrija 2011;66(2):211–21.

[61] Jokić A, Grahovac J, Dodić J, Zavargo Z, Dodić S, Popov S, et al. Interpreting the natural network for prediction of fermentation of thick juice from sugar beet processing. Acta Period Technologica 2011;42:241–9.

[62] Meaney M., Wilson S. Corporate transformation under pressure, McKinsey quarterly report, London, UK, April 2009.

[63] Mackechnie C, Maskell L, Norton L, Roy D. The role of the "Big society" in monitoring the state of the natural environment. J Env Monitor 2011;13(10):2687–91.

[64] Munitlak Ivanović O, Golusin M, Dodić S, Dodić J. Perspectives of sustainable development in countries of Southeastern Europe. Renew Sust Energy Rev 2009;13(8):2000–179.

[65] Munitlak Ivanović O, Golušin M. Strateško koncipiranje odnosa ekonomije i ekologije, SM 2005; X Internacionalni naucni simpozijum, Zbornik abstrakata, Ekonomski fakultet Subotica, Subotica; Serbia, 2005.

[66] Munitlak Ivanović, O. Ecological aspects of sustainable development – international and regional comparison. Ph D Thesis, Faculty of Economics, Subotica; 2005.

[67] Munitlak Ivanović O. Strategijski menadžment. Sremska Kamenica: Univerzitet Educons; 2009.

[68] Naseem M. Energy Law in India. India: Kluver Law International; 2010.

[69] National Rural Electrification Policies (NREP), 2006.

[70] Popov S, Dodić S, Vućurović D, Dodić J, Grahovac J, Tadijan, I. Modelling the process and costs of bioethanol production from corn stover. In: Proceedings of the 39th international conference of Slovak society of chemical engineering, Slovakia: Tatranské Matliare; 2012. pp. 654–62.

[71] Ohmae K. The Next Global Stage: Challenges and Opportunities in Our Borderless World. Wharton School Publishing; 2005.

[72] Popov S, Ranković J, Dodić J, Dodić S, Jokić A. Bioethanol production from raw juice as intermediate of sugar beet processing: a response surface methodology approach. Food Technol Biotech 2010;48(3):376–83.

[73] Ranković J, Dodić J, Dodić S, Popov S. Bioethanol production from intermediate products of sugar beet processing with different types of Saccharomyces cerevisiae. Chem Ind Chem Eng Quart 2009;15(1):13–16.

[74] Ranković J, Dodić J, Popov S, Dodić S, Zavargo Z, Jokić A. Bioethanol production from raw juice as intermediate product of sugar beet processing. International conference on science and technique in the agri-food business, 5–6 November, Segedin, Hungary; 2008.

[75] Recalde M, Ramos-Martin J. Going beyond energy intensity to understand the energy metabolism of nations: the case of Argentina. Energy 2012;37:1.

[76] Repetto R. The forest for the Trees?. Washington DC, USA: World Resource Institute; 1988.

[77] Sadrorsky P. Renewable energy consumption and income in emerging economies. Energy Policy October 2009;37(10):4021–8.

[78] Simpson AP, Edwards C F. An energy-based framework for evaluating environmental impact. Energy 2011;36(3):5824–31.

[79] Sorda G, Banse M, Kemfert C. An overview of biofuel policies across the world. Energy Policy 2010;38:11.

[80] Sustainable development: critical issues, OECD Publishing, 2001.

[81] Suzuki D. Earth time. USA: Stoddart; 1998.

[82] Thomas LW, Hunger JD. Strategic management and business policy. Pearson Education Inc.; 2001.

[83] Todorović J, Đuričin D, Janoševic S. Strategijski menadžment. Beograd: Ekonomski fakultet; 2003.

[84] Vućurović D, Dodić S, Popov S, Dodić J, Grahovac J, Tadijan I. Zero-emission process model of bioethanol production from sugar beet thick juice. Proceedings of the 39th international conference of Slovak society of chemical engineering. Tatranské Matliare, Slovakia; 2012. pp. 25–33.

[85] Vućurović D, Dodić S, Popov S, Dodić J, Grahovac J, Tadijan I, et al. Bioethanol as transport fuel. Tractors Power Mach 2011;16(3):113–20.

[86] Vućurović D, Dodić S, Popov S, Dodić J, Grahovac J, Tadijan I. Small-scale bioethanol production plants on farms. Tractors Power Mach 2011;16(3):129–36.

[87] Vućurović D, Dodić S, Popov S, Dodić J, Grahovac J. Process model and economic analysis of ethanol production from sugar beet raw juice as part of the cleaner production concept. Bioresource Technol 2012;104:367–72.

[88] Wayne CT. Energy management handbook. CRC Press; 2006.

[89] World Bank, Annual Report, 2011.

[90] Wustenhagen R, Menichetti E. Strategic choices for renewable energy investments: conceptual framework and opportunities for further research. Energy Policy 2012;40.

[91] Zavargo Z, Dodić S, Jokić A, Dodić J, Prodanic B, Popov S. Possibilities for bioethanol production in Vojvodina. 6th biennial intenational workshop – advances in energy studies, proceedings, June 29 to July 2, University of Technology, Graz; 2008. pp. 231–36.

[92] Zavodov K. Renewable energy investment and the clean development mechanism. Energy Policy 2012;40.

Index

Printed in the United States
By Bookmasters